ELECTRON MICROSCOPY

: 39 EXERCISES

: A RESEARCH PROTOCOL

: INTENSIVE SHORT–COURSE

AND WORKSHOP LAYOUT

Edited by

BETTY RUTH JONES, B.S., M.S., Ph. D.

Library Research Associates, Monroe, New York 1983

FRONT COVER: Pancreatic Acinar Cell, courtesy of Dr. Thomas L. Lentz, Dept. Of Anatomy, Yale University School of Medicine in Hartford, Connecticut. CELL FINE STRUCTURE by Thomas L. Lentz (1971) W.B.Saunders Co.,Phil.,Pa., p. 195

Printed in the United States of America.

Jones, Betty Ruth, 1951-
Electron microscopy: 39 exercises, a research protocol, an intensive short-course and workshop layout.
547 p. illus., micrographs

1.Electron microscopy--technique I. Title
ISBN: 0-912526-34-3 578.45

ELECTRON MICROSCOPY

Dedicated To

Jerry Washington Jones

and

Dr. Barnett Frissel Smith

Preface

Electron microscopy is a valuable teaching and research tool with vast applications in many branches of science and technology. EM techniques have particularly become indispensable when used in combination with biochemical, biophysical and immunocytochemical methods to eludicate the function of cellular components. The significance of data obtained from an EM study is greatly influenced by the technique utilized in preparation of the material. It is essential that every precaution be taken to obtain the best preservation of materials under normal conditions.

Despite rapid expansion of EM techniques, certain procedures are basic and standard. It is important that the novice electron microscopist be introduced properly to the preparation of materials for EM as well as to the operation and use of instruments.

The primary purpose of this manual is to introduce the students to basic principles and techniques of EM through laboratory exercises and to supplement basic knowledge with advanced EM applications. This manual will also aid the experienced microscopist who wishes to review basic techniques and to utilize advanced applications to meet research needs. It should bridge the gap between textbooks, handbooks and journals and fill the requirements of most college courses in electron microscopic techniques. This manual was developed as a result of teaching a course in electron microscopy in the Department of Biology, Morehouse College, Atlanta, Georgia to senior undergraduate and graduate students in the Atlanta University Center.

Section One provides twenty-three basic introductory exercises which have been tested on five groups of senior undergraduate and graduate students in the Atlanta University Center in Atlanta and technicians at Harvard Medical School, Boston, Massachusetts. The arrangement of exercises in Section One is based on the most efficient manner of introducing EM techniques. For example, if resin blocks are available in numbers, it is good practice to begin rough trimming and precision facing of specimen blocks by hand. This can be followed by making glass knives and the use of the ultramicrotome. When a degree of proficiency has been achieved using these techniques, tissues can be prepared for fixation and embedding.

Section Two consists of sixteen advanced exercises written by experts in the field of electron microscopy. Section Three includes key points for developing a research protocol using EM. Section Four provides a scheduled protocol and information for implementing intensive short courses and workshops in EM. Appendix A is a list of recipes, a list of tissue culture suppliers is included in Appendix B and C and Appendix D is a bibliography of reference EM textbooks.

To improve the use of this manual, I encourage suggestions and comments regarding revisions, additions or deletions.

June 1983 Betty Ruth Jones, Ph.D.

Acknowledgments

I am indebted to the Office of Naval Research (ONR) in Washington, D.C. through the auspices of the Mary Ingraham Bunting Institute of Radcliffe College, Cambridge, Massachusetts for providing supplementary financial support for the completion of this project (Grant Number 00014-80-G-0084). In this regard special thanks are extended to Dr. James Lester, Liaison Scientist of ONR in Boston, Massachusetts and Margaret McKenna, Director of the Bunting Institute and Administrative Vice-President for Program Planning at Radcliffe.

My gratitude is extended to the Department of Tropical Public Health, Harvard School of Public Health in Boston for providing office space and laboratory facilities to test exercises by carrying out extensive research in schistosomiasis. I am grateful to Dr. Steve Chia-Tung Pan, Director of EM, Dr. Gustave J. Dammin and Dr. Edward H. Michelson in the Department of Tropical Public Health for their critiques, suggestions and helpful comments on parts of this manual.

I am grateful also to Dr. Gustav A. Ofosu, Director of EM, Delaware State College, Dover, Delaware, Dr. Daniel Scheirer, Director of EM, Department of Biology, Northeastern University in Boston and Dr. Anita P. Hoffer, Department of Anatomy, Harvard Medical School in Boston for critically reading and making valuable suggestions and comments on the entire manual. Thanks are extended to Winston J. Peak, President of Ebony Industries, Woburn, Massachusetts for constructive critisms and profitable suggestions on the manuscript.

With sincere gratitude I am pleased to acknowledge the work of Dr. Reuven Miller, President of Research Graphics Company, Brookline,

Massachusetts and staff for excellent line drawings and illustrations throughout the manual.

I am indebted to my home institution, Morehouse College, Atlanta, Georgia, especially the Department of Biology for providing the opportunity to develop an extensive EM course for senior undergraduate and graduate students in the Atlanta University Center. I appreciate the encouragement and support from students who have taken my EM course to complete this laboratory manual and workbook.

Special thanks to the contributors who generously responded to my requests for providing exercises for this manual.

Appreciation is extended to Mrs. Allie Clift at Morehouse School of Medicine in Atlanta for word processing parts of the draft exercises of the manuscript. It is a pleasure to acknowledge Miss Patricia Thomason at Dana-Farber Cancer Institute in Boston for her patience, encouragement, editing, uniform accuracy and word processing of numerous draft copies and the final revisions of this manual. Acknowledgements are also extended to Mrs. Cheryl Carriere of Digital Equipment Corporation in Boston for assisting in the final revisions.

B.R.J.

CONTRIBUTORS

Dave P. Allison, Ph.D.
Biology Division
Oak Ridge National Laboratory
Oak Ridge, Tennessee 37830

G. Colin Budd, Ph.D.
Professor
Department of Physiology
Medical College of Ohio
Toledo, Ohio 43699

John P. Caulfield, M.D.
Associate Professor
Department of Pathology and Medicine
Harvard Medical School
and the Robert B. Brigham Hospital
Boston, Massachusetts 02115

Brenda R. Eisenberg, Ph.D.
Associate Professor
Department of Physiology
Rush Medical College
Chicago, Illinois 60612

Anthony J. Garrett-Reed, Ph.D.
Supervisor
of the Electron Microscopy
Center for Material Science and Engineering
Massachusetts Institute of Technology
Cambridge, Massachusetts 02139

G. William Gary, Jr., M.P.H., Dr. PH.
Chief
Division of Hepatitis and Viral Enteritis
Centers for Infectious Diseases
Centers for Disease Control
Phoenix, Arizona 85014

Frederic Giere, Ph.D.
Chairman and Professor
Department of Biology
Lake Forest College
Lake Forest, Illinois 60045

Raymond L. Hayes, Ph.D.
Professor
Department of Anatomy and
Director of Electron Microscopy
Morehouse School of Medicine
Atlanta, Georgia 30310

Tessa Hedley-Whyte, M.D.
Associate Professor
Department of Pathology
Harvard Medical School and
Massachusetts General Hospital
Boston, Massachusetts 02114

Patricianne M. Hurd, Ph.D.
Instructor and
Director of Electron Microscopy
Fernbank Science Center
Atlanta, Georgia 30307

Betty R. Jones, Ph.D.
Assistant Professor and
Director of Electron Microscopy
Department of Biology
Morehouse College
Atlanta, Georgia 30314

William J. Landis, Ph.D.
Assistant Professor
Department of Orthopedic Surgery
Laboratory for the Study of
Skeletal Disorders and Rehabilitation
Harvard Medical School and the
Children's Hospital Medical Center
Boston, Massahcusetts 02115

Morton D. Maser, Ph.D.
President
Woods Hole Educational Associates
Woods Hole, Massachusetts 02543

Sara E. Miller, Ph.D.
Associate Professor
Director of Electron Microscopy
Department of Microbiology and Immunology
Duke University Medical Center
Durham, North Carolina 27710

Hans Ris, Ph.D.
Professor
Department of Zoology and
Director of the Madison High Voltage
Electron Microscope Facility
Madison, Wisconsin 53706

John M. Robinson, Ph.D.
Assistant Professor
Department of Pathology
Harvard Medical School
Boston, Massachusetts 02115

Russell L. Steere, Ph.D.
Chief
Plant Virology Laboratory
Plant Protection Institute
U.S. Department of Agriculture
Beltsville Agriculture Research Center
Beltsville, Maryland 20705

Laboratory Safety

The prevention and control of accidents, injuries, fires and losses in any laboratory require safety practices that should be recognized in educational and research institutions, hospitals and industry. Students, investigators and technical staffs should be able to work in laboratory facilities without endangerment to health. It is essential that they recognize, accept and implement safety regulations and practices for their own protection as well as the protection of others.

Safety in the laboratory demands the same concern and attention as teaching and research. The utilization of techniques whether old, new or different and the use of chemical and equipment requires careful reading of instructions and supervision when necessary. The development of positive attitudes concerning laboratory safety is one step toward the prevention of laboratory accidents. There are certain dangerous hazards inherent in the laboratory which the student should learn and appreciate. The laboratory can be a safe environment to work and to learn if a high degree of emphasis is placed on safety regulations and practices.

This laboratory manual and workbook specifies safety precautions. Possible hazards concerning the use of chemicals and equipment have been emphasized in each exercise where necessary.

Strict precautions must be taken when using EM chemicals such as aldehydes, osmium tetroxide, buffers and propylene oxide. Embedding resins may cause inflammation of the skin and some accelerators are considered carcinogens.

Sharp objects such as razor blades, glass knives, needle probes and scalpels must be handled carefully. Speical precautions must be taken when using equipment such as critical point dryers and vacuum evaporators.

Users of this manual and workbook are urged to become familiar with the safety precautions suggested in the laboratory exercises.

GENERAL SAFETY GUIDELINES

1. Know the location, purpose and use of emergency laboratory first aid and safety equipment before starting laboratory work.

2. Wear protective equipment such as eye goggles, disposable gloves and a laboratory coat or apron.

3. Read each exercise carefully before coming to the laboratory so that you know what is to be done and the basic principles involved.

4. Do not begin work until a brief introduction is given by the instructor. Based on my personal experiences, I have found that good laboratory technique focuses on knowing what you are to do. Take notes on procedures and include any modifications for various exercises in your laboratory manual and workbook.

5. Always clean the bench top or work area before and after a laboratory exercise with a mild disinfectant solution. The same solution can be used for rinsing your hands after washing them with soap and water.

6. The instructor will introduce and demonstrate the use of equipment.

7. Use of any equipment in the laboratory should be checked by the instructor and approval obtained.

8. Carefully read all instructions on labels of chemicals and equipment prior to using them.

9. Label and store properly all chemicals and solutions.

10. Give strict attention to the care, handling and disposal of chemicals.

11. Primary attention must be given to the disposal of hazardous waste such as flammable liquids, toxic and corrosive materials, substances which may create an air or stream pollution problem, the handling and disposal of unstable materials and leaking containers of dangerous gases or liquids.

Contents

12. Use chemicals that are toxic and organisms that are pathogenic under a well ventilated laboratory hood.

13. Do not eat, drink or smoke in the laboratory.

14. Shoes must be worn at all times.

15. Notify instructor of all spills.

16. Place all chemicals back on the shelf in alphabetical order after use.

17. Replace all clean glassware.

18. Sign up for use of equipment in advance.

19. Complete designated log books after use of various equipment.

20. In summary, proceed with caution and care in carrying out the exercises in this workbook. Give special attention to toxic and chemical reactions, radiation, fire, biological, electrical and mechanical hazards.

SECTION II

SECTION III

SECTION IV

SECTION I

EXERCISE NUMBER 1

ROUGH TRIMMING AND PRECISION FACING OF SPECIMEN BLOCK BY HAND

Betty R. Jones, Ph.D.

OBJECTIVES:

1. To remove excess plastic embedding material from specimen block and face

2. To obtain a specimen size and shape suitable with the objective of the study

3. To obtain a uniform density of tissue within the block face

4. To obtain a trapezoid shape for the block face

5. To prepare the specimen block and face for thick and ultrathin sectioning

MATERIALS:

1. Untrimmed specimen blocks (embedded in beem and gelatin capsules, sizes 00, 0, 1,2,3,4,5)

2. Single and double edge razor blades

3. Acetone

4. Collect-type specimen holder (several sizes are available)

5. Stereoscopic dissecting microscope

NOTE: BLOCK MAY ALSO BE TRIMMED ON MICROTOME BY HAND DEPENDING ON TYPE AND MODEL.

6. A vise-type specimen holder can be used for flat embedding

RATIONALE AND DISCUSSION:

Rough trimming and precision facing of the specimen block is essential before sections can be cut for light and electron microscopy.

The specimen block must be trimmed to a suitable size and shape. The size is usually determined according to the objectives of the study. The final shape of the block face should be that of a trapezoid.

The purpose of rough trimming is to remove excess plastic embedding material from the specimen block and face. This is essential so that only a small additional amount of plastic need to be removed during precision or fine facing of the block.

This will result in a suitable size of the block face and uniform density with no part of the block face without tissue.

PROCEDURE:

1. Clean razor blades with acetone to remove the coating of oil which is used by the manufacturer to protect them from rust. Place several clean single and double edge razor blades on filter paper in separate dust-free clean glass petri dishes.

2. Remove specimen block from beem or gelatin capsule.

   To remove block from capsules, make two longitudinal cuts about 0.1 inches in diameter down the side of the capsule.

   Using forceps extend the cut slit from the top of the capsule to the end removing the specimen block.

   Some beem capsules can be purchased from EM supply houses that come prepared with a longitudinal slit guide that once fully extended will cause the specimen block to be removed easily (See Figure 1-a-e).

   A capsule beem press is a small apparatus specifically designed for removing the specimen block from the beem capsule and may be used if available.

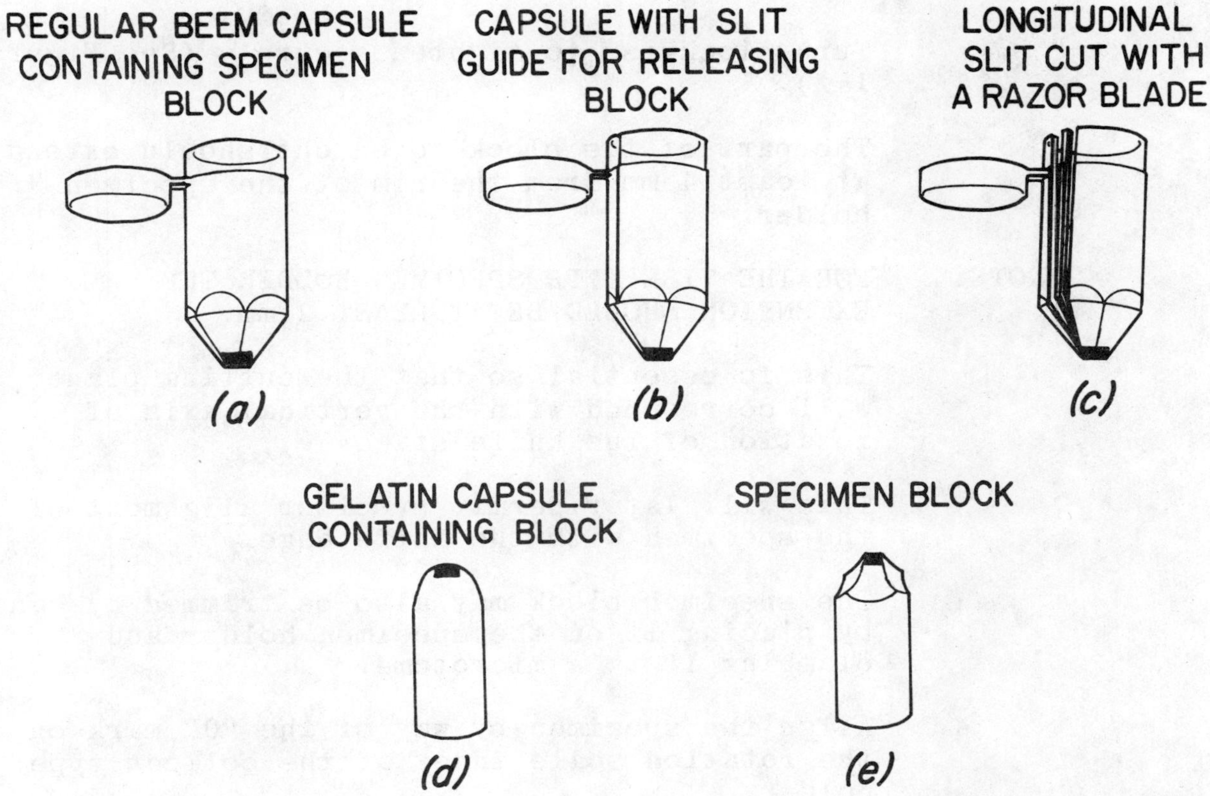

FIGURE 1-a-e. Appearance of specimen block in and out of capsule.

(a) Regular beem capsule containing specimen block.

(b) Beem capsule with longitudinal slit guide for removal of specimen block.

(c) Beem capsule containing specimen block showing how two longitudinal slits should be made with a razor blade.

(d) Gelatin capsule containing block.

(e) Specimen block removed from regular beem capsule.

3. For trimming of specimen block by hand under a dissecting microscope place the collect-type specimen holder in the trimming socket from the microtome.

   Place specimen block inside of specimen holder.

   Turn clockwise to tighten securely (See Figure 1-f).

   The part of the block to be cut should extend at least 4 mm from the rim of the specimen holder.

NOTE: FOR THE VISE-TYPE SPECIMEN HOLDER THE EXTENSION SHOULD BE AT LEAST 2 mm.

   This is essential so that the cutting plane will correspond with the vertical axis of rotation of the knife stage.

   This will later permit preciser alignment of the specimen with the knife edge.

   The specimen block may also be trimmed by hand by placing it on the specimen holder and clamping it to a microtome.

4. Align the specimen by way of the "0" mark on the rotation scale index of the collect-type holder.

   Make sure the specimen is tightened firmly in the holder and does not move.

5. Obtain the proper illumination from the light source of the dissecting microscope or microtome.

   Focus so that the face of the specimen block can be seen clearly (See Figure 1-g) for detailed trimming procedure.

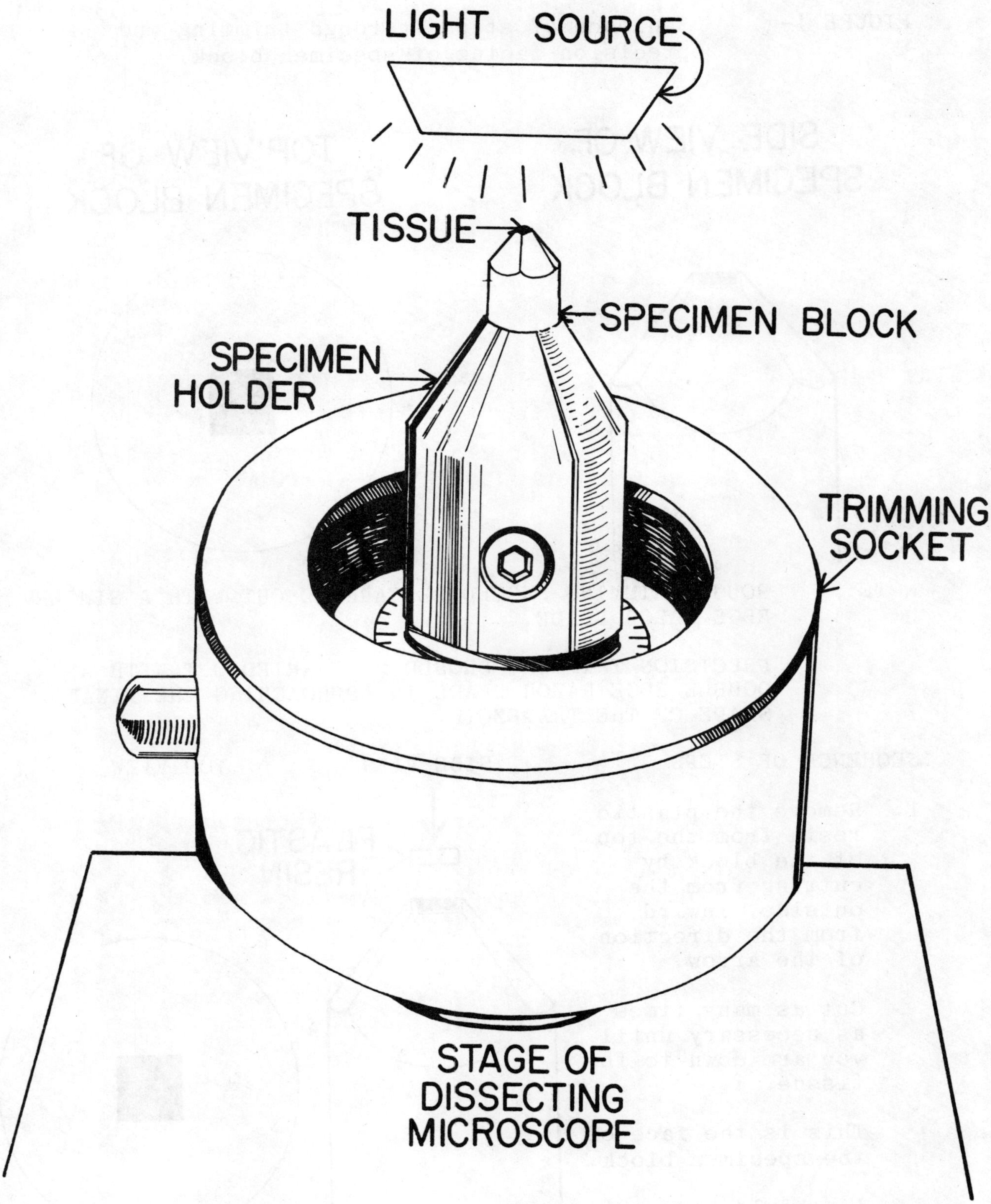

FIGURE 1-f. Specimen block mounted in specimen holder and trimming socket on the stage of a dissecting microscope.

FIGURE 1-g. Sequence of steps in rough trimming and precision facing of specimen block.

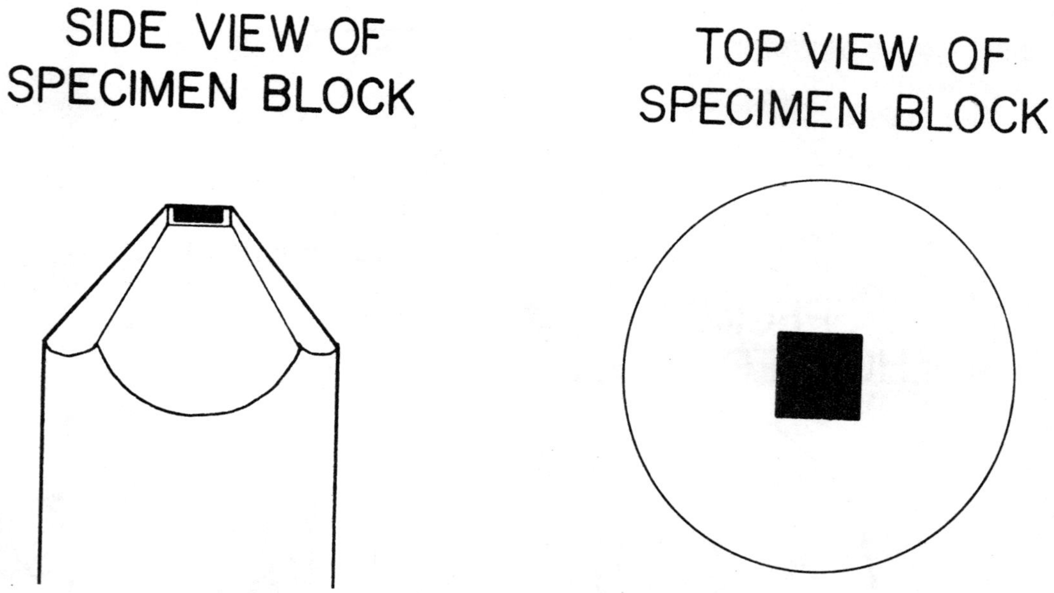

NOTE: ROUGH TRIMMING SHOULD BE CARRIED OUT WITH A SINGLE EDGE RAZOR BLADE.

PRECISION TRIMMING SHOULD BE CARRIED OUT WITH A DOUBLE EDGE RAZOR BLADE IN APPROACHING THE FINAL SHAPE OF THE TRAPEZOID.

SEQUENCE OF STEPS

SIDE VIEW TOP VIEW

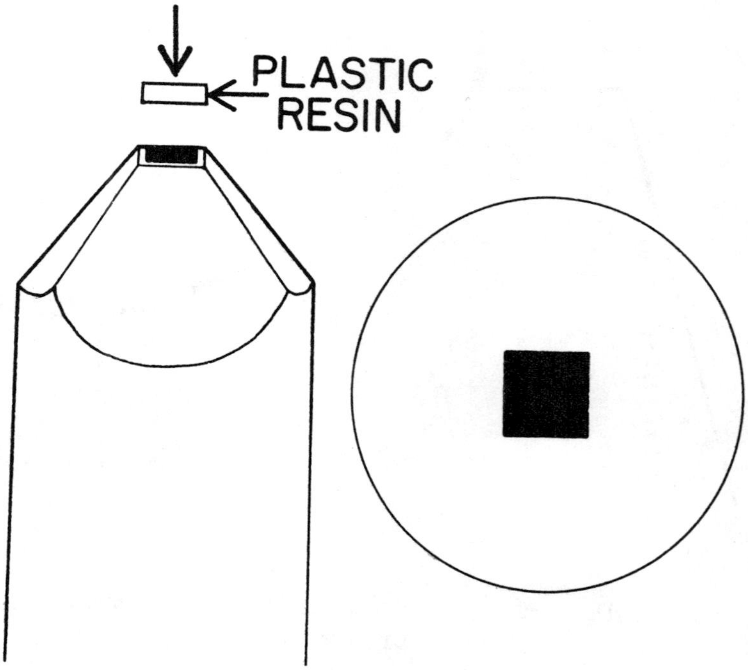

1. Remove the plastic resin from the top of the block by cutting from the outside, inward, from the direction of the arrow.

   Cut as many times as necessary until you are down to the tissue.

   This is the face of the specimen block.

   It should have a parallel top, parallel bottom edge, and smooth surface.

2. From the outside to the inside, make a veritcal cut on the left side of tissue.

NOTE: DO NOT HACK OR CHIP THE BLOCK.

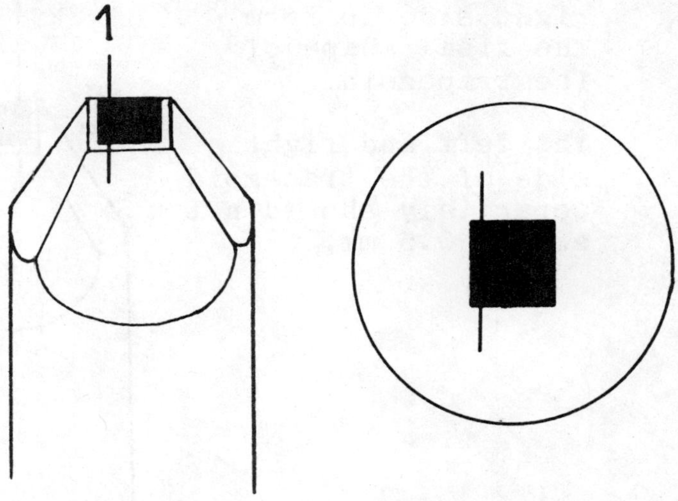

3. From the outside to the inside, make the left and right sides of the tissue parallel to each other.

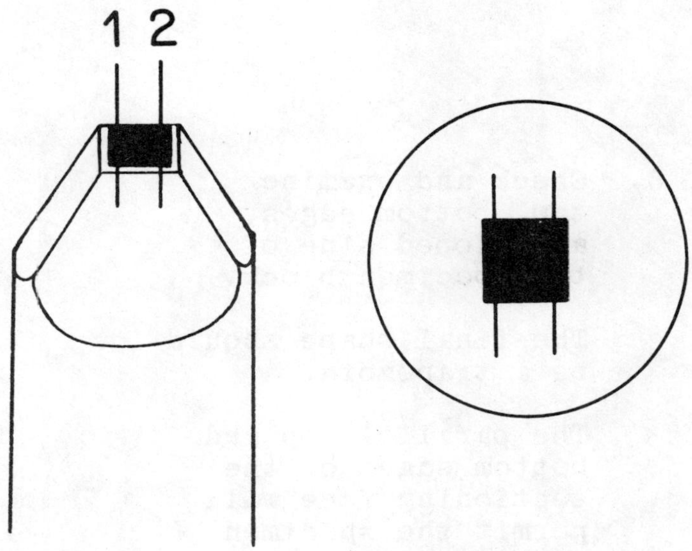

4. Rotate the entire specimen holder clockwise to the left side of tissue.

This side has not been trimmed using a double edge razor blade.

Cut one side of the trapezoid by sloping or slanting the side of the tissue.

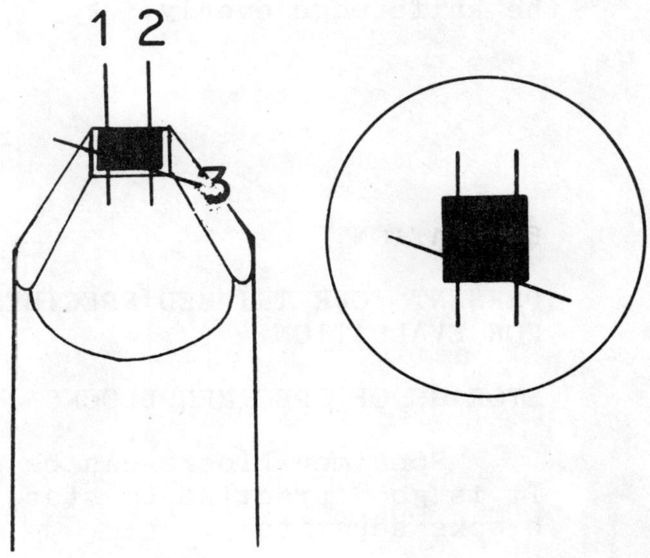

5. Cut the tissue on the right side to form the final shape of the trapezoid.

The left and right side of the trapezoid separately should not exceed 0.5 mm.

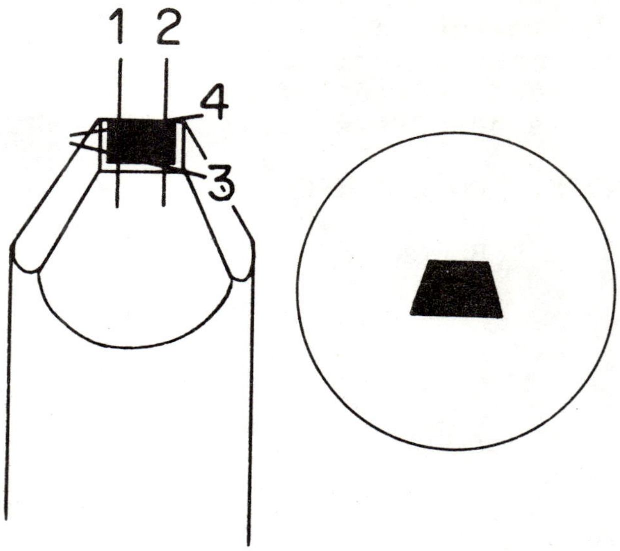

6. Check and examine top, bottom edges, and sloped side of the specimen block.

The final shape should be a trapezoid.

The parallel top and bottom edges of the sectioning face will permit the specimen to contact and leave the knife edge evenly.

This allows a straight ribbon of sections to be formed.

The sloping side edges cause each section as they are cut to move the previous section away from the knife edge.

This aids in obtaining good ribbon formation.

EVALUATION:

PRESENT FOUR TRIMMED SPECIMEN BLOCKS TO THE INSTRUCTOR FOR EVALUATION.

STORAGE OF SPECIMEN BLOCKS

Specimen blocks can be stored in small pill boxes. It is good practice to store trimmed and untrimmed blocks separately.

QUESTIONS:

1. By definition, differentiate between rough trimming and precision facing of a specimen block.

2. Why are single and double edged razor blades used in the trimming process?

3. Why is it necessary to cut away excess plastic from the specimen block?

4. What should be the final appearance of a trimmed specimen block and why is this essential?

5. Should razor blades be used for trimming blocks that are very hard? If not, why and what other tools are available for trimming?

6. What are the advantages of precision facing a specimen block?

7. What is the size range for the block face?

8. What are some disadvantages of a block face that is trimmed too large?

9. Why should the upper and lower sides of the block face be made parallel to each other?

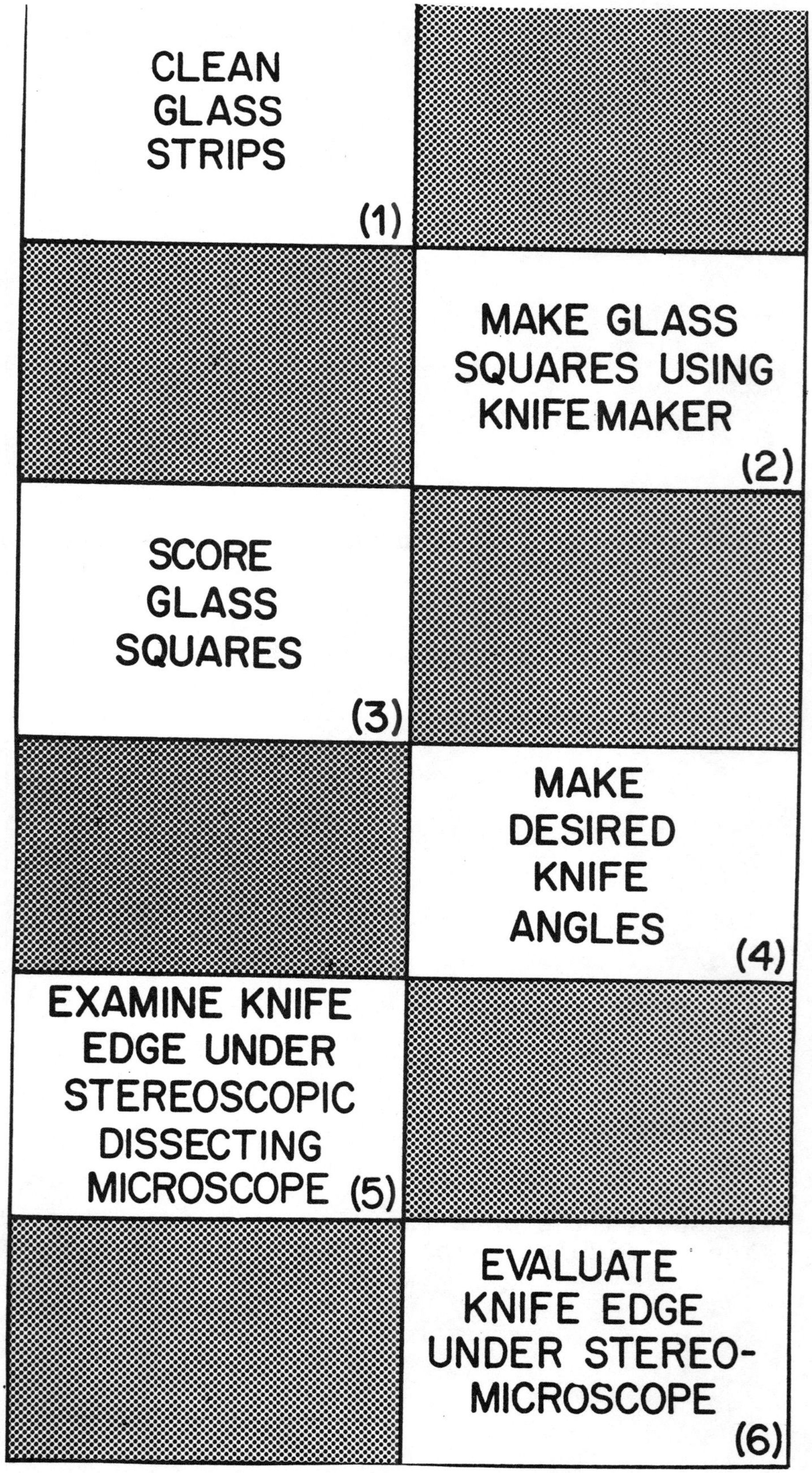

FIGURE 2-a. Major steps in preparing glass knives.

EXERCISE NUMBER 2

PREPARATION OF GLASS KNIVES WITH A KNIFEMAKER MACHINE

Betty R. Jones, Ph.D.

OBJECTIVES:

1. To break a strip of glass into squares
2. To make a diagonal score line across the center of the glass square
3. To break the glass square into two pieces
4. To produce two knives with thin and sharp knife angles suitable for cutting thick and thin tissue sections

MATERIALS:

1. Knifemaker machine
2. Glass strips (16 inches long, 1/4 inch thick, 1 inch wide)
3. Mild detergent solution
4. Warm running tap water
5. Distilled water
6. 95% alcohol
7. Lent-free paper tissues

RATIONALE AND DISCUSSION:

A knifemaker machine in the electron microscope laboratory provides a rapid and more productive means of making glass knives. Figure 2-a shows the major steps in preparing glass knives.

However, if one is not available, knives can be made by hand (See Exercise Number 3). The procedures in this exercise apply to the LKB knifemaker machine.

NOTE: THE KNIFEMAKER WIILL BE ADJUSTED AND DEMONSTRATED BY THE INSTRUCTOR BEFORE ANY STUDENT USES IT.

DO NOT CHANGE ANY OF THE SETTINGS (SEE FIGURE 2-d).

ANY FURTHER ADJUSTMENTS REQUIRED SHOULD BE CALLED TO THE ATTENTION OF THE INSTRUCTOR.

A = + 2¾	C = 9.5
B = + 1¾	D = 4.8

NOTE: DO NOT CHANGE ANY SETTING WITHOUT PERMISSION FROM THE INSTRUCTOR.

FIGURE 2-d. Knife setting for 45° angle knives.

PROCEDURE:

The instructor will demonstrate how to use the knifemaker and discuss the working parts and functions.

Instructions for use of the machine are stored on a shelf at the base of the instrument.

NOTE: STUDENTS SHOULD USE SCRAP GLASS TO PRACTICE GETTING ACQUAINTED WITH THE WORKING PARTS OF THE MACHINE (SEE FIGURES 2-b-c).

Cleaning of Glass Strip

1. Clean glass strip under warm running tap water from faucet with mild detergent solution.

2. Rinse with distilled water and then 95% alcohol for final rinsing.

3. Allow glass strip to drain dry in a closed dust-free area.

   If time is a factor dry the glass with several pieces of lint-free paper tissues.

NOTE: NEVER TOUCH GLASS THAT HAS BEEN CLEANED DIRECTLY WITH FINGERS. WEAR DISPOSABLE LABORATORY GLOVES. HANDLE GLASS WITH LENT-FREE TISSUES. ALWAYS KEEP IN MIND THAT IF A KNIFE IS MADE FROM DIRTY GLASS IT CANNOT BE CLEANED, IT CAN ONLY BE DISCARDED.

POSITIONING AND BREAKING GLASS STRIP

1. Place clean glass strip on knifemaker.

2. Place needle edge of glass strip in downward position to obtain 45° angle knives. For 50° angle knives needle edge should be up (See Figure 2-e).

3. Line right end of glass strip up with dot on right side of the instrument table top.

   The dot permits positioning of the 16 inch glass strip in order to break it exactly in half.

   Score and break glass as demonstrated below.

   The resulting product here is two 8 inch glass strips.

   Remove one strip and store for later use.

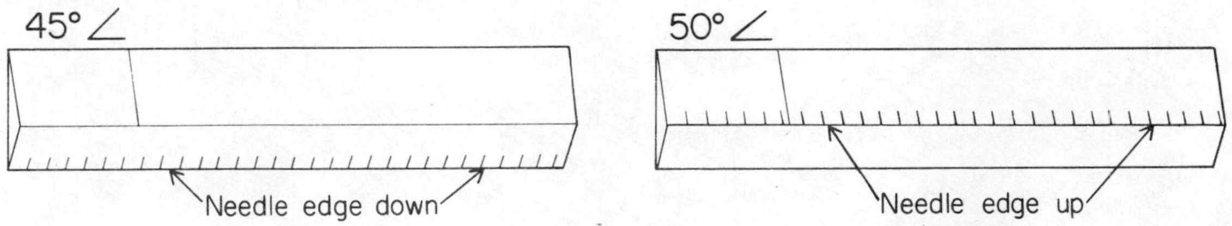

FIGURE 2-e. Appearance of glass strip positions under knife-maker machine for obtaining 45° and 50° knife angles.

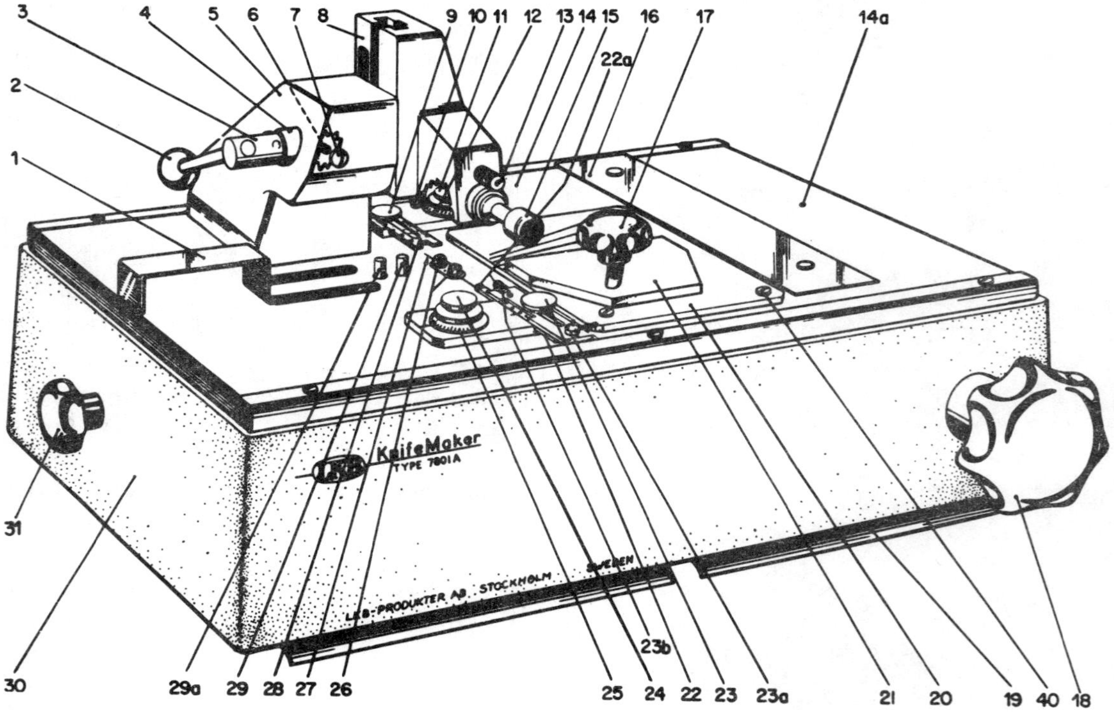

FIGURE 2-b. LKB knifemaker machine and parts (courtesy of LKB Instruments, Inc.).

1. Fork used to remove the glass knives from the apparatus
2. Locking lever
3. Shaft
4. Adjusting sleeve for (8)
5. Bracket
6. Set screw for (4)
7. Cover screw for (6)
8. Clamping head
9. Locking screw for (28)
10. Support studs
11. Locking screw for (12)
12. Rear dial
13. Score selector
14. Cover plate
14a. Dot (marking 200 mm strip length)
15. Scoring shaft with cutter wheel
16. Support plate for glass strip
17. Locking knob for (21)
18. Breaking knob
19. Operation Instructions
20. Angle setting plate
21. Guide plate
22. Front glass holder
23. Locking screw for (22)
23a. DPA-lever (lever for Damping Pressure Adjustment)
23b. Damping pad
24. Locking screw for (25)
25. Front dial
26. Breaking pins
27. Centre line
28. Rear glass holder
29. Arresting stud for 25 mm glass strips width
29a. Arresting stud for 38 mm glass strips width
30. Foundation block (silumingamma)
31. Disengage knob

FIGURE 2-c. List of parts for LKB knifemaker machine (courtesy of LKB Instruments, Inc.).

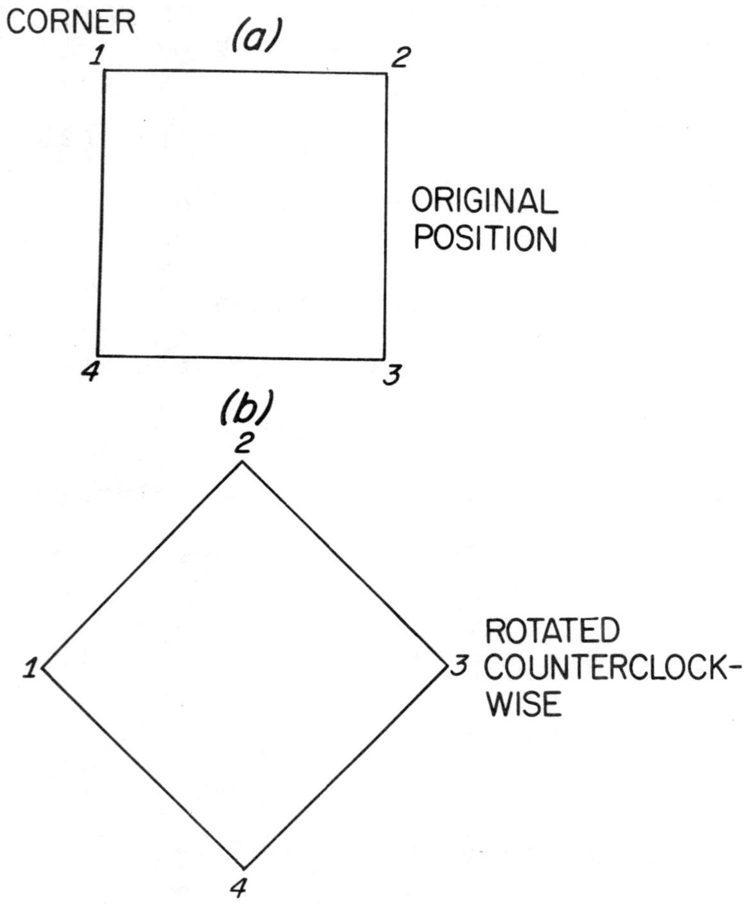

FIGURE 2-f
Glass knife square under knifemaker machine.

(a) Before rotating.

(b) After rotating.

FIGURE 2-g
Positions where glass knives should be handled during breaking.

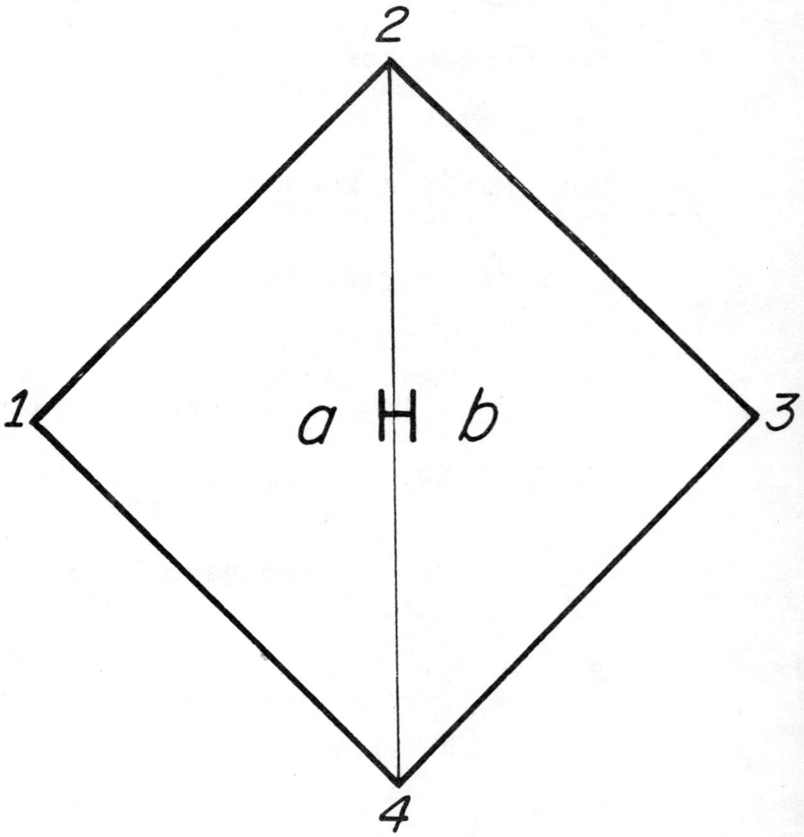

Making Glass Squares

4. Place locking lever (2) in rear position.

   Push scoring shaft (15) in and turn breaking knob all the way counter-clockwise.

   Turn disengage knob (31) all the way counter-clockwise and pull out to lock it in this position.

5. Slide glass strip to arresting stub (29) allowing it to rest against guide plate (21).

6. Hold glass strip lower then clamping head (8) with locking lever (2) until it rests on the glass.

   Remove hand from glass and gently swing locking lever all the way down, clamping strip.

   DO NOT USE UNDUE FORCE IN PRESSING LOCKING LEVER DOWNWARD.

7. Place score selector (13) in position and fork (1) under left end of strip.

   Slowly pull scoring shaft (15) forward until it stops.

   Turn breaking knob (18) clockwise until glass breaks.

   Then turn knob counterclockwise until it stops.

8. Lift up scoring shaft (15) and swing locking lever to rear position.

   Push scoring shaft (15) to rear position.

   Remove square with fork.

NOTE: THE FORWARD LEFTHAND CORNER OF GLASS SQUARE IS CALLED CORNER 1 (SEE FIGURE 2-F).

9. Repeat the above process until several glass squares are made.

Making Glass knife Diagonals

10. Place working parts in same position as Number 4.

Set score selector on 25 and glass square on breaking pins (26) with corner 1 in knotch in front of glass holder (22).

Push disengage knob (31) in and turn clockwise.

This is done to move rear glass holder (28) forward to corner 3 (See Figure 2-g).

Swing locking lever (2) forward to locked position.

11. Place fork (1) under glass square and pull score shaft (15) forward until it stops.

    Slowly turn breaking knob (18) clockwise to break glass.

    After breaking return to counter-clockwise stop.

12. Lift score shaft (15) and at the same time swing locking lever (2) to rear position.

    Push scoring shaft to rear position and slowly turn counterclockwise.

    Pull out disengage knob (31) to locked position.

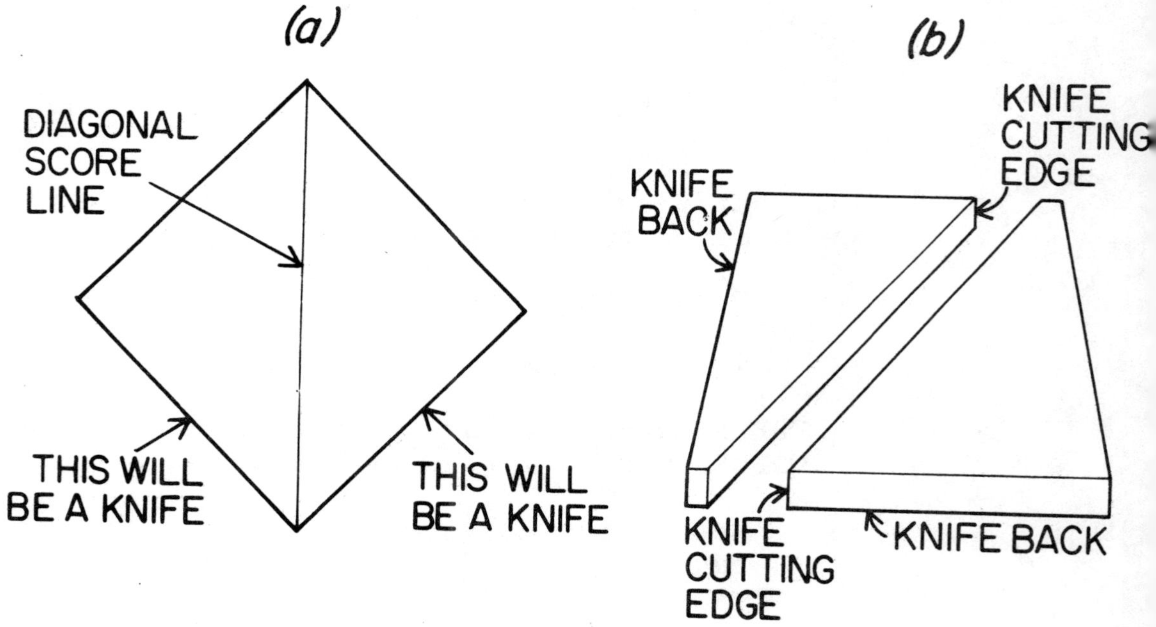

FIGURE 2-h. (a) Position of diagonal score line.
(b) End-product after break; two usable knives. NOTE KNIFE CUTTING ED

<u>REMOVE KNIVES CAREFULLY WITH FORK.</u>

Handling of Glass Knives

13. Grasp both halves of glass knives with the thumb and middle finger of the right hand in the area marked H (See Figure 2-g).

14. With the left hand remove area a handling it at corner (1) (See Figure 2-g).

    NOTE: USUALLY THE LEFT KNIFE IS THE BETTER ONE.

    KNIVES SHOULD BE HANDLED WITH CARE IN ORDER NOT TO DAMAGE THE CUTTING EDGE (See Figure 2-h).

    Several other knife angles can be obtained using a knifemaker machine (See Figure 2-i).

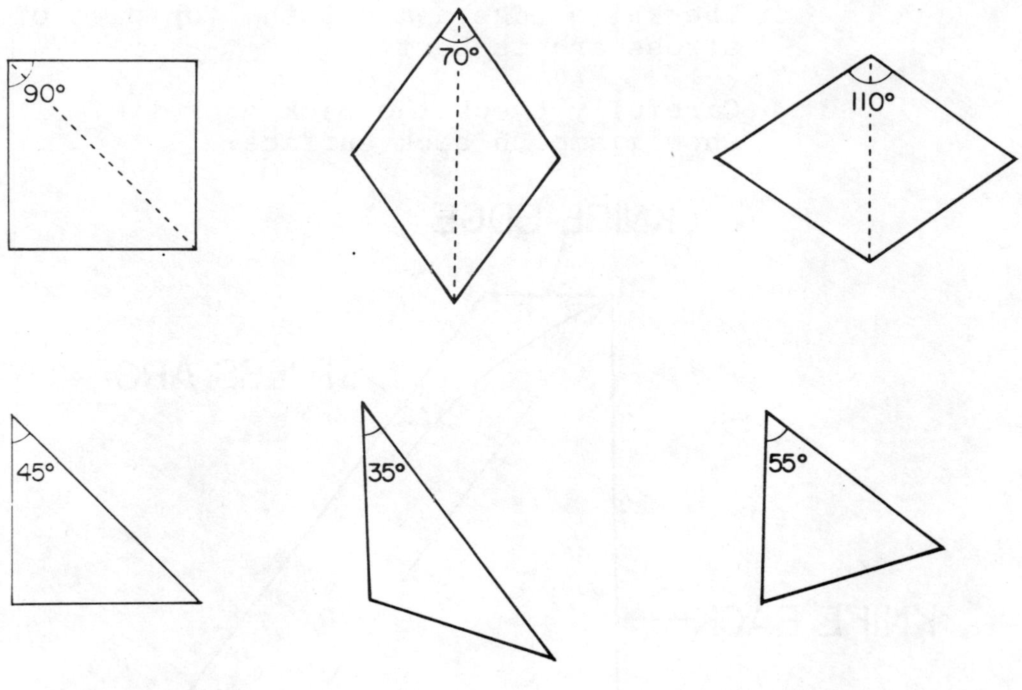

FIGURE 2-i. Various knife angles that can be obtained with a knifemaker machine (courtesy of R.L.Griffin, <u>Ultramicrotomy</u>, 1972).

RESULTS

<u>Examination of Glass Knife Edge</u>

1. Using a stereoscopic dissecting microscope examine the upper edge of knife corner.

   Focus on the cutting edge and direct the light source onto the edge until it appears to form a bright line.

   The angle of the knife will vary if the beam of light is not parallel to the knife edge.

   <u>A GOOD KNIFE SHOULD BE SHARP, THIN, STRAIGHT, AND WITHOUT NICKS OR IRREGULARITIES (See Figure 2-j).</u>

2. Examine the back face of knife for any defects.

   Use stress arc to determine the best side of knife edge for thin sectioning.

   The knife edge nearest the top part of the stress arc is best.

3. Carefully check the back of knife. It should show a smooth back surface.

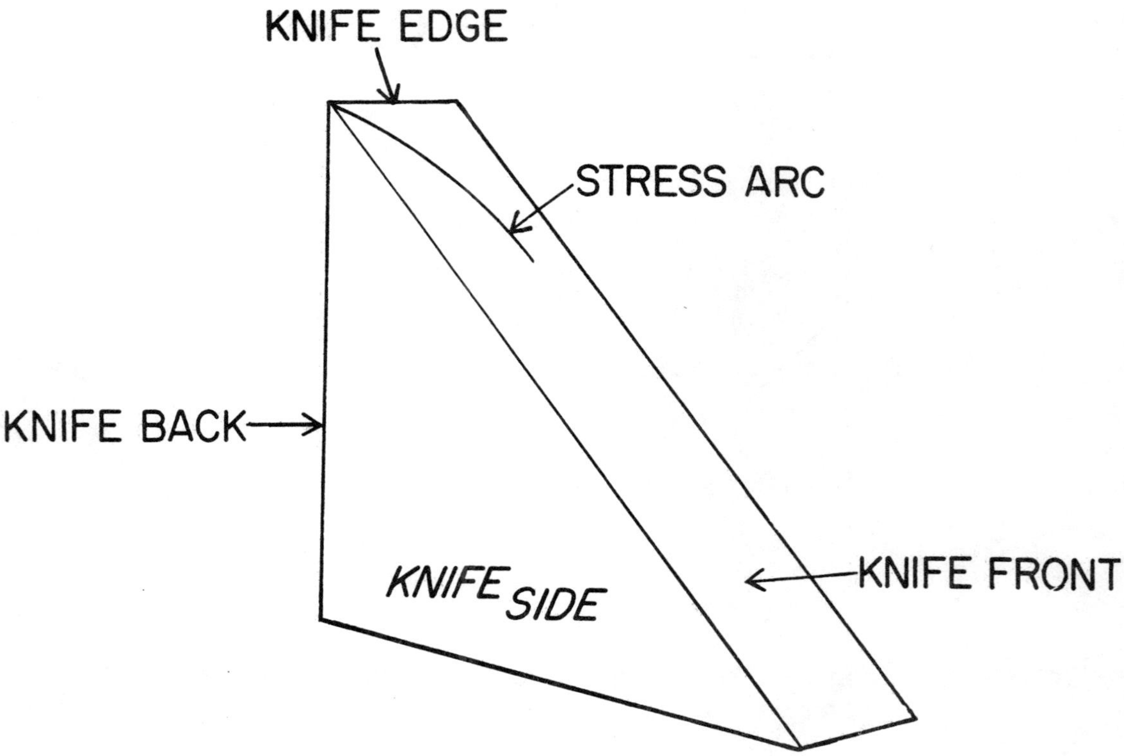

FIGURE 2-j. Parts of a glass knife.

If there are whirl marks on the back of the knife the glass square has been rotated clockwise instead of counterclockwise (See Figure 2-k).

This is a poor knife. Repeat the procedure following directions carefully.

4. A poor knife will not show a direct straight line of bright light on the edge when viewed under the dissecting microscope.

   Whiskers on back face of knife indicates glass chips and a poor cutting edge (See Figure 2-l).

   Poor knives may be saved for rough trimming of specimen blocks.

5. The length of a knife satisfactory for sectioning is dependent upon the shape of the cutting edge of the glass (See Figure 2-m).

   Various knife defects can be observed in examining poor knives (See Figure 2-n).

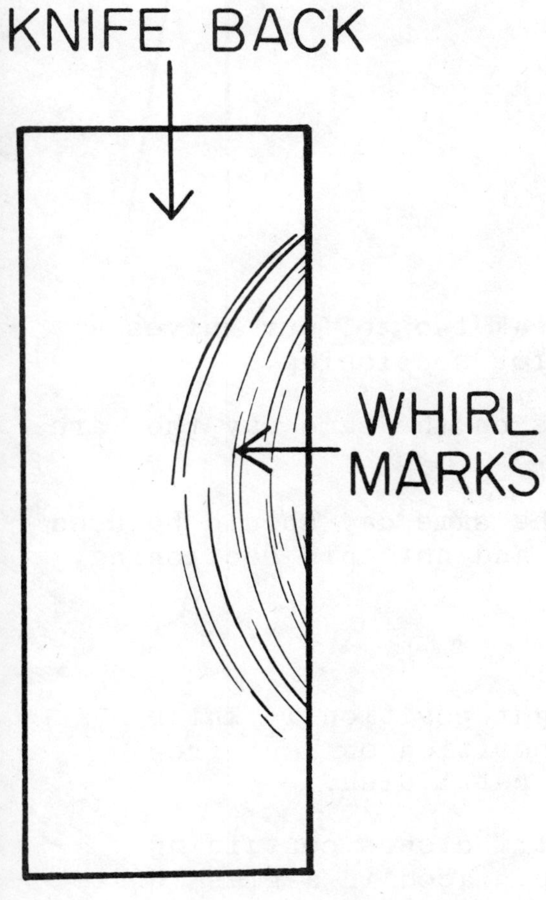

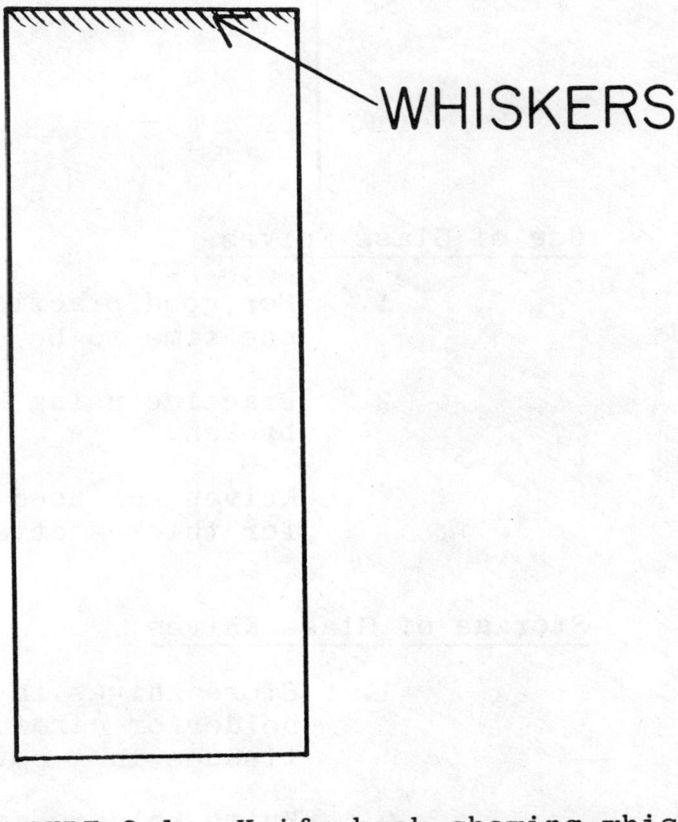

FIGURE 2-l. Knife back showing whiskers.

FIGURE 2-k. Knife back showing whirl marks.

FIGURE 2-m. Knife edge showing areas that can be used for sectioning.

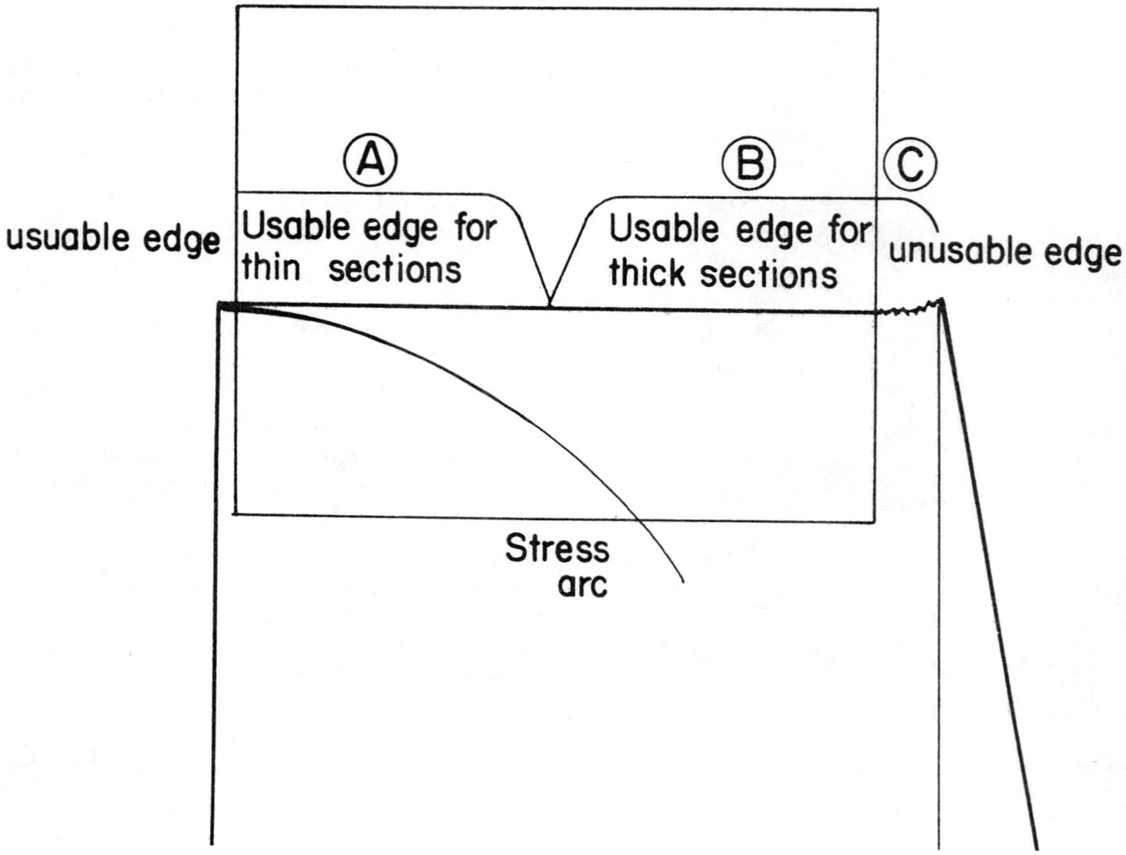

Use of Glass Knives

1. For good practice break two to four knives at one time to be used for sectioning.

2. Practice using knives on the same day they are broken.

   Knives not used on the same day should be used for thick sectioning and not thin sectioning.

Storage of Glass Knives

1. Store knives in upright position in knife holder or carefully position on lent-free tissues in a covered petri dish.

2. Knife holders and petri dishes containing knives may further be placed in a glass dust-free storage cabinet.

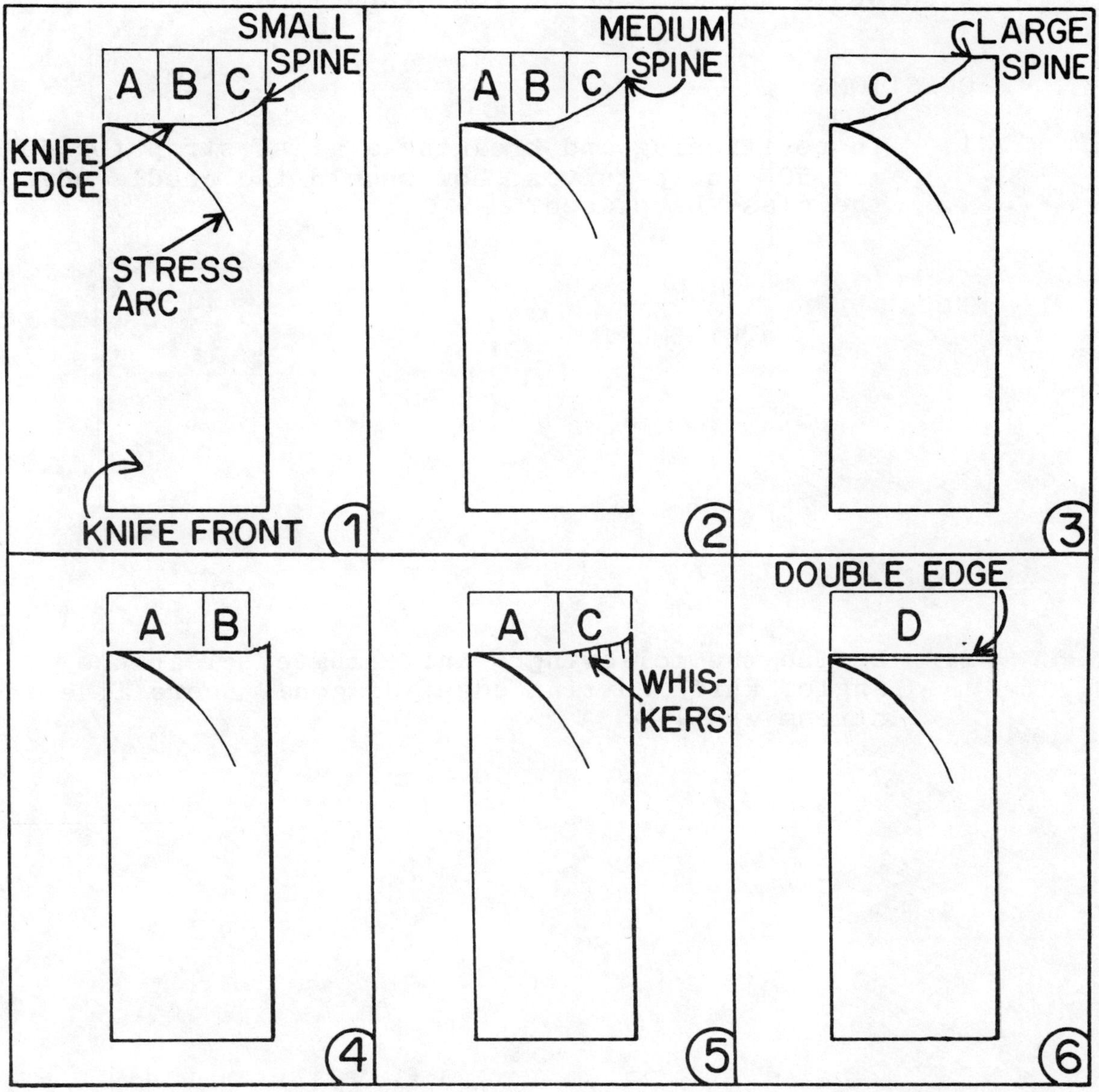

FIGURE 2-n. Various knife defects and usable knife parts for sectioning.

(A) Portion of knife edge for thin sectioning.

(B) Portion of knife edge for thick sectioning.

(C) Unusable spine region that cannot be used for sectioning.

(D) Double knife edge that cannot be used for sectioning.

E. EVALUATION OF GLASS KNIVES:

PRESENT FOUR KNIVES PREPARED WITH THE KNIFE MAKER MACHINE TO THE INSTRUCTOR FOR EVALUATION.

F. QUESTIONS:

1. In positioning and breaking a glass strip for 45° and 50° angle knives, how should the needle edge of the glass be placed?

2. Define the following: Knife angle, clearance angle, knife cutting edge, diagonal score line and stress arc.

3. Briefly discuss the appearance of a good glass knife and defects of a poor knife.

4. Who first introduced glass knives and why should they be used within a short time after being made?

5. What are the advantages of a knifemaker machine?

6. What dimensions of a strip of glass are commonly used for making glass knives?

7. How should glass strips be cleaned and handled?

8. What are the major steps in making glass knives with a knifemaker machine?

9. List what you think are the five most important parts of a knifemaker machine.

10. What part of the knife edge is usable for thin sections and what part is usable for thick sections?

EXERCISE NUMBER 3

PREPARATION OF GLASS KNIVES BY HAND

Betty R. Jones, Ph.D.

OBJECTIVES:

1. To prepare glass knives without the use of a knifemaker machine

2. To break a sheet of glass into adequate sized strips from which knives will be produced

3. To produce knives with a sharp cutting edge free from defects for thick and thin sectioning of tissues

MATERIALS:

1. 1 sheet of plate glass (0.25 inches thick, 4 inches long, 4 inches wide)

2. Pair of round nosed glazier pliers

3. Wheel-type glass scorer

4. Scotch double stick tape

5. Ruler or metal straight edge

6. 1 sheet of white cardboard

7. Lent-free paper EM towels (at least 2 square feet)

RATIONALE AND DISCUSSION:

If a knifemaker machine is not available, knives can be made by hand. This takes patience and practice and most likely the degree of failure will be high at first.

However this method of making glass knives requires inexpensive supplies and is important to be familiar with in instances where a knifemaker machine is not available.

A method learned from scratch sometimes results in better appreciation of more sophisticated methods. Several methods have been devised for breaking glass knives. The best practice is achieved by selecting the method of your choice and sticking with it. Only in this way can one gain experience, thereby increasing the percentage of good knives.

A key feature in making knives from plate glass is the selection and quality of the glass. Selection of the glass is important because different types of glass may yield a different quality of knives.

Unpolished sheets of glass are probably far superior to those made smooth by a polishing process. Polishing may cause stress in the glass.

Thickness is also essential in selection of plate glass. Plate glass or glass strips with a thickness of 0.25 inches should be used. Glass of this thickness is fairly easy to break.

Special pliers and glass scorers may be purchased from EM supply companies (See Figures 3-a and 3-b). Glazier pliers can also be modified with electrical strips of tape. The tape is applied to the pressure point ends.

FIGURE 3-a. Pair of modified glazier pliers.

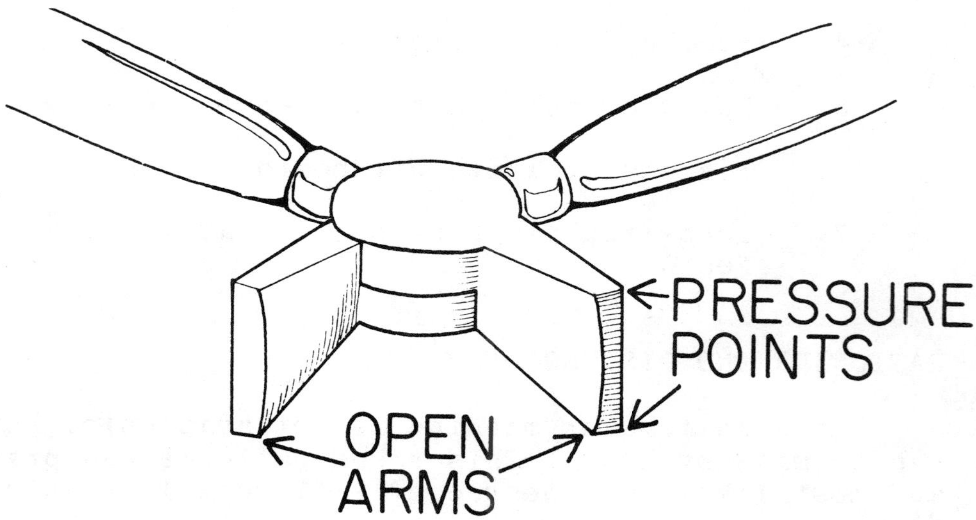

FIGURE 3-b. Wheel-type glass scorer.

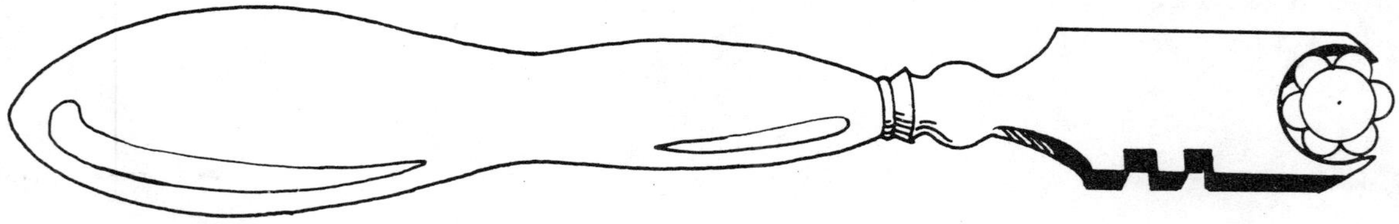

PROCEDURE:

APPLICATION OF THE FREE BREAK METHOD

1. Clean glass according to procedures described in Exercise Number 2 of manual.

2. Place clean white cardboard on a flat and firm table surface to serve as cutting guide and table cover.

3. In order to score the glass at the desired knife angle design and draw a chart illustrating angles that may be used for making glass knives (See Figure 3-c).

NOTE: MOST KNIFE ANGLES MUST BE 45° OR LESS.

THIS DEPENDS ON THE TYPE OF SPECIMEN BEING CUT AND TYPE OF EMBEDDING RESIN USED.

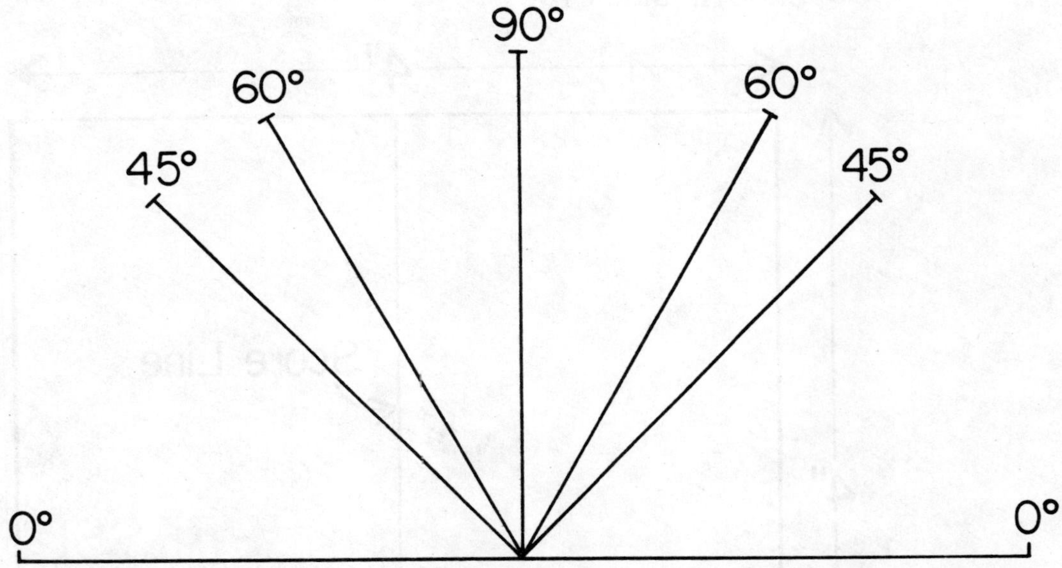

FIGURE 3-c. Illustration of knife angles.

4. Tape securely one lent-free paper EM towel to table top surface.

5. Place one clean sheet of plate glass on top of EM paper towel.

   Score down the center of glass sheet with wheel-type scorer.

6. Apply special ready made glazier pliers about 3/8" from edge of glass sheet.

   Line up index line of the arm of pliers with score line on glass.

   Break the glass into two halves by applying equal pressure on the edge of the score line (See Figure 3-d).

   One half of glass may be stored for later use.

NOTE: DO NOT SCORE GLASS MORE THAN ONCE TO INSURE SMOOTH BREAKING.

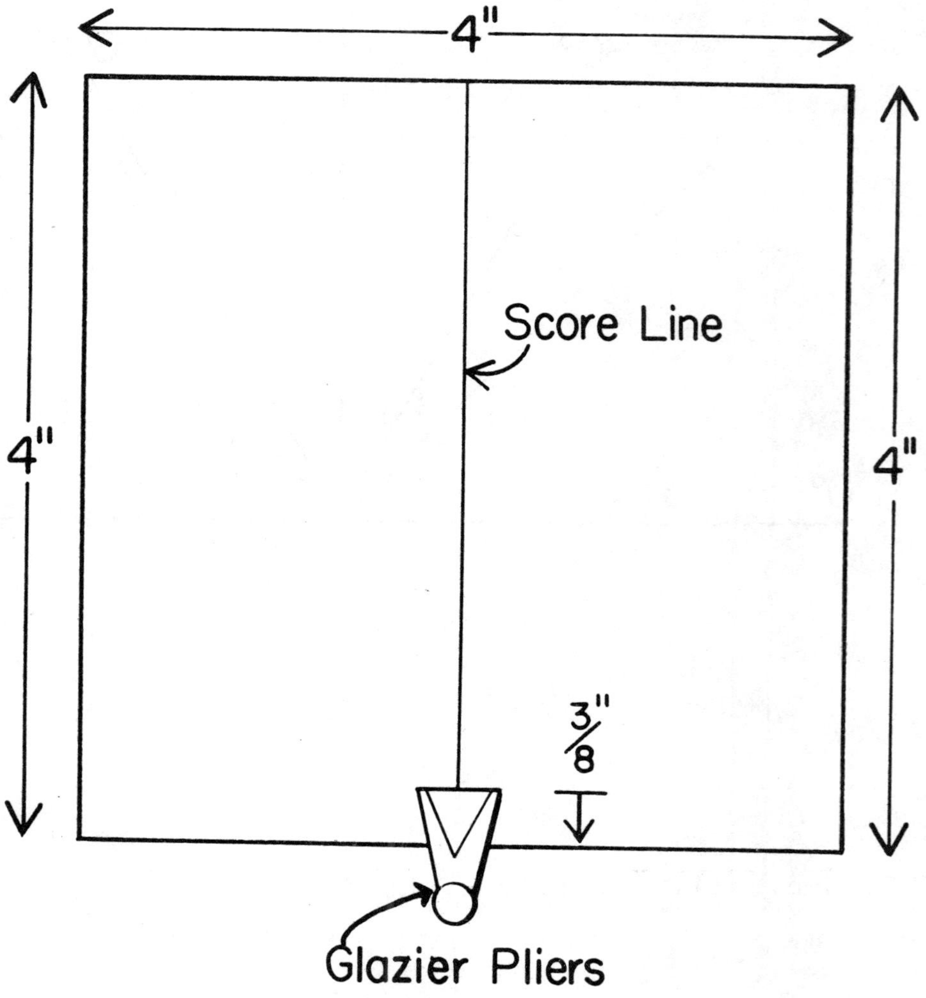

FIGURE 3-d. 1st scoring and breaking of 4X4 inch glass sheet.

7. Using one of the 2" glass sheets make a partial score at the center.

   Start 3/8's of an inch from the lower edge and continue.

   Apply pressure using pliers and execute a straight free break along the midline center.

   You now have two 1" glass strips (See Figure 3-e).

8. Take one of the 2X1" glass sheets and place in vice position.

   Make a partial score with wheel-scorer and break with pliers.

   You now have two 1" squares.

   Repeat the same procedure for the other 2X1" glass sheet.

   A total of four 1" glass squares will be obtained (See Figures 3-f-g).

9. To obtain proper 45° angle measure glass square on white cardboard (See Figure 3-h).

10. Using scoring wheel make a score opposite the free break corner.

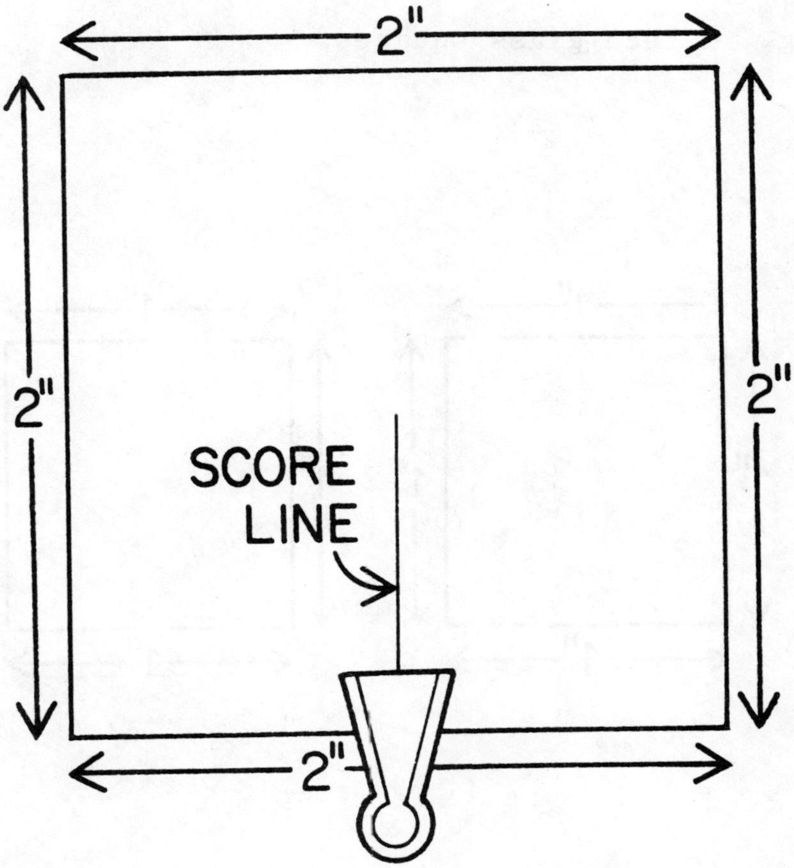

FIGURE 3-e. 2nd scoring and breaking of 2 inch glass sheet.

FIGURE 3-f. 3rd scoring and breaking of 2X1 inch glass sheet.

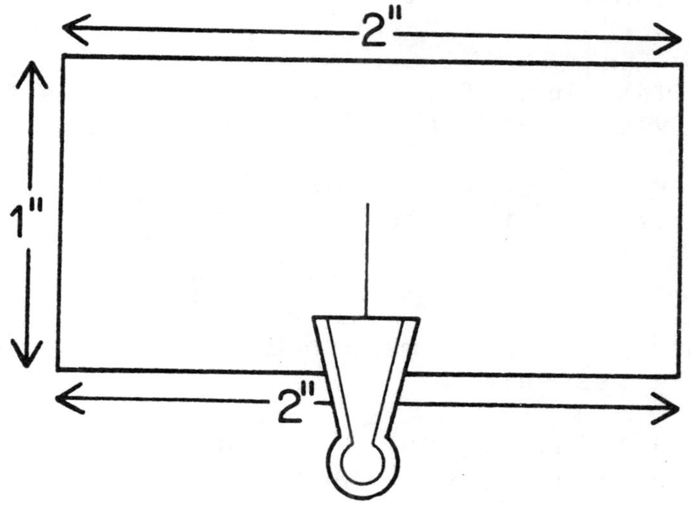

FIGURE 3-g. 1 inch glass squares.

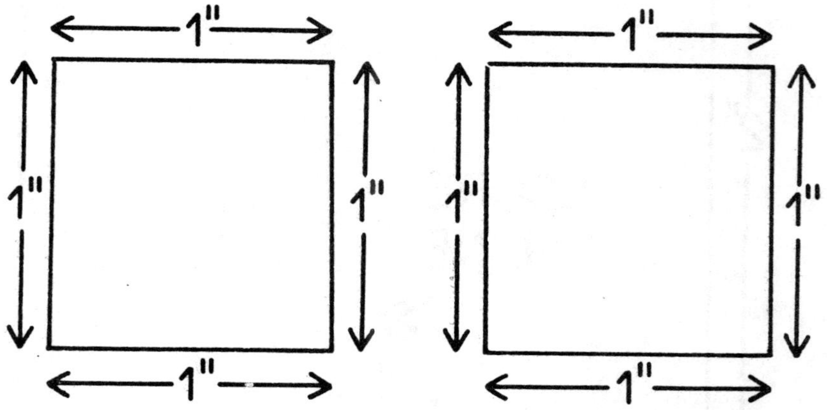

Set the square so that the 45° line passes just in front of the corner (See Figure 3-h).

11. Make score starting at the edge of the free break corner and continuing at 45° to the lower left edge of square (See Figure 3-i).

12. Place pliers at the opposite end of free break area.

    Apply pressure slowly and steady so that the break travels along the score line and breaks freely between the end of the score mark and the edge of the glass.

NOTE: THIS FREE BREAK WILL FORM THE KNIFE EDGE.

    This will result in one or two good knives (See Figure 3-j).

13. You now have two diagonal knife halves (See Figure 3-k).

    Examine knives under stereoscopic dissecting microscope.

NOTE: USUALLY ONLY ONE KNIFE IS GOOD.

14. Repeat the described process until four good knives are obtained.

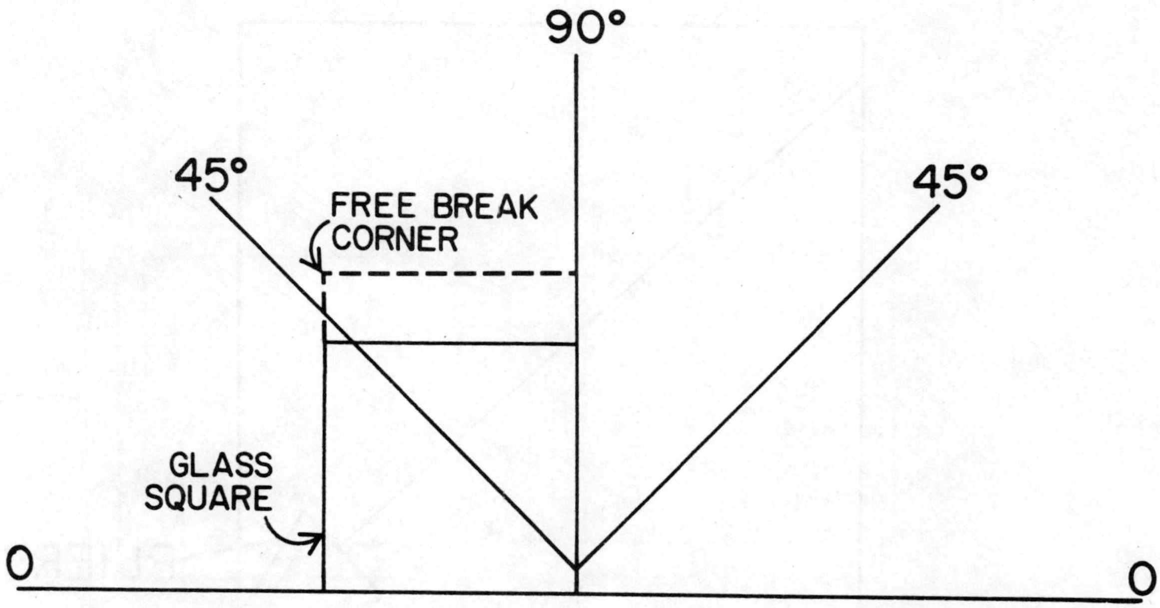

FIGURE 3-h. Positioning of glass square on angled cardboard.

FIGURE 3-i. Diagonal score line.

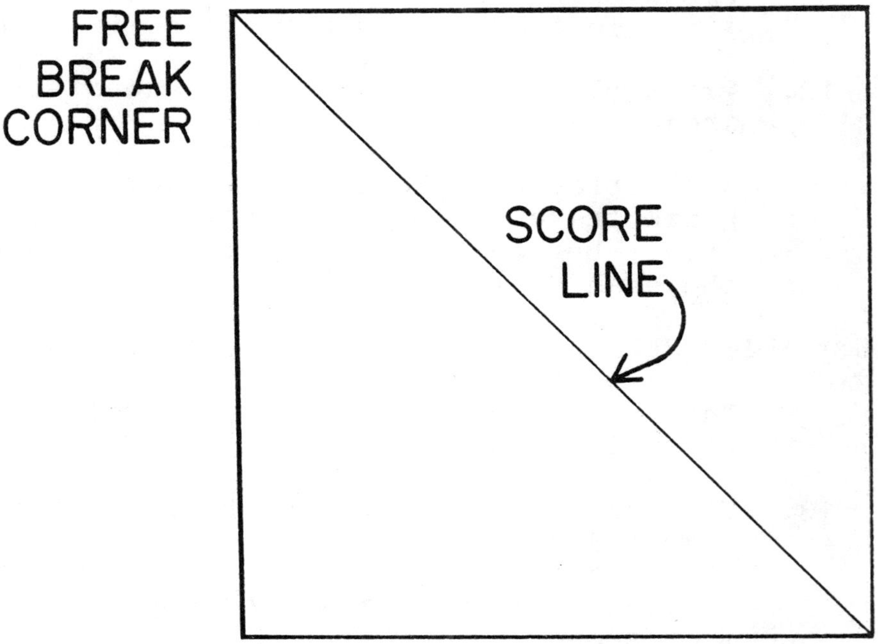

FIGURE 3-j. Breaking of diagonal square to obtain glass knives.

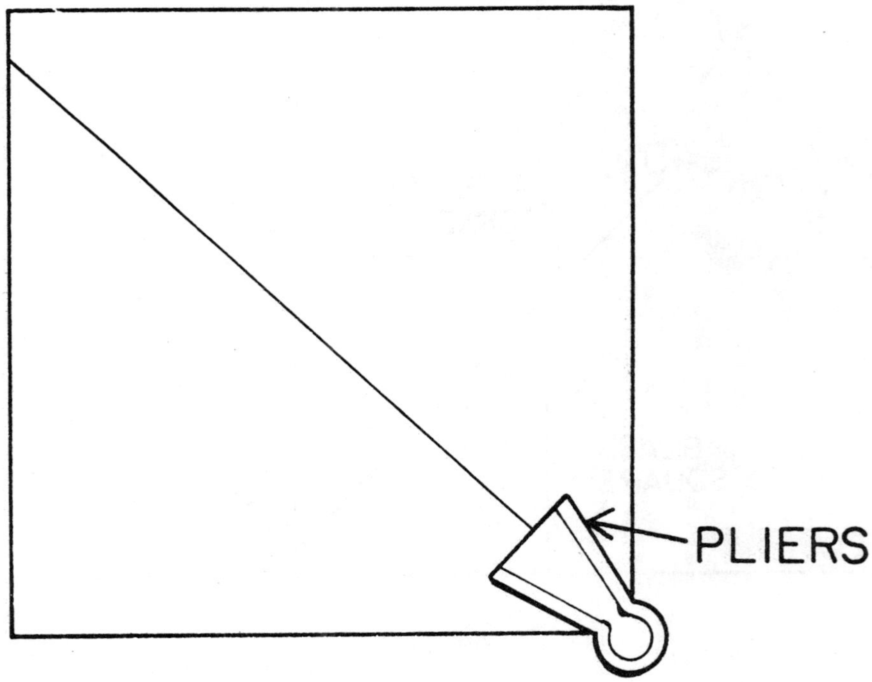

EVALUATION:

PRESENT TWO KNIVES PREPARED BY HAND TO THE INSTRUCTOR FOR EVALUATION.

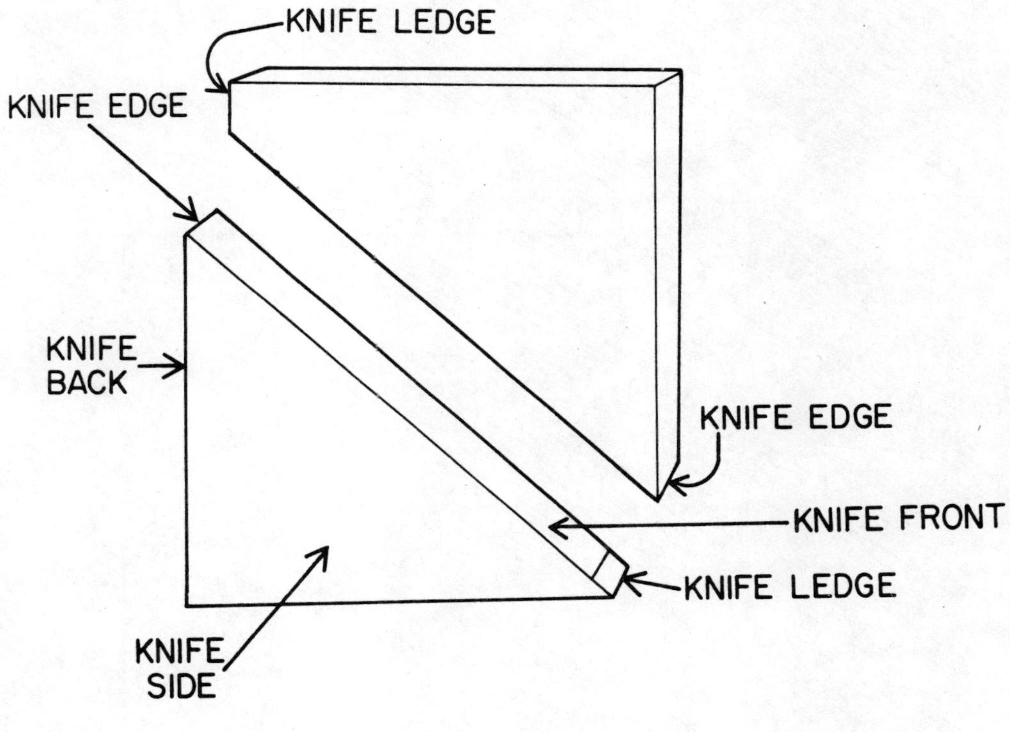

FIGURE 3-k. Final product; two glass knives.

QUESTIONS:

1. What are two important features one should keep in mind in selection of plate glass? Why?

2. What are the advantages and disadvantages of preparing glass knives by hand?

EXERCISE NUMBER 4A

DRESSING OF GLASS KNIVES

PREPARATION OF GLASS KNIFE RESERVOIRS WITH PRESHAPED METAL

Betty R. Jones, Ph.D.

OBJECTIVES:

1. To make a reservoir or trough using preshaped metal for dressing glass knives
2. To provide a greater surface area for floatation fluid and sections

MATERIALS:

1. Preshaped metal reservoirs or troughs
2. 1 sheet of dental wax
3. Alcohol torch
4. Glass knives

RATIONALE AND DISCUSSION:

The use of metal reservoirs for the dressing of glass knives provides a much faster and easier procedure. It allows a greater surface area for sections thereby preventing sections from crashing or sticking to the walls of the reservoir. The tape method is also satisfactory. Students should experience using methods and then decide on the one that works best for them.

NOTE: PRESHAPED METAL RESERVOIRS MAY BE PURCHASED FROM LKB COMPANIES IN DIFFERENT SIZES.

THE SIZE OF THE RESERVOIR WILL DEPEND ON THE SIZE OF THE GLASS. DISPOSABLE RESERVOIRS ARE ALSO AVAILABLE.

PROCEDURE:

1. Carefully place the metal reservoir around the knife edge. Make sure metal does not touch the cutting edge of knife.

   Metal reservoirs are malleable and can be made to fit the knife properly (See Figure 4-a).

2. Melt dental wax in covered glass petri dish on hot plate at a low temperature.

3. Heat end of metal spatula in flame from alcohol torch.

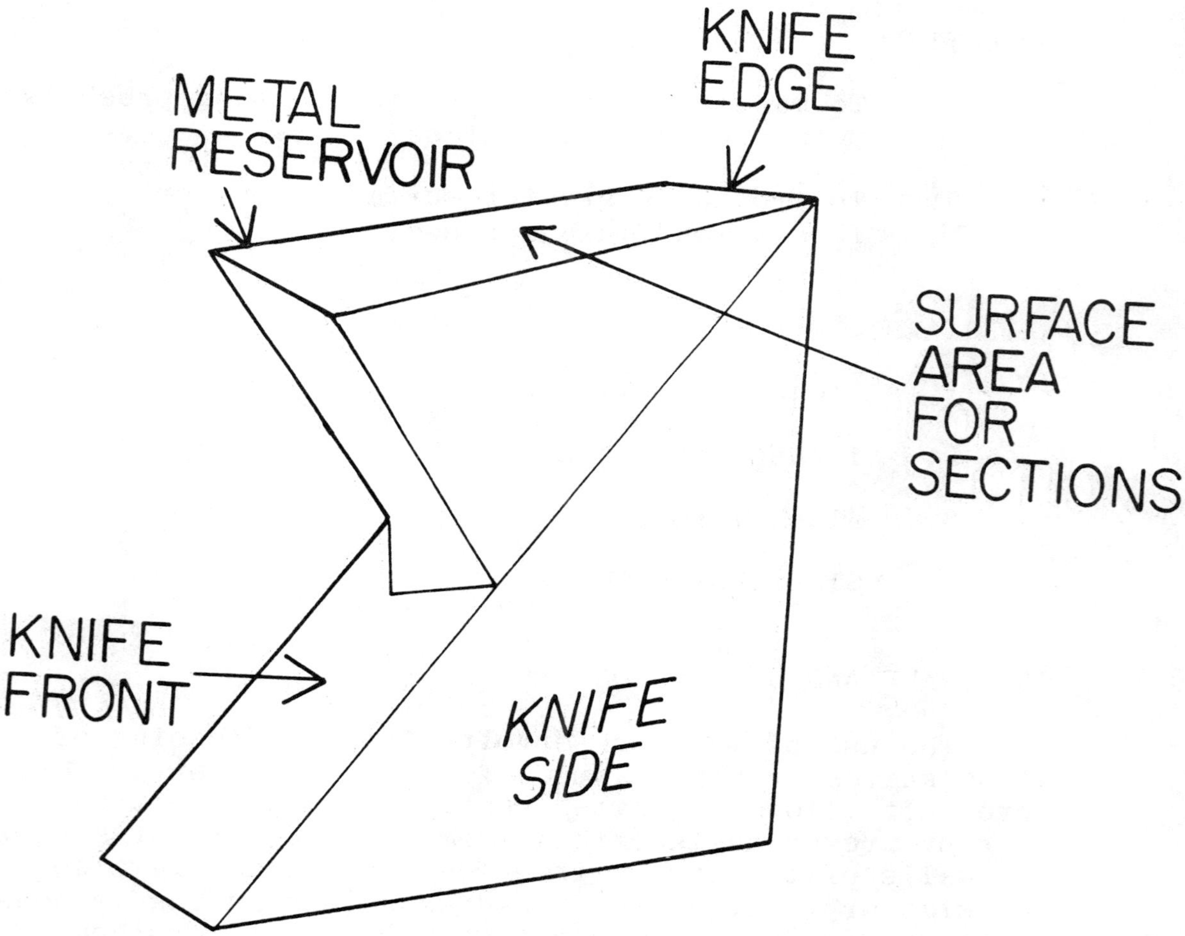

FIGURE 4-a. Positioning of metal reservoir around knife.

4. Place spatula in melted dental wax and transfer small amount around the metal reservoir.

5. Begin placing wax against the metal reservoir at the front of the knife then the sides (See Figure 4-b).

6. Heat spatula each time wax is applied to the reservoir.

NOTE: USE THE ALCOHOL TORCH WITH CARE.

EXAMINATION OF METAL RESERVOIR

7. Examine the dressed knife under a dissecting microscope.

EVALUATION OF METAL RESERVOIR:

Fill two knives half-full with distilled water.

PRESENT TO INSTRUCTOR FOR EVALUATION.

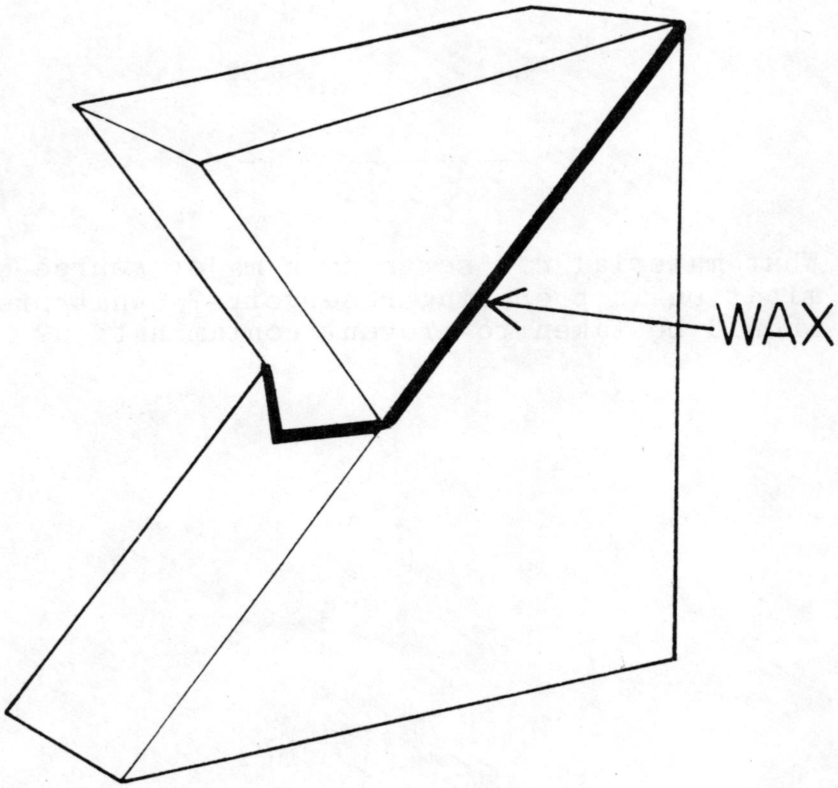

FIGURE 4-b. Glass knife completely dressed.

QUESTIONS:

1. What are the advantages of dressing glass knives with preshaped metal?

2. What role does a reservoir play in obtaining an optimal water meniscus level?

3. What material can serve as a major source of contamination in preparing reservoirs? What precautions should be taken to prevent contamination?

EXERCISE NUMBER 4B

DRESSING OF GLASS KNIVES

Preparation of Glass Knife Reservoirs With Electrical Tape

Betty R. Jones, Ph.D.

OBJECTIVES:

1. To provide a leak-free dressed knife
2. To provide a surface area for the floatation fluid
3. To obtain a flat meniscus level with the floatation fluid
4. To provide an area for section preservation and storage for a short period of time

MATERIALS:

1. Black electrical tape, number 33 or silver polyester film tape, number 850
2. Dental wax, beeswax, paraffin or nail lacquer

NOTE: DENTAL WAX IS PREFERRED OVER BEESWAX OR PARAFFIN.

IT IS CLEANER AND LESS BRITTLE.

NAIL LACQUER IS EASY TO APPLY BUT TAKES LONGER TO DRY.

HOWEVER, IF A MULTIPLATE APPARATUS IS AVAILABLE, SUPPLIES AND EQUIPMENT SUCH AS HOT PLATE, WAX, SPATULA AND SO FORTH ARE NOT NEEDED.

THE MULTIPLATE APPARATUS CAN BE PURCHASED FROM LKB SUPPLIERS.

3. Single edge razor blades
4. Hot plate
5. Small spatula, brush, or needle probe

FIGURE 4-c. Positioning of adhesive tape to corner 1 of knife edge.

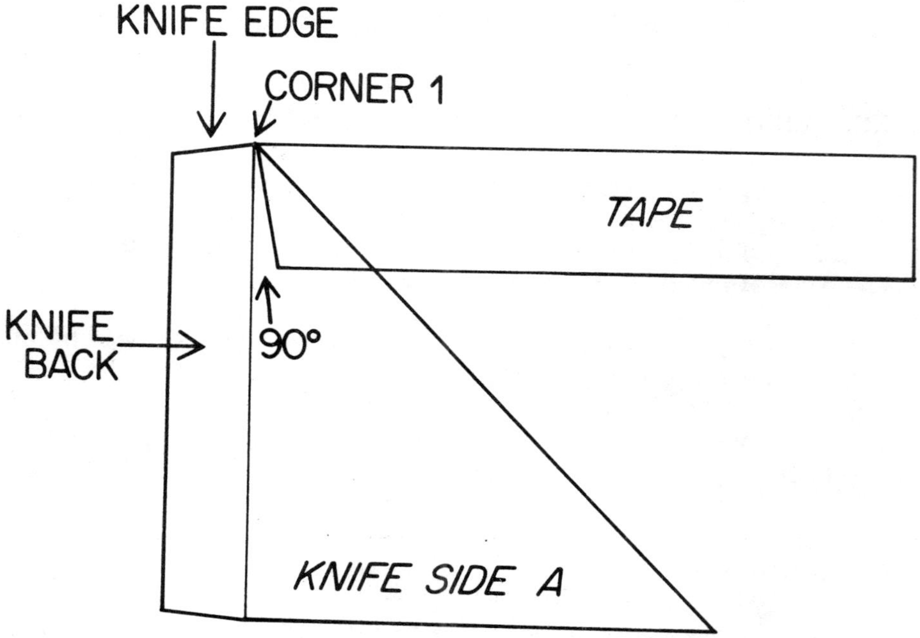

FIGURE 4-d. Positioning of adhesive tape to corner 2 of knife edge.

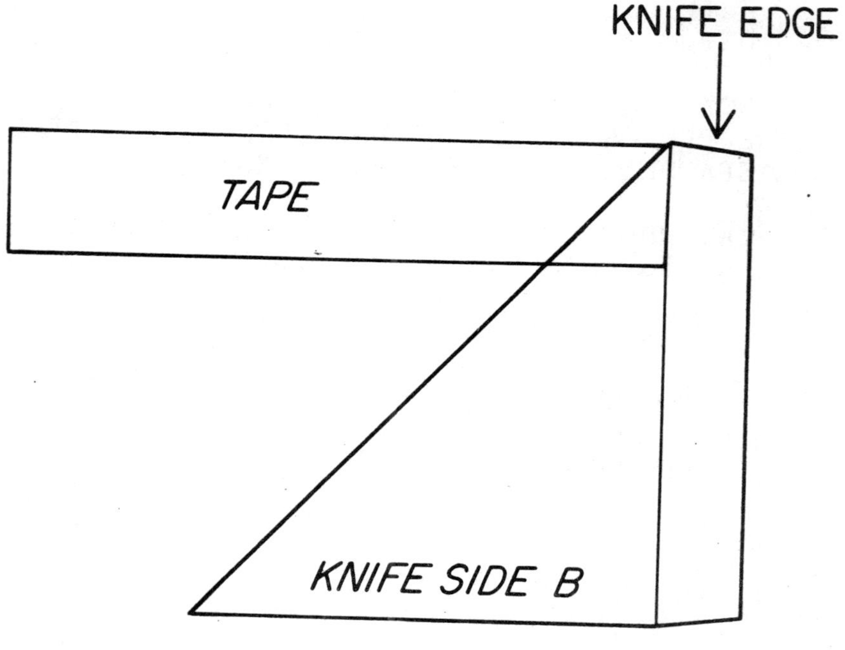

6. Glass knives

7. Glass petri dish

RATIONALE AND DISCUSSION:

The reservoir of a glass knife is important in several ways. It provides support for holding the floatation fluid. This allows sections to float on the fluid surface for collection. A knife is completely dressed when it is adequately sealed with a sealant which prevents leakage of the floatation fluid.

A good reservoir is necessary to maintain the correct fluid meniscus level. This allows sections to float away from the knife cutting edge and to show interference colors for judging section thickness.

NOTE: IT IS IMPORTANT TO USE THE CORRECT TYPE OF TAPE.

SOME TAPES MAY FORM A CONTAMINATION FILM ON THE SURFACE OF THE FLUID.

PROCEDURE:

1. Using a single edge razor blade cut a piece of tape (1 1/4" to 1 3/8") depending on size of glass knife.

2. Place one end of tape firmly onto one side of knife corner.

NOTE: BE CAREFUL NOT TO TOUCH KNIFE EDGE WITH FINGERS.

THE TOP EDGE OF TAPE SHOULD FORM AN ANGLE OF 90° WITH THE KNIFE EDGE OF CORNER 1 AND NOT EXTEND BEYOND IT (SEE FIGURE 4-c).

3. Slowly and carefully construct reservoir by extending tape to the opposite side of knife edge of corner 2 (See Figure 4-d).

   Make sure boundary line of tape is lined up with knife back edge.

   Cut off excess tape with razor blade (See Figure 4-e).

4. Melt dental wax in small glass dish on a hotplate or inside an oven.

FIGURE 4-e. Appearance of knife reservoir dressed with tape.

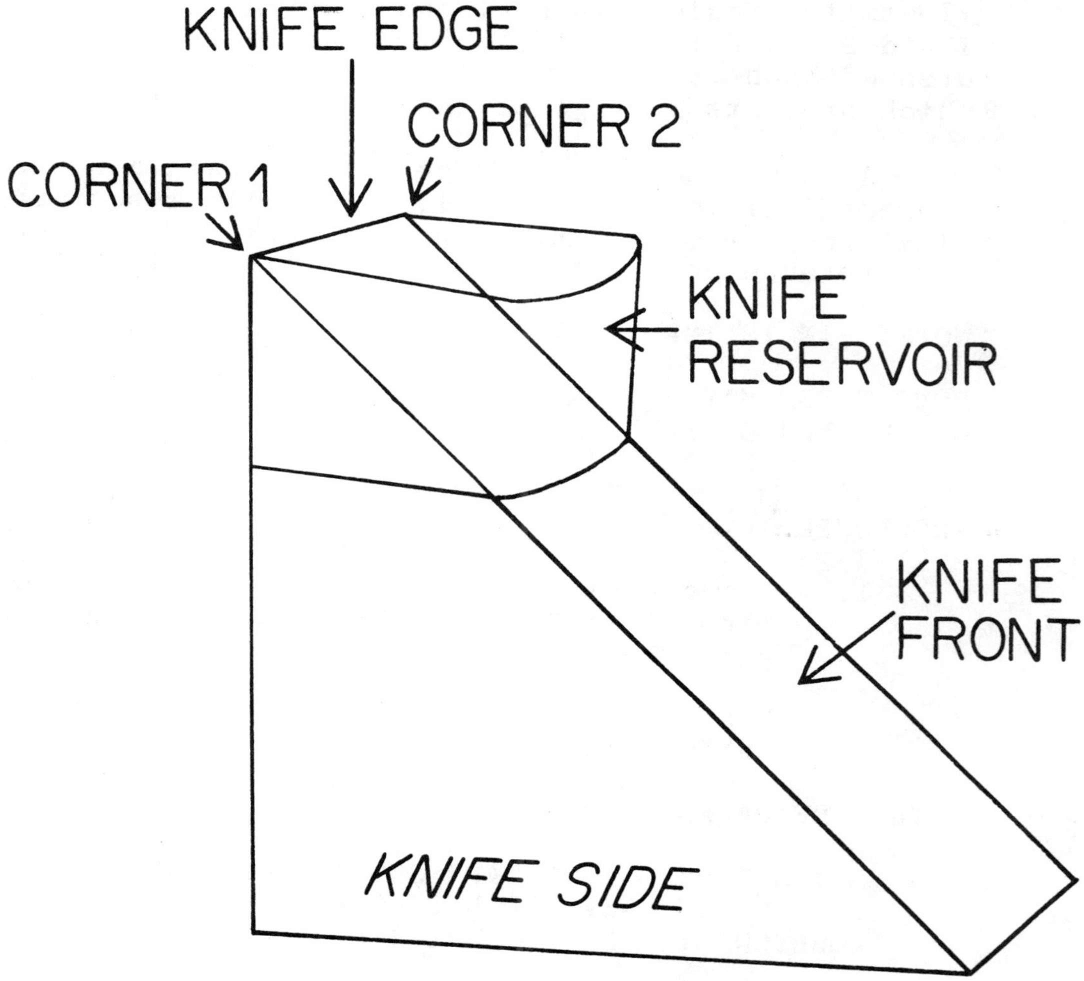

5. Using a spatula, small brush or needle probe, seal boundary lines, sides, and front of knife.

   Be careful not to touch knife edge with wax (See Figure 4-f).

NOTE: IF DENTAL WAX IS NOT AVAILABLE AND PARAFFIN OR BEESWAX ARE USED PRECAUTIONS SHOULD BE TAKEN.

FUMES FROM THE HEATING OF THESE WAXES ARE CONSIDERED CARCINOGENIC.

6. Examine dressed knife reservoir under a stereoscopic dissecting microscope.

FIGURE 4-f. Appearance of a dressed and sealed knife.

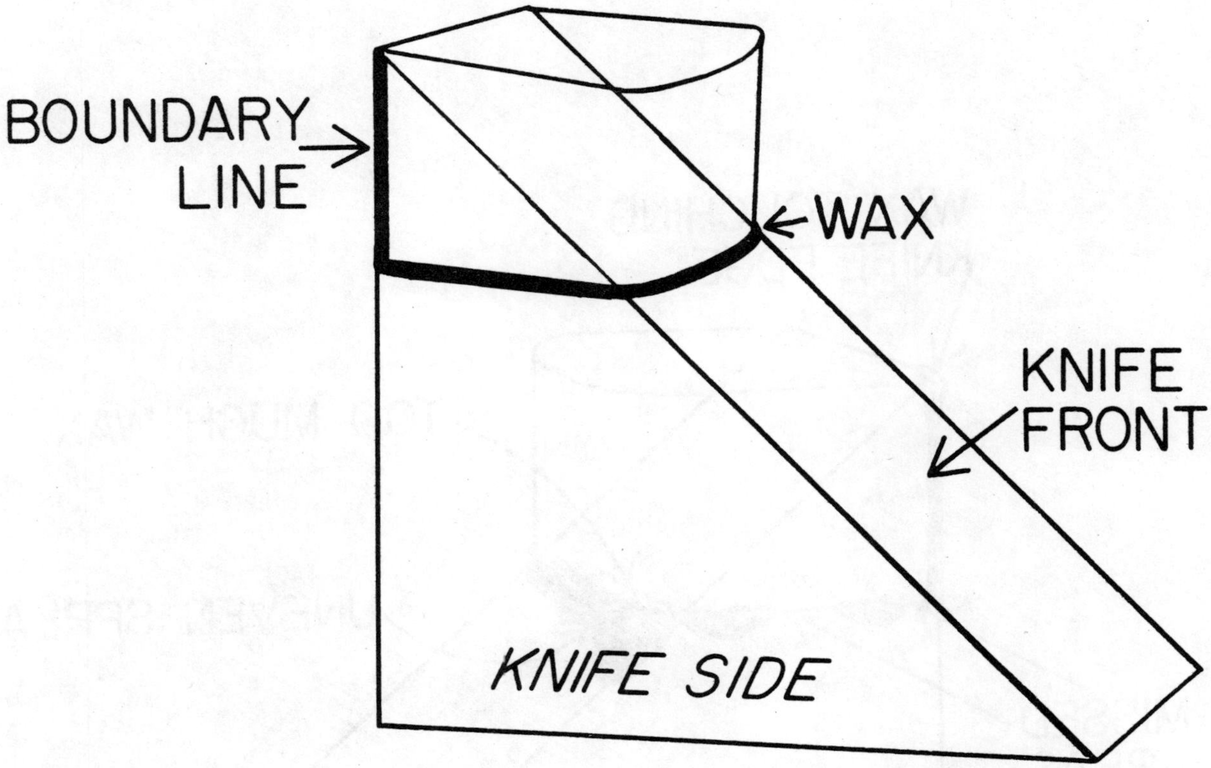

7. Check all parts of the knife that have been sealed with wax for any defects such as holes in wax, uneven spread or missed areas. (See Figure 4-g).

   If defects such as these are present repeat the sealing process on the problem areas.

NOTE: IF CARE IS TAKEN IN SEALING RESERVOIRS THESE PROBLEMS CAN BE PREVENTED.

EVALUATION OF KNIFE RESERVOIRS:

FILL TWO KNIVES HALF FULL WITH DISTILLED WATER.

PRESENT TO INSTRUCTOR FOR EVALUATION.

NOTE: IF RESERVOIRS LEAK SEAL AGAIN WITH A LIGHT COAT OF WAX.

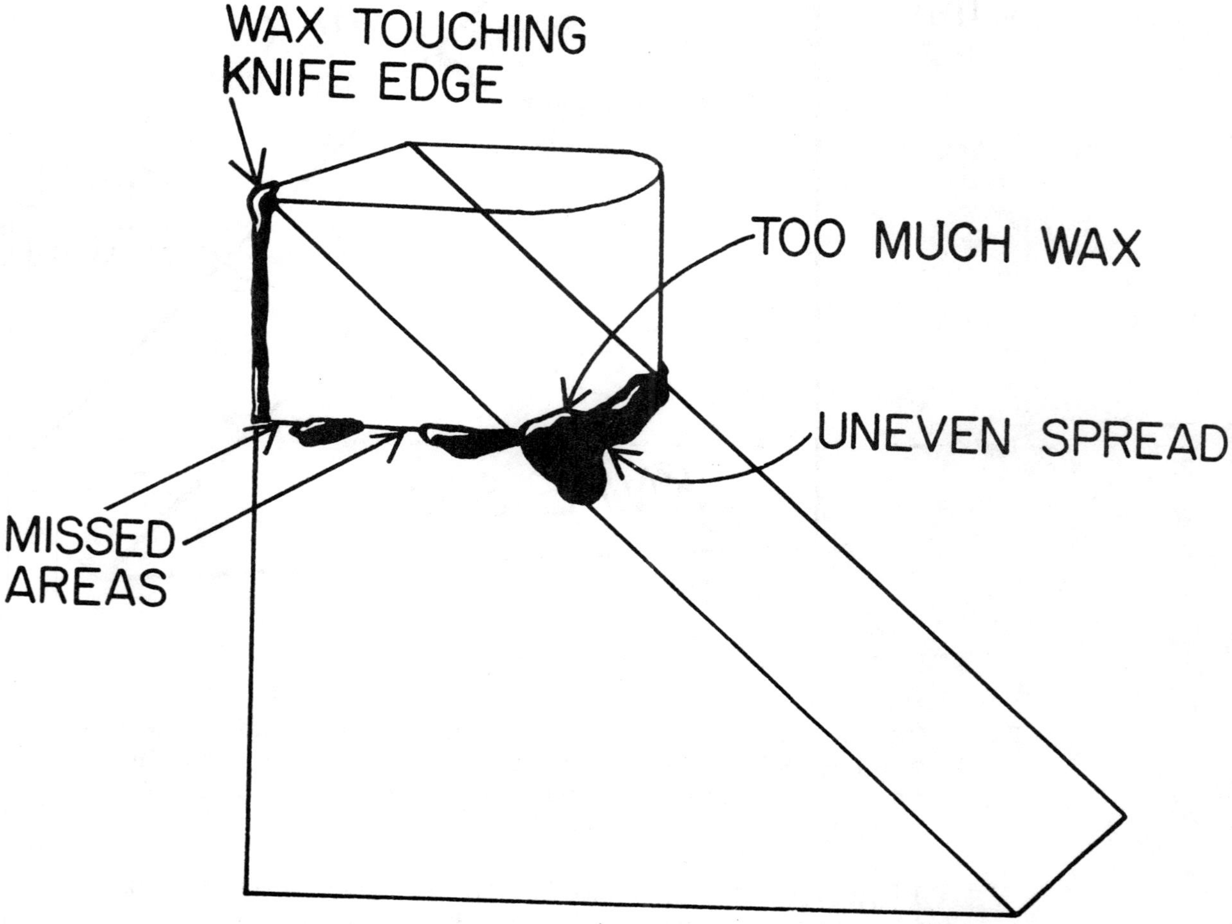

FIGURE 4-g. Some common problems in sealing a knife reservoir.

QUESTIONS:

1. Compare the advantages and disadvantages of preparing reservoirs with preshaped metal versus adhesive tape.

2. What persons are given credit for first developing reservoirs?

3. Why is black adhesive tape preferred over other colors for preparing reservoirs?

NOTES

EXERCISE NUMBER 5

USE OF DIAMOND KNIVES

Betty R. Jones, Ph.D.

OBJECTIVES:

1. To be used for cutting ultrathin sections
2. To cut tissues that are too hard to be cut with glass knives
3. To be used when a long series of sections are required
4. To be used by an experienced microscopist

MATERIALS:

1. Diamond knife (size may vary)
2. Squirt bottle containing distilled water
3. Sodium hydroxide
4. Soft wood stick
5. Acetone or alcohol
6. Glass covered dish

NOTE: ACETONE, ALCOHOL AND SODIUM HYDROXIDE SHOULD ONLY BE USED FOR KNIFE EDGES THAT ARE EXTREMELY DIRTY.

RATIONALE AND DISCUSSION:

Diamond knives should not be used by beginners in electron microscopy. Beginners should first learn how to prepare and use glass knives. Diamond knives are expensive and without previous experience in sectioning one can easily damage the cutting edge.

Users of diamond knives should take great care in mounting and removing it from the microtome to prevent chipping or nicking the edge.

Diamond knives can be used repeatedly without affecting the quality of sections. Knives may be obtained from EM suppliers and other manufacturers (See Appendix for List of Suppliers).

PROCEDURE:

Examination of Diamond Knife

1. As soon as a diamond knife is received from a company, check the knife edge under the dissecting microscope for defects.

   An illustrative diagram of a diamond knife is shown in Figure 5-a.

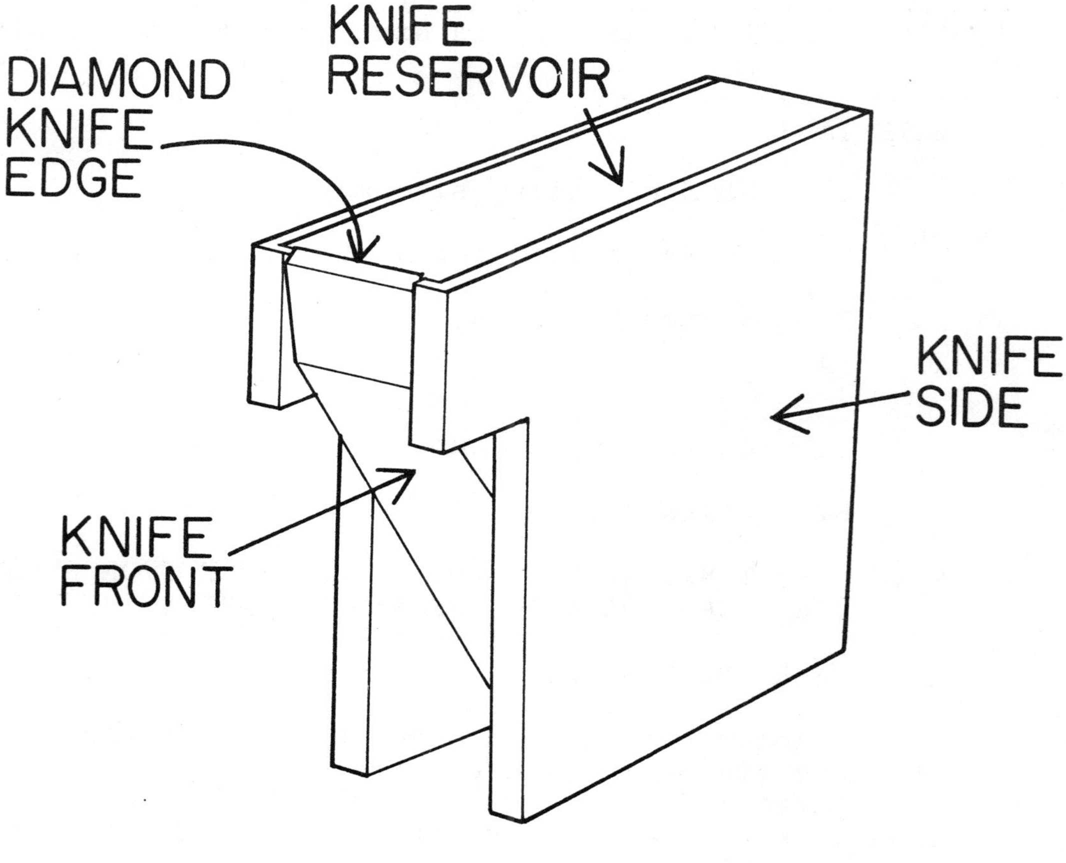

FIGURE 5-a. Diamond knife.

2. Cut a few sections and examine them under the electron microscope.

   This is important in determining whether the knife should be returned to the manufacturer for sharpening or polishing.

Mounting of Diamond Knife

3. The diamond edge of the knife is very fragile.

NOTE: EXTREME CARE SHOULD BE USED IN MOUNTING THE KNIFE ON THE MICROTOME.

   Carefully place knife into microtome holder and set knife angle to the desired sectioning angle. Diamond knife angles vary between 40° and 60°.

   Generally, smaller angles are used for soft biological materials and larger angles used for hard materials.

4. Obtain the correct clearance angle by test sectioning or refer to the manufacturer's instructions.

   To determine clearance angle by test sectioning, start with a low angle of $0.1\text{-}0.9^{o}$ and then increase angle until best results are obtained.

NOTE: DIAMOND KNIVES SHOULD NEVER BE USED FOR TRIMMING SPECIMEN BLOCKS EXCEPT IF THEY HAVE BEEN DISCARDED FROM NORMAL USE.

AVOID CUTTING SECTIONS THICKER THAN 1 µm TO PREVENT DAMAGING KNIFE EDGE.

Cleaning of Diamond Knives

5. Submerge diamond knife in a covered glass container filled with distilled water for 2-4 hours or overnight if prolonged sectioning has occurred.

6. Remove knife from container and place under a dissecting microscope.

7. Overfill knife reservoir with distilled water.

8. Bring knife edge and reservoir into focus.

9. Gently move wet soft wooden stick over knife edge.

   Practice moving the stick in the same direction each time the knife is cleaned preferably from left to right (See Figure 5-b).

10. Move stick gently over knife edge one or two times.

11. Using a squirt bottle containing distilled water flood knife edge several times.

12. Place knife back in original container for drying.

    If time is a factor knife may be dried under a warm air dryer.

13. For excessive dirty knives place concentrated sodium hydroxide on wooden stick and clean knife edge as previously described.

    This should be followed by rinsing knife edge several times with distilled water.

NOTE: SODIUM HYDROXIDE SHOULD BE USED IF ACETONE OR ALCOHOL FAILS TO CLEAN THE KNIFE EDGE.

ACIDS SHOULD NEVER BE USED FOR CLEANING DIAMOND KNIVES.

14. If after repeating this process several times and the knife is still dirty, send it back to the manufacturer for resharpening.

NOTE: PRACTICE CLEANING DIAMOND KNIFE AFTER IT IS USED.

DIAMOND KNIFE CLEANING KITS ARE ALSO AVAILABLE FROM EM SUPPLIERS.

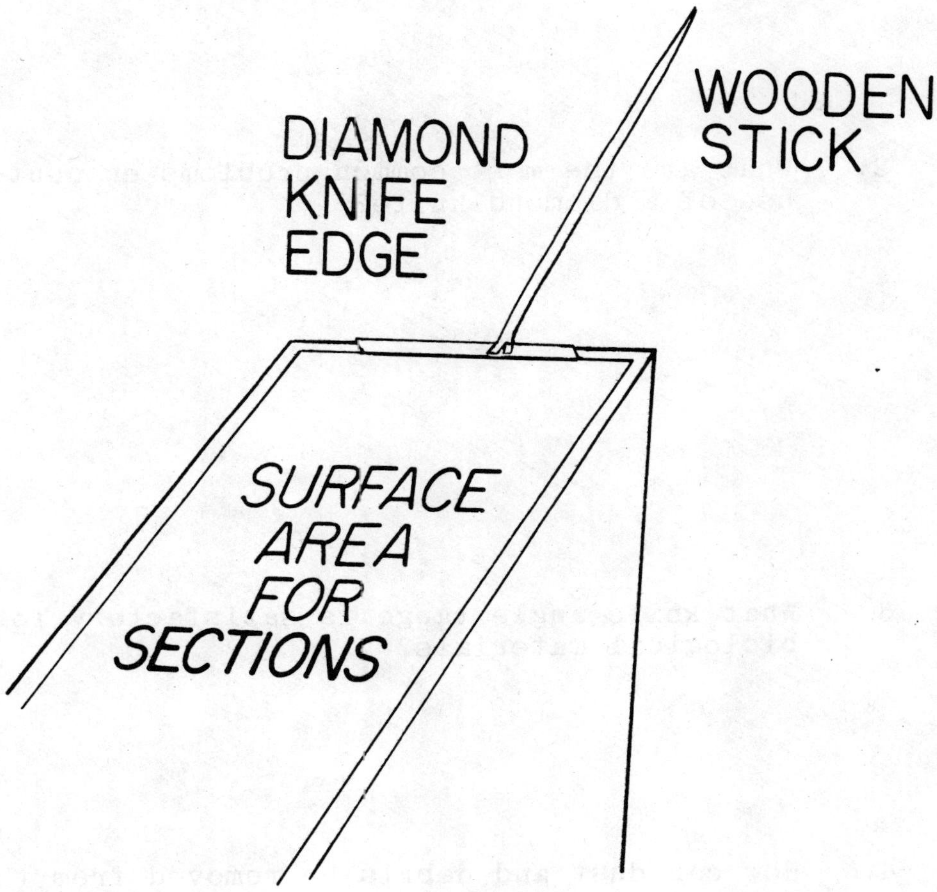

FIGURE 5-b. Cleaning a diamond knife.

E. QUESTIONS:

1. Briefly discuss the advantages and disadvantages of diamond knives to that of glass knives.

2. What are the most common problems encountered with use of a diamond knife?

3. What knife angle range is satisfactory for cutting biological materials?

4. How can dust and debris be removed from the cutting edge of a diamond knife?

EXERCISE NUMBER 6

CLEANING COPPER GRIDS

Betty R. Jones, Ph.D.

A. OBJECTIVES:

1. To prevent section contamination
2. To prevent contamination of microscope grid holder

NOTE: ONCE A SECTION IS CONTAMINATED IT CANNOT BE DECONTAMINATED.

B. MATERIALS:

1. 5 small, clean 10 ml beakers
2. 20 ml double distilled water
3. 10 ml 10% formic acid
4. 10 ml chloroform
5. New and used copper grids
6. Tweezers
7. Saturated solution of sodium hydroxide in methanol
8. Butyl acetate

C. RATIONALE AND DISCUSSION:

The resultant end-product of electron microscopy work are high quality electron micrographs preferably free of any defects from the chemical procedures and techniques. In order to work toward obtaining the best possible micrographs all precautions must be taken to avoid any type of contamination.

An ultrathin section is subject to contamination from the time it is cut until the time electron micrographs are made. The majority of the time contamination occurs as a result of the postsectioning process as opposed to presectioning.

One can perhaps better appreciate the importance of cleanliness in sectioning and microscopy once a large dirt particle is found in the area of a section of much interest.

The only defect might be a dirt particle obliterating part of the structures in the tissue. The use of clean grids is only one step, but an important one, in carrying out good electron microscopy.

D. PROCEDURE:

Cleaning New Grids

1. On a clean flat surface line up 4 small beakers.

   Label as follows: 10% formic acid double distilled water, and chloroform (See Figure 6-a).

2. Clean from 10-15 grids.

   Using a pair of tweezers, pick up the grid at the rim from the glass petri dish and gently lower it into 10% formic acid for 2 seconds.

NOTE: BE CAREFUL NOT TO LET THE GRID FALL FROM THE TWEEZERS.

3. Rinse in double distilled water for 2 seconds.

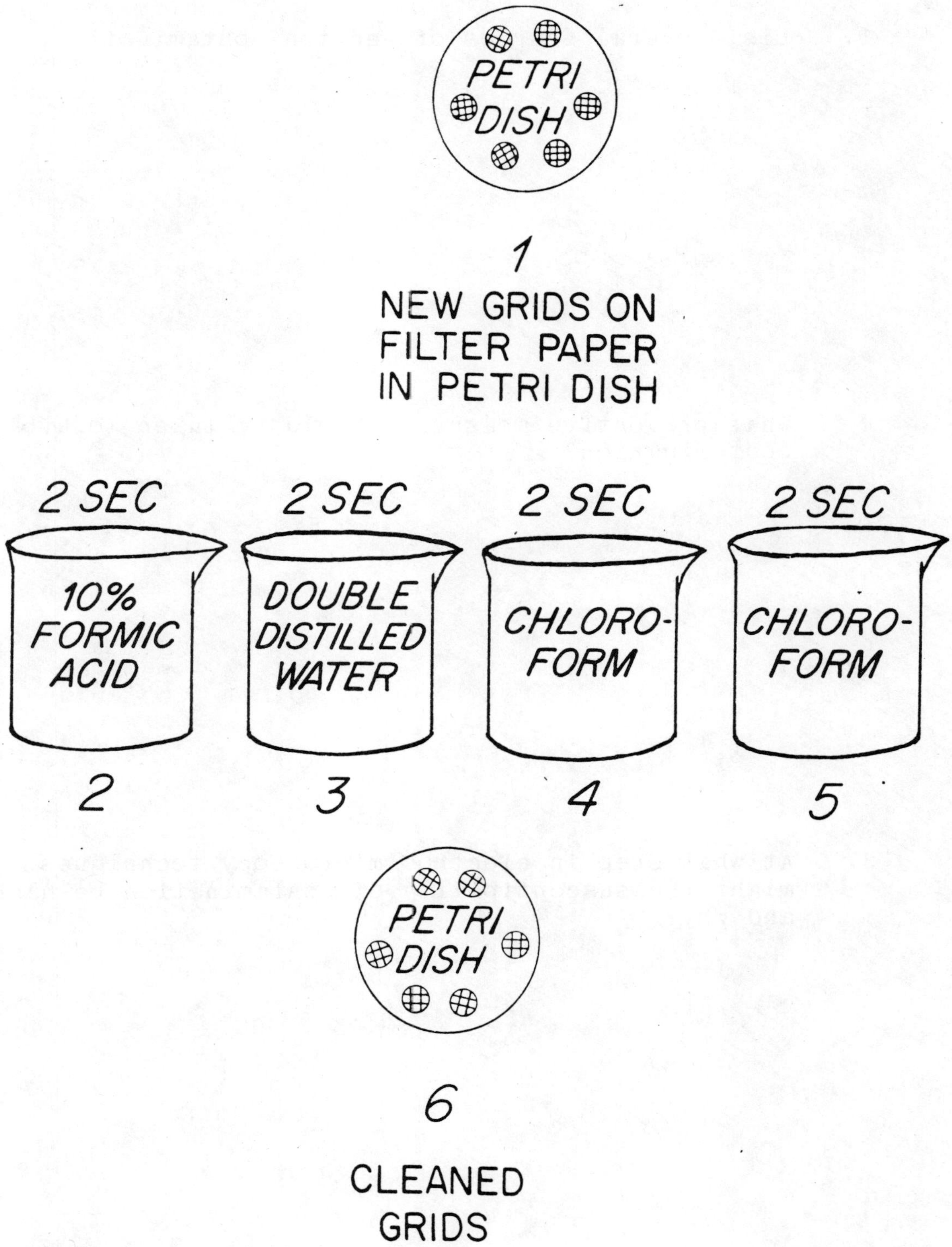

FIGURE 6-a. Cleaning grids.

E. QUESTIONS:

1. List several sources of section contamination.

2. What preventive measures should be taken to avoid contamination?

3. At what step in electron microscopy techniques might the susceptibility of contamination be higher and why?

EXERCISE NUMBER 7

ROUGH TRIMMING AND PRECISION FACING OF SPECIMEN BLOCK BY ULTRAMICROTOME

Betty R. Jones, Ph.D.

A. OBJECTIVES:

1. To introduce the student to the operation of the ultramicrotome for rough trimming and precision facing of specimen blocks

2. To obtain a more precise shaped block face

3. To obtain a block face with a smooth shiny-like finish

4. To prepare the face of the block for thick and ultrathin sectioning

NOTE: PRECISION TRIMMING OF BLOCK FACE MAY BE REFERRED TO AS PRECISION FACING OR FINE FACING.

B. MATERIALS:

1. Ultramicrotome

2. Eight glass knives

3. Four untrimmed specimen blocks

4. Small size water color brush with fine hairs

5. 1 can of EM refrigerant

C. RATIONALE AND DISCUSSION:

Trimming of the specimen block by microtome will always give a more precise shaped block face than the most skillful hand trimming. However, this method requires a little more time than hand trimming and much more practice.

It is my opinion that beginners should practice trimming first by hand and then by microtome. Both methods are rewarding in that a trapezoid block face can be produced.

The advantage of trimming by microtome is that a smoother shiny-like finish is produced from the block face. This type of surface reflects the knife edge enabling one to see clearer the closeness of the knife edge to the specimen.

A smoother surface will also retain and save the cutting edge of the knife. The smoother the surface of the block face the less knife marks encountered.

D. PROCEDURE:

1. Identify the following parts and controls of your ultramicrotome. The various parts of ultramicrotomes may have slightly different names depending on various models.

   However the functions are the same. Be able to recognize and identify the parts of the ultramicrotome listed below that are essential in microtomy.

TABLE I

MAJOR PARTS AND FUNCTIONS OF PARTS OF THE ULTRAMICROTOME.

PARTS	FUNCTIONS
a. Specimen Trimming Socket (or specimen stand)	a. Supports the specimen holder which contains the specimen block. Permits orientation of the specimen.
b. Specimen Trimming Socket Holder	b. Holds the specimen block.
c. Knife Holder	c. Holds knife firmly. May have attachment for adjusting the

	height of knife. Can be adjusted to obtain the desired clearance angles.
d. Knife Stage Assembly	d. Holds knife holder. Allows independent rotation of the knife around both vertical and horizontal axis through the point of contact with the specimen. This permits critical alignment of the knife close to the specimen.
e. Knife Holder Clamp	e. Locking lever located on the knife stage that clamps the knife in place. Scale on the knife holder indicates clearance angle of the knife in units of 2°.
f. Horizontal Movements of	f. Knob on the knife stage that allows the knife to move from left to right.
g. Macro Feed Wheel (coarse knife advance)	g. Knob clibrated in increments of 10 microns. It brings the knife closer to the specimen in very small units.
h. Micro Feed Wheel	h. Knob calibrated in increments of 1 micron. It brings the knife closer to the specimen in very small units.

i. Specimen Advance System	i. Knob to advance the specimen block toward the knife in varying preset micrometers.
j. Panel Controls	j. Knobs located on the front panel of the ultramicrotome or on a separate control unit that comes with the microtome. These controls may include the following: button or lever type switches for focused and diffused light or lever type switches for focused and diffused light sources thickness control cutting speed motor drives position of cutting strokes duration of cutting strokes and etc.
k. Viewing System	k. Steromicroscope permits focusing of the specimen block.

---

2. Remove specimen block from beem capsule.

   Insert untrimmed block into specimen trimming socket or specimen chuck.

   Fasten the trimming socket firmly and tightly around the block.

3. Make sure that the specimen block is arranged in a manner that the majority of the tissue is facing you.

If the tissue is off-centered in the block the specimen trimming socket must be adjusted to the appropriate angle (See Figure 7-a-c) for the appearance of centered and off-centered specimens.

4. Mount glass knife on knife stage and adjust it to a 4° clearance angle.

   A clearance angle of 4° is recommended for glass knives (See Figure 7-d).

   When trimming with a diamond knife the clearance angel must be determined by trial and error if not indicated by the supplier.

5. Carefully advance the knife stage assembly toward the trimming socket containing the specimen block.

   The binoculars need not be used for this advancement.

   Use the naked eye to bring the knife stage assembly within 2-3 mm of the specimen face.

6. Using the binoculars on the steromicroscope of the microtome carefully bring the knife edge and specimen face into sharp focus.

   Advance the knife stage assembly toward the block face using the macro feed wheel or coarse knife advance.

   The original position of the knife stage should be at 0° or straight forward.

   Slowly and carefully continue to advance the knife stage until large amounts of plastic are cut from the block face.

   Continue to cut plastic from block face until 1-3 mm of tissue is removed along with the plastic.

   NOTE: USE EXTREME CARE AT THIS POINT TO AVOID CUTTING LARGE CHUNCKS OF PLASTIC.

7. Precision trim the specimen block until a smooth shiny-like surface can clearly be seen from the tissue of the block face.

FIGURE 7-a-c. Appearance of tissue centered and off-centered in specimen block.

NOTE: CAREFUL POSITIONING OF TISSUE IN BEEM CAPSULE DURING EMBEDDING WILL PREVENT OFF-CENTERING.

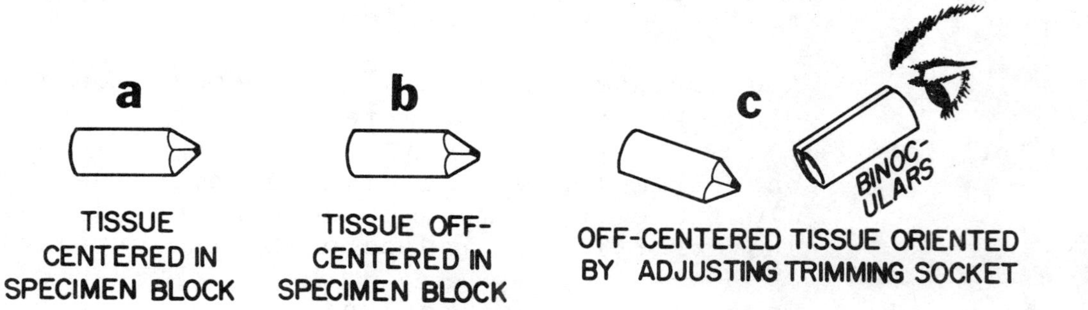

FIGURE 7-d. Clearance angle of 4° for glass knife used in trimming specimen block.

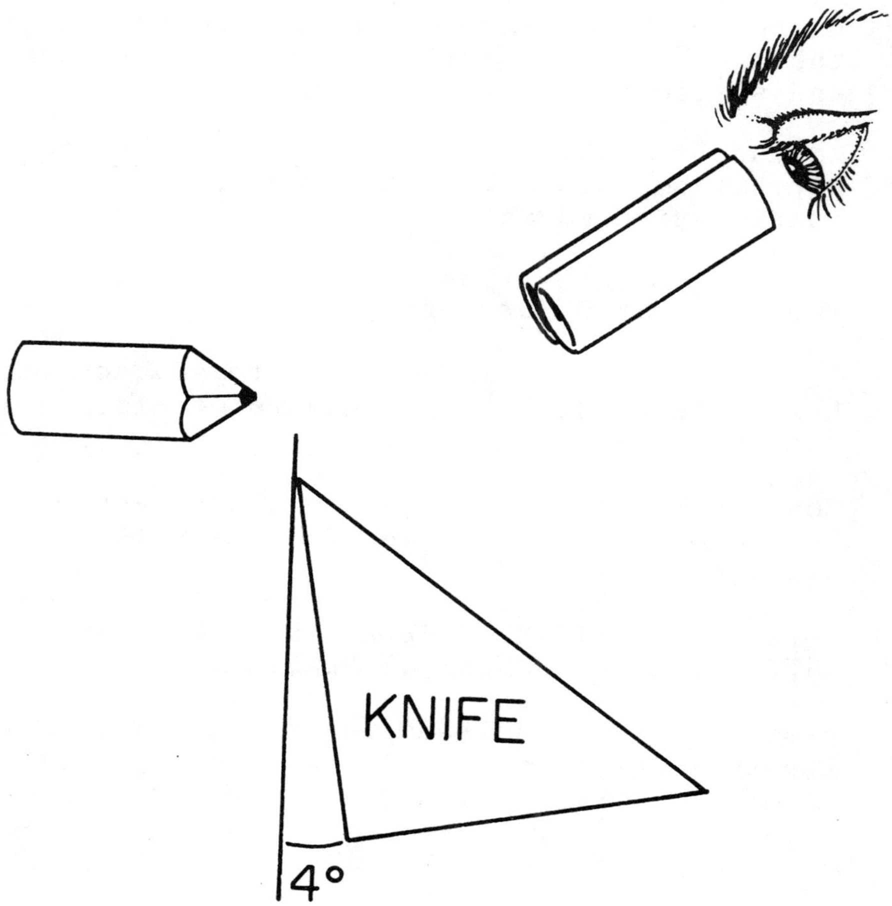

8. Remove excess plastics that have been cut with a fine hair brush or by spraying with any type of EM refrigerant.

NOTE: IF A REFRIGERANT IS USED MAKE SURE THE CAN IS HELD IN AN UPRIGHT HORIZONTAL POSITION.

9. Having obtained a smooth shiny surface of the block face with 1-2 mm of tissue exposed carefully retract or back up the macro feel wheel.

Move the knife stage assembly 30° to the left side of the block.

NOTE: TAKE PRECAUTION AT THIS STAGE TO MAKE SURE YOU RETRACT THE MACRO FEED WHEEL.

If this is not done you will damage the knife edge and tissue specimen (See Figures 7-e-i) for knife stage and specimen rotation positions.

The arrow in the first two figures is a starting point to assist in the following rotation of the specimen from step to step.

10. Retract the macro feed wheel and move the knife stage assembly 30° to the right side of the block.

You now have a precision trimmed face with a pair of parallel sides 1 and 2 (See Figure 7-e).

11. Retract the macro feed wheel and rotate the specimen trimming socket holder 90°.

The cut parallel sides 1 and 2 should now be the top and bottom parts of the trapezoid.

Slowly move specimen holder an additional 5° to the right.

Make sure that the knife stage assembly has been left at 30° to the right side of the specimen (See number 10, Figure 7-f).

Now cut the right side of the block. This will be cut 3.

Cut plastic until about 0.8-1 mm of tissue is removed from that side of the specimen block **(See figure 7-g)**

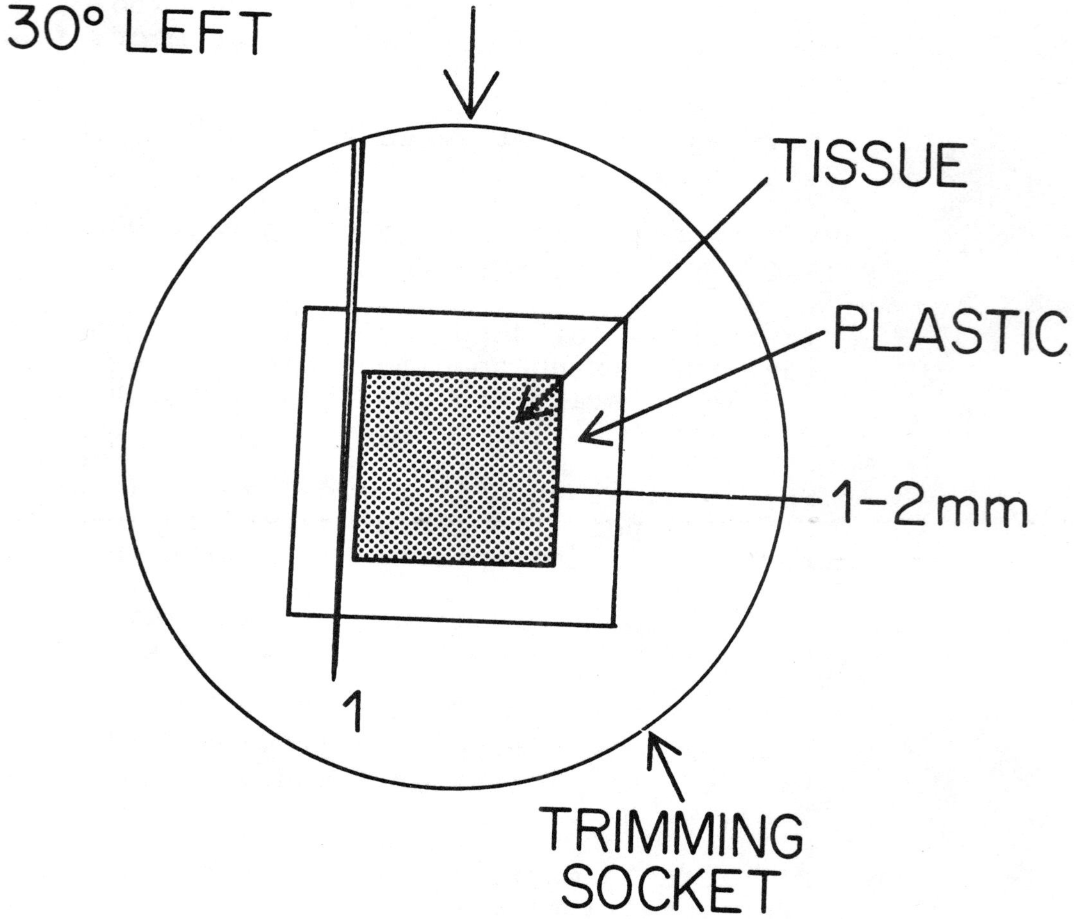

FIGURE 7-e.

FIGURE 7-e-i. Knife stage and specimen rotation positions.

NOTE: Parallel sides 1 and 2 will become the parallel top and bottom of the block face once the entire precision trimming procedure is completed.

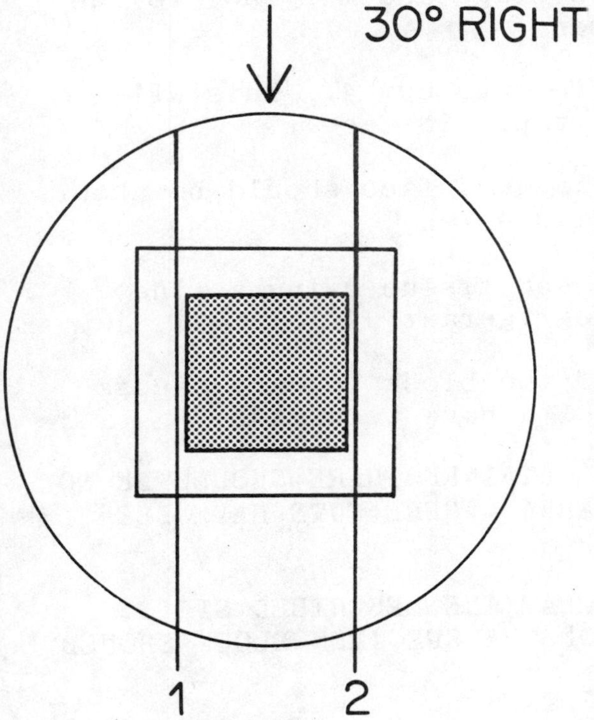

FIGURE 7-f

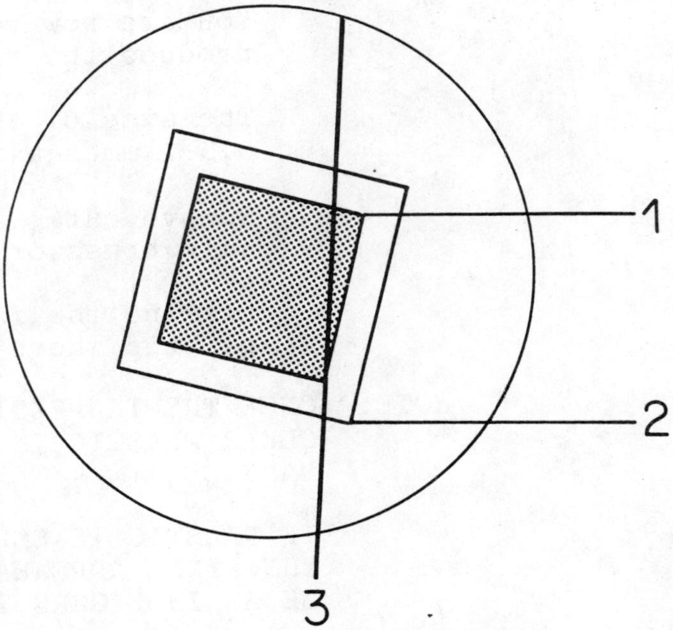

FIGURE 7-g

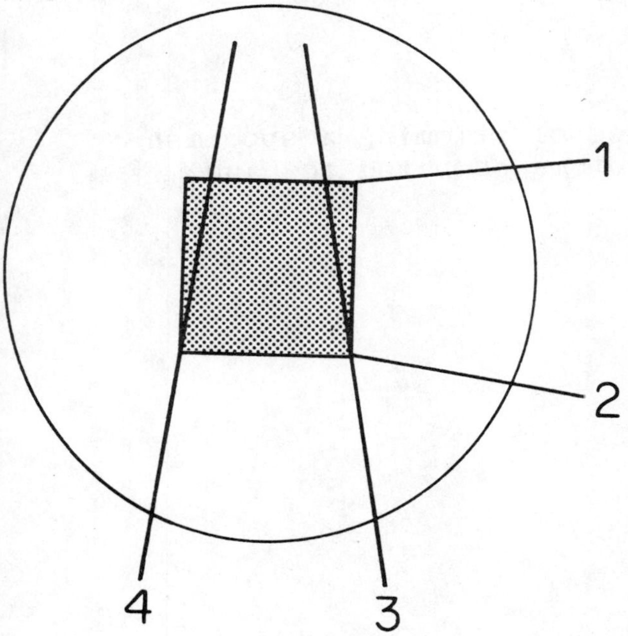

NOTE: see 12
Size 0.5-1 mm

FIGURE 7-h

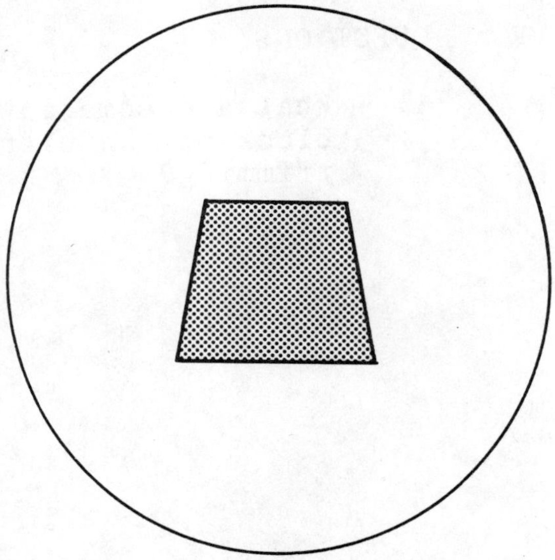

FIGURE 7-i

12. Retract the macro feed wheel and tip the specimen holder 10° to the left.

Then move the knife stage assembly 30° to the left of the specimen.

You are now ready to make cut 4. This will produce the final trapezoid face.

The size of the trapezoid face should be about 0.5-1 mm square (Figure 7-h).

13. Remove cut plastic and tissue using a fine hair brush or EM regrigerant.

14. Examine the face of the block for any excess plastic where cuts 1-4 have been made.

NOTE: ONCE THE TRAPEZOID IS OBTAINED THERE SHOULD BE NO CLEAR PLASTIC IN THE AREA WHERE CUTS HAVE BEEN MADE.

IF PLASTIC IS SEEN REPEAT THE DESCRIBED STEPS. THE FINAL APPEARANCE OF THE SPECIMEN BLOCK SHOULD BE AS IN FIGURE 7-i.

E. EVALUATION:

PRESENT TWO TRIMMED SPECIMEN BLOCKS TO THE INSTRUCTOR FOR EXAMINATION.

F. QUESTIONS:

1. What are some advantages of trimming a specimen block on the ultramicrotome compared to hand trimming?

2. Why is proper mounting of the specimen block important?

3. What defects might occur in sections due to a loose block?

4. Briefly discuss sectioning tissue culture monolayers with reference to specimen block alignment, cutting edge of knife and limited thickness.

5. List several reasons why sections might not come off in a straight ribbon.

NOTES

EXERCISE NUMBER 8

FLOATATION FLUIDS

Betty R. Jones, Ph.D.

A. OBJECTIVES:

1. To fill a glass knife reservoir with fluid
2. To obtain a flat fluid meniscus
3. To release sections from the knife edge
4. To provide a liquid surface for sections allowing them to float
5. To cause no damage to cell ultrastructure

B. MATERIALS:

1. Distilled water
2. Dressed glass knives
3. Syringe
4. Microtome
5. Xylol or chloroform solution

C. RATIONALE AND DISCUSSION:

Fresh distilled water is an excellent floatation fluid. It allows sections to float for any length of time without being at risk to any dissolving action that might be caused by other solvents. Solvents such as acetone and ethanol may be used depending on the type of embedding resin.

Xylol or chloroform vapors may be used if necessary to prevent section compression. Exposure to the vapor results in flattening of sections on the fluid surface. If sections are being cut with a diamond knife xylol or chloroform vapors are not necessary.

D. PROCEDURE:

1. Mount and adjust glass knife on microtome.

2. Using a clean syringe filled with fresh distilled water and looking through the stereomicroscope, inject water into the surface area of knife reservoir.

NOTE: DO NOT TOUCH KNIFE EDGE OR RESERVOIR WITH SYRINGE.

3. Fill knife reservoir from end opposite knife edge (See Figure 8-a).

NOTE: DO NOT OVERFLOW KNIFE EDGE WITH WATER.

4. Fill the reservoir with fluid until a flat meniscus is obtained (See Figure 8-b).

Examination of Floatation Fluid

5. Using stereomicroscope of microtome check water surface for correct water meniscus, cleanliness of water surface and possible signs of leakage.

NOTE: A WATER MENISCUS CURVED INWARD MAY CAUSE CRUMPLING OF THE SECTIONS.

A WATER MENISCUS CURVED OUTWARD MAY CAUSE WETTING OF BLOCK FACE (SEE FIGURES 8-c-d).

6. Make sure knife reservoir has a <u>flat</u> fluid meniscus.

E. EVALUATION OF FLOATATION FLUID:

The knife reservoir containing floatation fluid will be checked by the instructor.

FIGURE 8-a. Filling reservoir with floatation fluid.

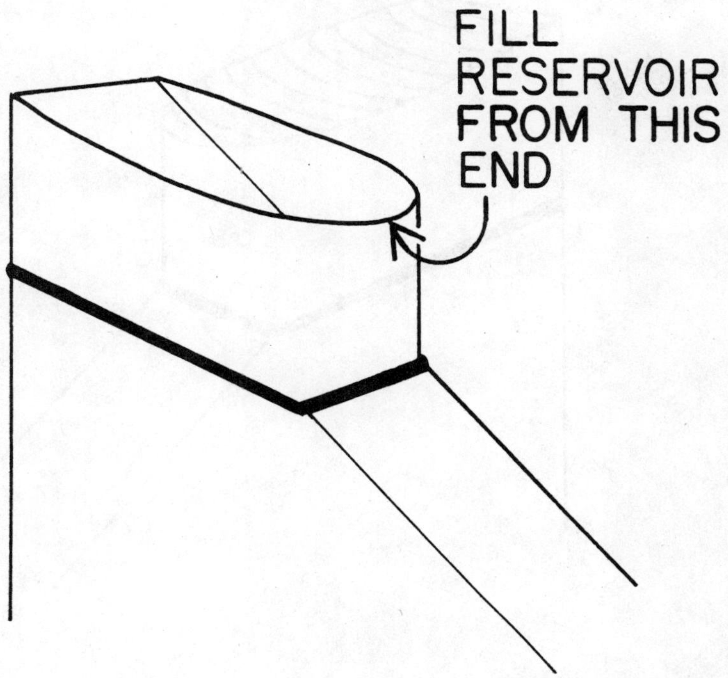

FIGURE 8-b. Flat water meniscus.

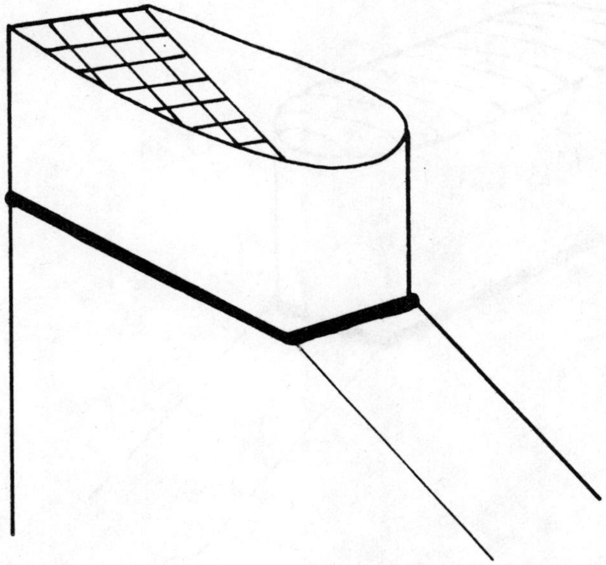

FIGURE 8-c. Appearance of incorrect fluid meniscus. Fluid level curved inward.

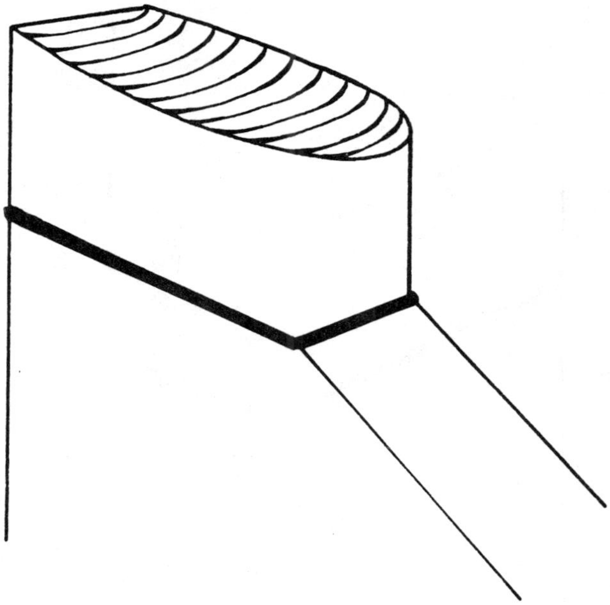

FIGURE 8-d. Appearance of incorrect fluid meniscus. Fluid level curved outward.

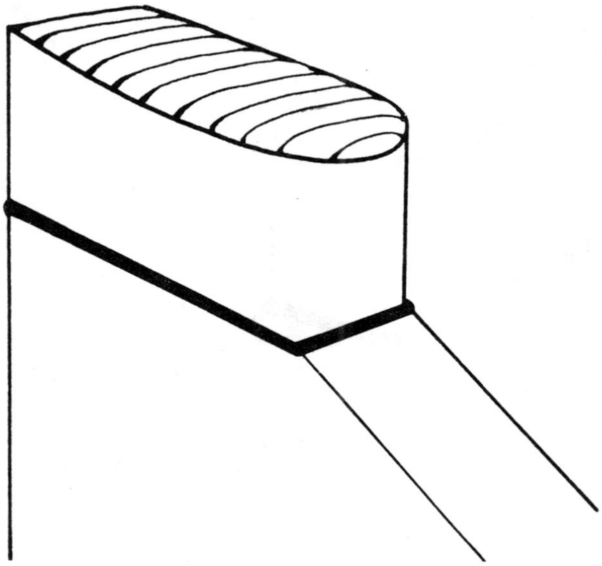

F. QUESTIONS:

1. Discuss several general characteristics for a floatation fluid.

2. Discuss the use of distilled water as a satisfactory floatation fluid compared to other solvents.

3. Briefly discuss the following: concave meniscus, convex meniscus, and flat meniscus. Indicate the meniscus level satisfactory for section floatation.

4. What conditions are essential in obtaining the release of intact sections from the cutting edge of the knife?

EXERCISE NUMBER 9

CUTTING AND STAINING THICK SECTIONS FOR LIGHT MICROSCOPY

Betty R. Jones, Ph.D.

A. OBJECTIVES:

1. To cut sections 0.2-1 micron in thickness
2. To produce thick sections thinner than those cut from paraffin blocks
3. To obtain precise orientation and location of a desired or suitable area in the tissue before ultrathin sectioning
4. To stain thick sections with a dye that will penetrate the embedding resin and color the tissues to permit viewing in a light microscope
5. To retrim the specimen Block once the area of interest has been located under the light microscope
6. To examine the quality of fixation and cytological detail of the tissue
7. To be able to correlate thick sections for light microscopy with ultrathin sections for electron microscopy

NOTE: THICK SECTIONS MAY ALSO BE VIEWED UNDER A PHASE CONTRAST MICROSCOPE WITHOUT THE USE OF A DIFFERENTIAL STAIN.

B. MATERIALS:

1. Precision trimmed specimen block
2. Dressed knives

3. Fine wire loop or paint brush for transferring sections

4. Fine eyelash rod for section manipulation

5. Precleaned glass slides

6. Precleaned glass coverslips

7. Floatation fluid (distilled water) for glass knife reservoir

8. Glass syringe for filling knife reservoir

9. Toluidine Blue 0 Stain (0.5% toluidine blue 0 in 1% sodium borate)

10. Periodic Acid (1%)

11. Mounting medium

12. Electric hot plate

C. RATIONALE AND DISCUSSION:

It may be construed as true that techniques in electron microscopy are time consuming but the beginner microscopist must develop patience, have confidence and acquire a certain finesse for carrying out various procedures and more importantly understand the principles accompanying certain techniques.

Cutting and staining thick sections for light microscopy is an important intermediate step preceding ultrathin sectioning. It allows one to save more time and effort simply by examining sections at lower magnifications in a light microscope.

This will permit specific areas to be located, the orientation of the section can be determined (i.e. transverse, longitudinal, sagittal, tangential, oblique or etc.) and the quality of fixation can be studied.

Areas of uninterest in the tissue or poorly fixed material may be cut away once the block is pretrimmed. It is much easier to cut thick sections than ultrathin sections.

Useful information on cytological structure can be obtained when thick sections are stained and examined in a light microscope. Unstained sections may be examined using a phase contrast microscope.

For the beginner electron microscopist it is sometimes difficult to relate what is seen at high magnifications under the electron microscope to material observed with the naked eye.

Examination of thick sections under the light microscope will provide an understanding of the material as a whole. It is therefore important to correlate light microscopy with electron microscopy.

A wide variety of differential stains are used in light microscopy but only a few are adequate for embedded synthetic resins of electron microscopy. This is primarily due to the poor uptake of various dyes and penetration into the resin.

Toluidine blue 0 stain is probably the most widely used stain for observing thick sections from synthetic resins under the light microscope. A histology atlas should be used for correct cytological interpretation of tissue structures.

D. PROCEDURE:

1. Before using the ultramicrotome make sure the macro and micro feed wheel and knife stage assembly are at the start of their cycle and not near the end.

   The former has reference to LKB microtomes and the latter for Sorval microtomes.

2. Insert the precision trimmed specimen block into the trimming socket and its holder.

   Make sure the smallest part of the trapezoid face is turned upward and that 1-2 mm of the block projects from the holder.

   The block should be fastened securely to the microtome.

3. Insert the dressed knife into the knife stage assembly. Adjust the knife clearance angle to 4°.

4. Slowly and carefully move the entire knife stage assembly up toward the face of the specimen block being careful not to touch it.

   The knife edge should be left at a distance of

2-3 mm from the face of the block.

Make sure the knife edge is parallel to the face of the block before fastening the knife assembly in place.

Movement of the knife stage assembly is not necessary for LKB microtomes.

5. Looking through the binoculars of the stereomicroscope on the microtome check the alignment of the knife to the face of the block.

   Also check the knife edge for nicks and dust particles.

   Dust particles can be removed by spraying with an EM refrigerant.

6. Withdraw the floatation fluid from a stoppered glass bottle using a glass syringe.

   Fill the dressed knife with distilled water until a flat meniscus is obtained and a silver-white reflection can be seen on the surface of the water.

   Make sure the water is in contact with the edge of the knife.

7. If it is difficult to obtain the correct silver-white reflection on the water surface adjust the light source or the knife clearance angle.

NOTE: BE CAREFUL NOT TO OVERFILL THE KNIFE RESERVOIR TO AVOID WETTING THE KNIFE EDGE AND THE FACE OF THE BLOCK.

8. Carefully advance the macro feed wheel or coarse advance toward the knife edge.

   In case of sorvall microtomes advance the coarse knife advance toward the face of the specimen block.

   Proceed cautiously until the knife edge is approximately 1 mm away from the block face.

9. Adjust the light source to illuminate the block face and knife edge. A reflection line should appear on the smooth shiny block face.

This reflection line represents the knife edge on the smooth surface of the block face. Once the line is seen the final alignment of the knife edge with the block face can be checked.

It is of utmost importance to identify this line so that the block face is parallel to the knife edge.

10. The instructor will check to see if the reflection line is on the face of the block.

11. The reflection line should be a straight line.

    If it is not straight it can be adjusted by moving the knife assembly until it is aligned parallel to the face of the block.

12. Slowly and cautiously advance the macro feed moving the block face up and down manually or mechanically until contact is made between the knife edge and the block.

NOTE: EXERT EXTREME CARE AND PATIENCE AT THIS POINT.

THE ART OF CONQUERING THIS PROCEDURE WILL PROVE WORTHY NOT ONLY FOR THICK SECTIONS BUT ULTRATHIN SECTIONS.

13. Once contact is made between the knife edge and the block, the reflection line will no longer be seen.

    Gently cut pieces of plastic and tissue from the block face.

    Continue to cut the face of the block until trapezoid sections of the entire block face are removed.

NOTE: THE MICRO ADVANCE WHEEL MIGHT NEED TO BE ADJUSTED SLIGHTLY IF SECTIONS DO NOT COME OFF.

14. Make sure you are familiar with the settings on the microtome for cutting micrometer or micron sections.

    Refer to the instruction manual of the microtome.

15. Cut three to four 0.2-1 micron thick sections.

NOTE: SELECTION OF A SPECIFIC AREA OF THE ULTRATHIN SECTION CAN BE MADE AFTER OBSERVING A FEW THICK SECTIONS UNDER THE MICROSCOPE (SEE Figure 9-a-c).

16. Collect the sections from the floatation fluid by placing the fine wire loop or fine paint brush underneath the section.

    Transfer one section at a time to a clean glass slide containing one drop of distilled water.

    If sections are to close together and are coming off in ribbons in the floatation fluid, use the eyelash rod to manipulate or move them apart (See Figure 9-d-g).

17. For the convenience of locating the section on the glass slide circle the designated area underneath the slide with a glass marking pencil or a wax pencil.

18. Rapidly check the section under the light microscope for defects such as wrinkling, curling and folding.

    Make sure the drop of water does not dry up. Cut more sections if needed.

Staining

19. Place the slides on a hot plate at 40°C for drying.

    Allow the water to evaporate. The heat causes the sections to adhere to the slide.

20. After drying remove slides from hot plate and place 1 drop of 1% periodic acid solution on sections.

    Leave on for 5 minutes only.

21. Rinse slides for 1 minute in distilled water.

22. Place slides back on hot plate for drying or air dry.

    Place one drop of 0.5% toluidine blue O stain on the sections.

    Stain for 2 minutes.

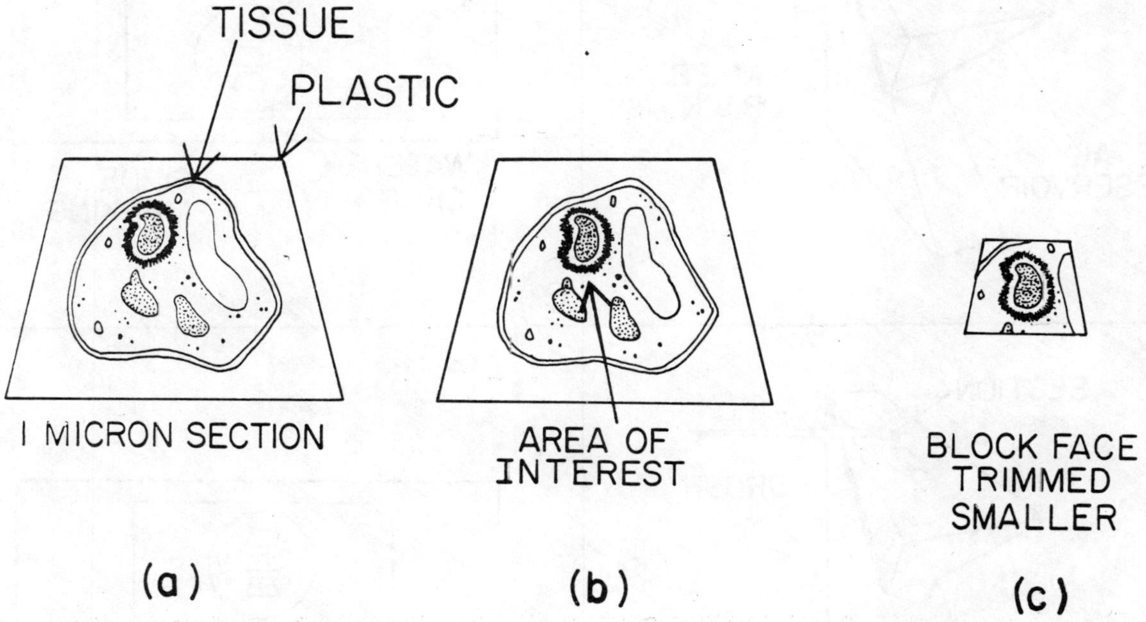

FIGURE 9-a-c. Selection of a specific area for ultrathin sectioning.

a. 1 micron thick section stained with toluidine blue.

b. Area of interest selected from block face.

c. Block face trimmed smaller to area of interest.

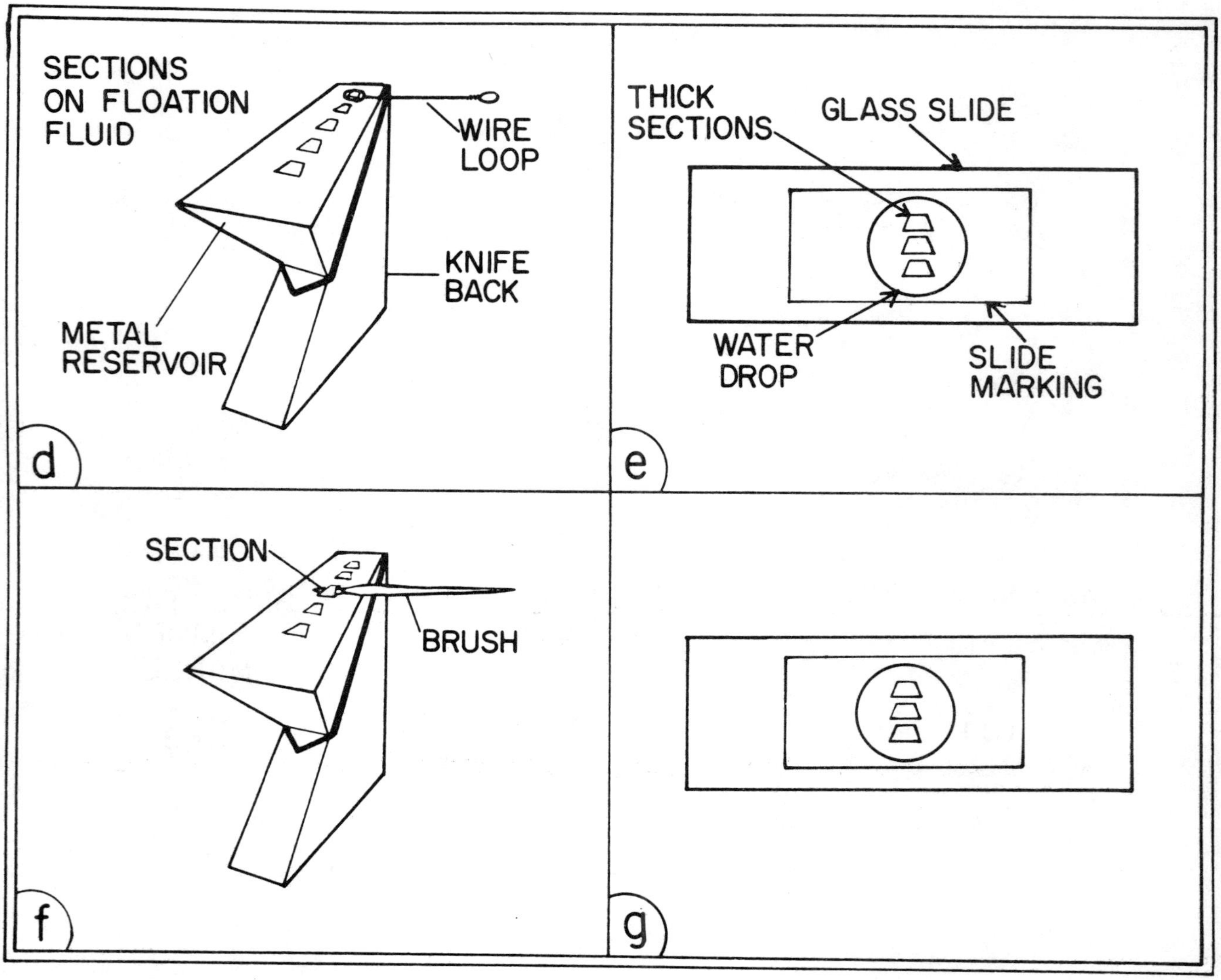

FIGURE 9-d-g. Picking up thick sections.

d. Picking up sections with a wire loop.

e. Transfer of sections to glass slide.

f. Picking up sections with a brush.

g. Transfer of sections to glass slide.

NOTE: DO NOT LET THE STAIN DRY UP OR EVAPORATE.

23. Rinse slides gently in distilled water for 1 minute to remove excess stain.

24. Examine slides under light microscope to check staining quality.
If stain is too pale repeat step 22.

25. Place one drop of synthetic mounting medium on section and lower a coverslip on it.

NOTE: BE CAREFUL NOT TO INTRODUCE AIR BUBBLES.

26. Properly label and observe under the light microscope.

E. RESULTS:

Toluidine blue 0 stain will produce various shades of blue making identification of tissue structures such as nuclei, membranes, and other cytoplasmic organelles easy. It provides various degrees of contrast and specific structures are not stained selectively.

F. EVALUATION:

NOTE: PRESENT TWO SLIDES CONTAINING THICK SECTIONS TO THE INSTRUCTOR FOR EVALUATION.

Storage of Glass Slides

Glass slides should be properly labeled and stored in slide boxes for future use.

F. QUESTIONS:

1. What are the advantages of cutting thick sections from resin blocks prior to beginning ultrathin sections?

2. What is the thickness range for sections prepared from blocks for light microscopy?

3. Briefly describe how thick sections are collected from the knife reservoir and stained.

4. Discuss three common stains used for staining thick sections cut from resin blocks.

5. Why are only a few differential stains available for staining embedded resin material for light microscopy?

EXERCISE NUMBER 10

CUTTING AND COLLECTING ULTRATHIN SECTIONS FOR ELECTRON MICROSCOPY

Betty R. Jones, Ph.D.

A. OBJECTIVES:

1. To obtain ultrathin sections from the same area or smaller area from which thick sections have been cut
2. To obtain silver-gray sections ranging in thickness from 500 to 600 Å; and silver and gold sections ranging in thickness from 600-900 Å
3. To obtain sections in ribbon formation floating on the surface of the floatation fluid
4. To obtain sections free of defects such as chatter, wrinkles, breaks, folds and contamination
5. To collect sections on mesh grids
6. To provide two-dimensional data from good quality electron micrographs resulting from nonserial or serial sectioning
7. To provide data that will meet the objectives of the study

B. MATERIALS:

1. Trimmed specimen block (face of the block should be a trapezoid less than 1 mm^2)

NOTE: USE SAME BLOCK FROM WHICH THICK SECTIONS WERE OBTAINED.

2. Several dressed glass knives

3. Glass syringe with long needle

4. Distilled water in glass stoppered bottle

5. 1 pair of fine tweezers

6. A supply of precleaned grids (200 or 300 mesh) on Whatman #2 filter paper in a petri dish

7. Eyelash rod

8. 1 can of EM refrigerant

9. 2 cotton swabs

10. Chloroform (10 ml) in tightly sealed container

C. RATIONALE AND DISCUSSION:

Ultrathin sections are necessary in order to examine biological tissues in the electron microscope. The image of the specimen is formed by an electron beam focused through a series of magnetic coils or lenses.

As the electron beam is emitted from the filament through the lenses and specimen, electron scattering or electron deflection occurs. This scattering or deflection of electrons is due to dense components in the specimen.

If the scattering of electrons is great enough, electrons will fall outside of the objective apperture and thus be eliminated from the beam. Henceforth, a pattern will be formed on the flourescent screen or photographic plate beneath the specimen in the path of the transmitted beam.

The amount of scattering that occurs in a specimen depends upon the thickness of the section and the atomic weights of the atoms that constitute it. The section must be ultrathin so the electron beam can penetrate through it.

The process of cutting ultrathin sections is called ultramicrotomy and requires a standard methodical approach if reproducible results are to be obtained.

Section thickness is determined by the interference colors of the sections as they reflect light. One should aim for obtaining silver-gray sections which range from 500-600 A^o in thickness.

However, silver sections ranging from 600-900 A^o in thickness and pale gold sections ranging from 900 to 1200 A^o in thickness are adequate and usable but incapable of providing high resolution.

Sections showing interference colors such as deep gold, purple, pink, blue, green, brown, red, and violet are too thick for electron microscopy. Sections ranging in thickness from 500 A^o and below are gray.

Gray sections may be difficult for the beginning microtomist but with patience, concentration, time and effort, this task can be mastered.

After 8-10 sections have been cut and a ribbon is floating on the water surface the sections must be collected on precleaned mesh grids.

Always keep available a clean supply of grids on filter paper in a petri dish prior to beginning the sectioning procedure.

NOTE: REFER TO EXERCISE NUMBER 6 FOR CLEANING NEW AND USED GRIDS.

Collecting sections from the floatation fluid of the knife reservoir requires practice and skill.

NOTE: THE SECTIONS SHOULD BE IN THE CENTER OF THE GRID SO THEY WILL NOT FALL OUTSIDE THE FIELD OF VIEW IN THE ELECTRON MICROSCOPE.

THEY SHOULD NOT OVERLAP ONE ANOTHER. IF SECTIONS HAVE COME OFF IN RIBBON FORMATION THE PROBLEM OF OVERLAPPING SELDOM OCCURS.

If sections have not formed a ribbon and are scattered they must be gently grouped together using the eyelash rod toward the center of the reservoir.

If they are within an area of about 1 mm^2, 200, 300, and 400 mesh grids can be used without being coated.

NOTE: SEE EXERCISE NUMBER 31 FOR USE OF SUPPORT FILMS.

D. PROCEDURE:

NOTE: AT THIS POINT THE OPERATOR SHOULD BE TOTALLY

FAMILIAR WITH THE PARTS AND FUNCTIONAL OPERATIONS OF THE ULTRAMICROTOME BASED ON PREVIOUS EXERCISES.

THE FOLLOWING STEPWISE PROCEDURE SHOULD BE FOLLOWED CAREFULLY FOR OBTAINING ULTRATHIN SECTIONS.

THIS PROCEDURE HAS WORKED WELL IN THE AUTHOR'S LABORATORY.

1. Check trimmed specimen block in which thick sections were cut under a stereoscopic dissecting microscope.

   Determine if retrimming of the block or a smaller desired area is needed.

2. Reset the advance system of the microtome. Make sure the macro and micro advance and knife stage assembly are in their original positions.

   This is important in order to avoid hitting the knife and specimen block into each other and to ensure advancing the system the entire coarse if necessary during the sectioning process.

3. Insert the trimmed specimen block into the holder and fasten securely in place.

4. Insert the dressed knife into the knife stage assembly.

   Adjust the knife clearance angle between 2 and 5°.

5. Adjust the correct height of the knife according to the microtome setting.

6. Carefully remove any plastic debris, dust particles or contamination from the face of the block and sharp knife edge by using EM refrigerant spray.

NOTE: THE EM REFRIGERANT MUST BE HELD IN A HORIZONTAL POSITION WHEN SPRAYING.

7. Adjust the binoculars and the illumination system of the microtome to view the shiny face of the block and examine the knife edge.

8. With a syringe carefully fill the knife reservoir with distiilled water until a flat

water meniscus is obtained and a silver-white reflection can be seen on the surface of the water.

9. Identify the reflection line on the smooth shiny surface of the block face.

   The block face must be parallel to the knife edge and this can be determined according to the straightness of the reflection line.

   Make any adjustments needed for final alignment of the block face and knife edge.

10. Adjust the knife by a lateral knob on the knife holder so that sections can be cut by the best portion of the knife cutting edge.

11. While viewing through the binoculars, slowly and cautiously advance the macro and then micro adjustment until contact is made between the block face and knife edge and a section is cut.

12. Adjust the desired cutting speed and section thickness.

    The cutting speed may vary from 1-2 mm/sec and the section thickness may range from 700-800 A^{o} at first for the beginner microtomist.

    This range of thickness should produce sections that are pale gold to silver in color.

    One can practice obtaining thinner sections as they become more skilled in using the ultramicrotome.

13. If the block is properly trimmed the sections should come off in ribbons.

    You are now ready to section until a ribbon of sections are cut (See Figure 10-a).

NOTE: DURING THE SECTIONING PROCESS DO NOT ALLOW ANY PART OF YOUR BODY TO BE CONTACT WITH THE MICROTOME OR MICROTOME TABLE.

DO NOT MAKE NOISE, TALK OR BREATH IN THE CUTTING AREA.

MOVEMENTS, VIBRATIONS, DRAFTS OR TEMPERATURE CHANGES WILL AFFECT SECTIONS BEING CUT.

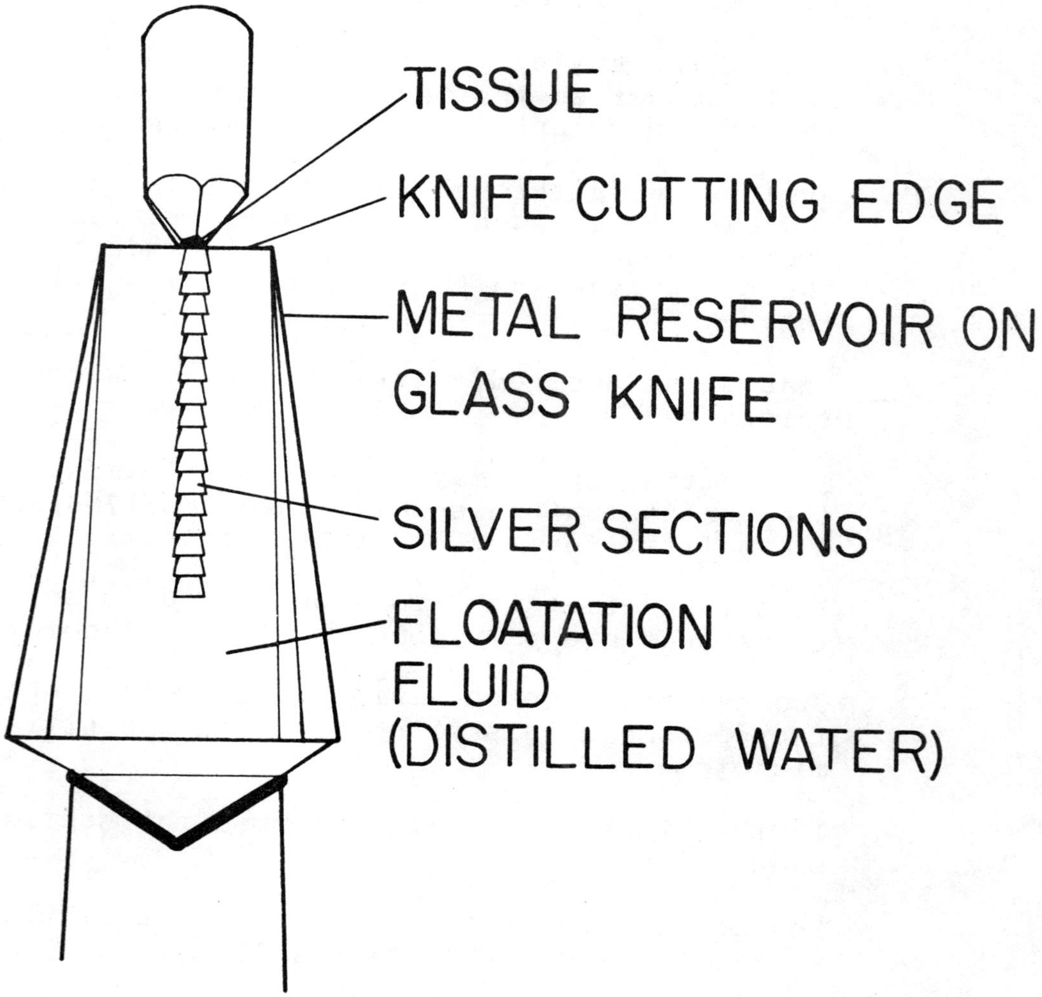

FIGURE 10-a. Appearance of ribbon formation cut from specimen block.

14. Carefully keep check on the quality of sections, the water menicus and proper illumination.

15. If you are not getting pale gold to silver sections or if the sections are not being cut every stroke of the cutting cycle, stop the automatic cycling control and check the following conditions:

    a. cutting speed (1-2 mm/sec)

    b. section thickness control (700-800 Å)

    c. proper illumination on microtome set to give white reflection on water surface

    d. floatation fluid in contact with the cutting edge of the knife

    e. floatation fluid not leaking from reservoir

    f. microtome controls sufficiently holding specimen and knife firmly

    If these conditions are accounted for sections should come off at each stroke of the cutting cycle and should float on the surface of the floatation fluid.

    However, the art of mastering the sectioning procedure requires an awareness, balance and correction of possible problems that might occur.

NOTE: REFER TO THE TABLE AT THE END OF THIS EXERCISE FOR THE MOST COMMON PROBLEMS THAT OCCUR DURING SECTIONING.

16. After cutting the desired number of sections stop the automatic advance control on the microtome.

17. Gently bring the sections away from the cutting edge using an eyelash rod.

18. Flatten the sections if necessary by holding a cotton swab that has been dipped in chloroform close to the sections.

NOTE: DO NOT BRING THE COTTON SWAB IN CONTACT WITH THE SECTIONS.

EXPOSE THE SECTIONS TO THE CHLOROFORM VAPOR FOR 5-6 SECONDS.

E. RESULTS:

Sections should range in interference colors from pale gold to silver.

F. EVALUATION:

Call instructor for observation of pale gold to silver sections in a ribbon formation floating on the water surface. Then prepare to go to the next step.

Collection of Ultrathin Sections for Electron Microscopy On A Mesh Grid

G. PROCEDURE:

1. Become familiar with the dull and shiny side of the grid, the grid mesh and rim (See Figure 10-b).

NOTE: THE AUTHOR PREFERS PLACING THE SPECIMEN ON THE DULL SIDE OF THE GRID BECAUSE THE SECTIONS CAN BE EASILY SEEN AGAINST A DULL BACKGROUND WITH THE NAKED EYE AND BETTER CONTACT IS MADE ON THE DULL SIDE.

2. If sections are in ribbon formation at least 8-10 sections can be picked up on one grid at one time.

   If the ribbon consists of about 20 sections (i.e. 10-20 sections), it should be separated into two ribbons using the eyelash rod.

   Each ribbon should be lined up side by side and picked up on one grid or separate grids.

3. If sections did not come off in ribbon formation and are scattered, gently group them together using an eyelash rod.

4. Gently pick up one grid at the rim using tweezers. The dull side of the grid should face up.

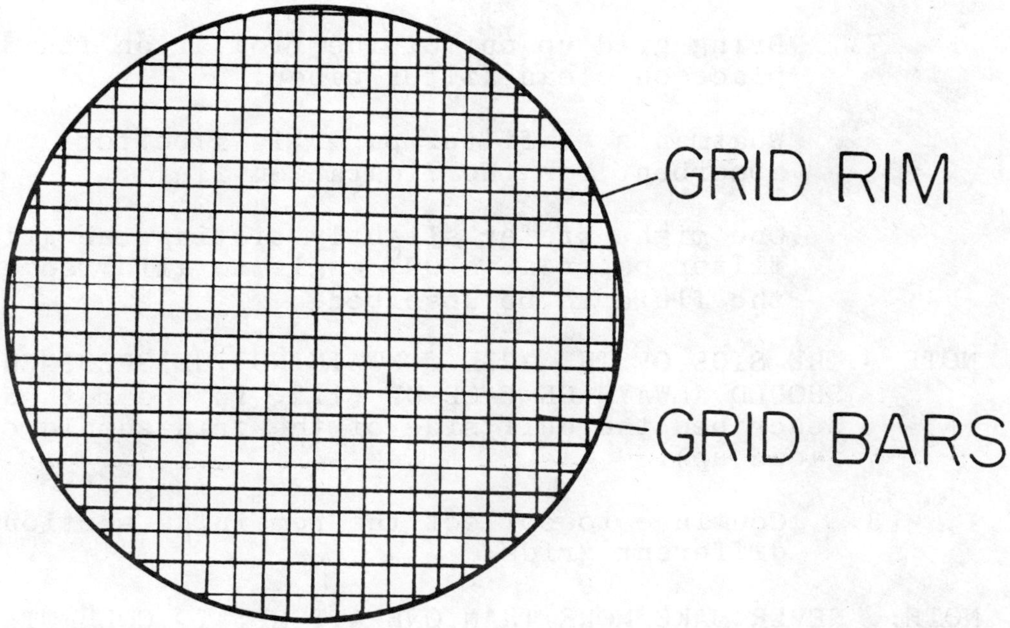

FIGURE 10-b. Mesh grid showing outer grid rim and grid bars.

NOTE: ALWAYS HANDLE GRIDS WITH TWEEZERS AT THE RIM.

5. While viewing through the binoculars of the microtome with the sections in full view carefully lower the dull side of the grid on top of the sections.

NOTE: THE RIBBON SHOULD BE COLLECTED IN THE CENTER OF THE GRID AND NOT NEAR THE RIM TO PREVENT BEING OUTSIDE THE FIELD OF VIEW IN THE ELECTRON MICROSCOPE.

6. After lowering the dull side of the grid on top of the sections continue to immerse the whole grid under the floatation fluid until it is completely wet.

7. Bring grid up out of the floatation fluid and place on clean filter paper.

   Whatman's #2 filter paper is a better absorbent for the floatation fluid.

   One might prefer slightly tilting the grid on filter paper at a 90° angle to allow some of the fluid to be absorbed.

NOTE: THE SIDE OF THE GRID CONTAINING THE SPECIMEN SHOULD ALWAYS BE FACE UP (i.e. in the method described the dull side of the grid should be faced up).

8. Continue to collect the remaining sections on different grids.

NOTE: NEVER MAKE MORE THAN ONE ATTEMPT TO COLLECT SECTIONS ON THE SAME GRID.

IF ONE IS UNCERTAIN AS TO WHETHER THE SECTIONS ARE ON THE GRID, EXAMINE THE GRID UNDER THE BINOCULARS OF THE MICROTOME, A DISSECTING MICROSCOPE OR WITH THE NAKED EYE.

IF THE GRID IS TILTED UNDER ILLUMINATION, TINY SMALL SPECKS CAN BE SEEN. THIS IS AN INDICATION OF THE PRESENCE OF TISSUE IN THE SECTION.

9. Another method of collecting sections is to dip the grid in the floatation fluid at an angle and pick up the sections from underneath.

10. Lower the grid under the fluid and gently raise it up collecting the sections.

This method however is not one of preferance by the author because the sections may tend to scatter and are not easily centered on the grid.

NOTE: FOR GOOD PRACTICE NEVER COLLECT SECTIONS ON MORE THAN FIVE GRIDS AT A TIME BEFORE OBSERVING UNDER THE ELECTRON MICROSCOPE.

11. Store grids in petri dish on filter paper until the next step.

NOTE: AFTER GRIDS HAVE BEEN VIEWED IN THE ELECTRON MICROSCOPE THEY SHOULD BE STORED WITH THE ORIGINAL BLOCK FOR FUTURE REFERENCE.

They can be placed in gelatin capsules, one grid to each capsule. The capsules should be labeled with the block identification number.

If grids are properly stored and maintained they can be re-examined ten years later.

H. EVALUATION:

PRESENT TO THE INSTRUCTOR 4 GRIDS CONTAINING SECTIONS.

Figure 10-c. Chart of some of the most common problems and probable causes encountered during the sectioning procedure. For possible prevention of some of these problems the reader should refer to basic EM texts.

In addition to common difficulties mentioned in the chart it is the author's experience that one can never overemphasize the importance of cleanliness in implementing electron microscopy techniques, especially sectioning.

A clean microtomy laboratory will enhance by 95% better quality work. Often it is easy for the beginning microtomist to site the problem as being the microtome itself but the majority of the time this is seldom true.

Thus one must have patience and be ready to meet and correct problems as they occur. This task is learning the art of microtomy.

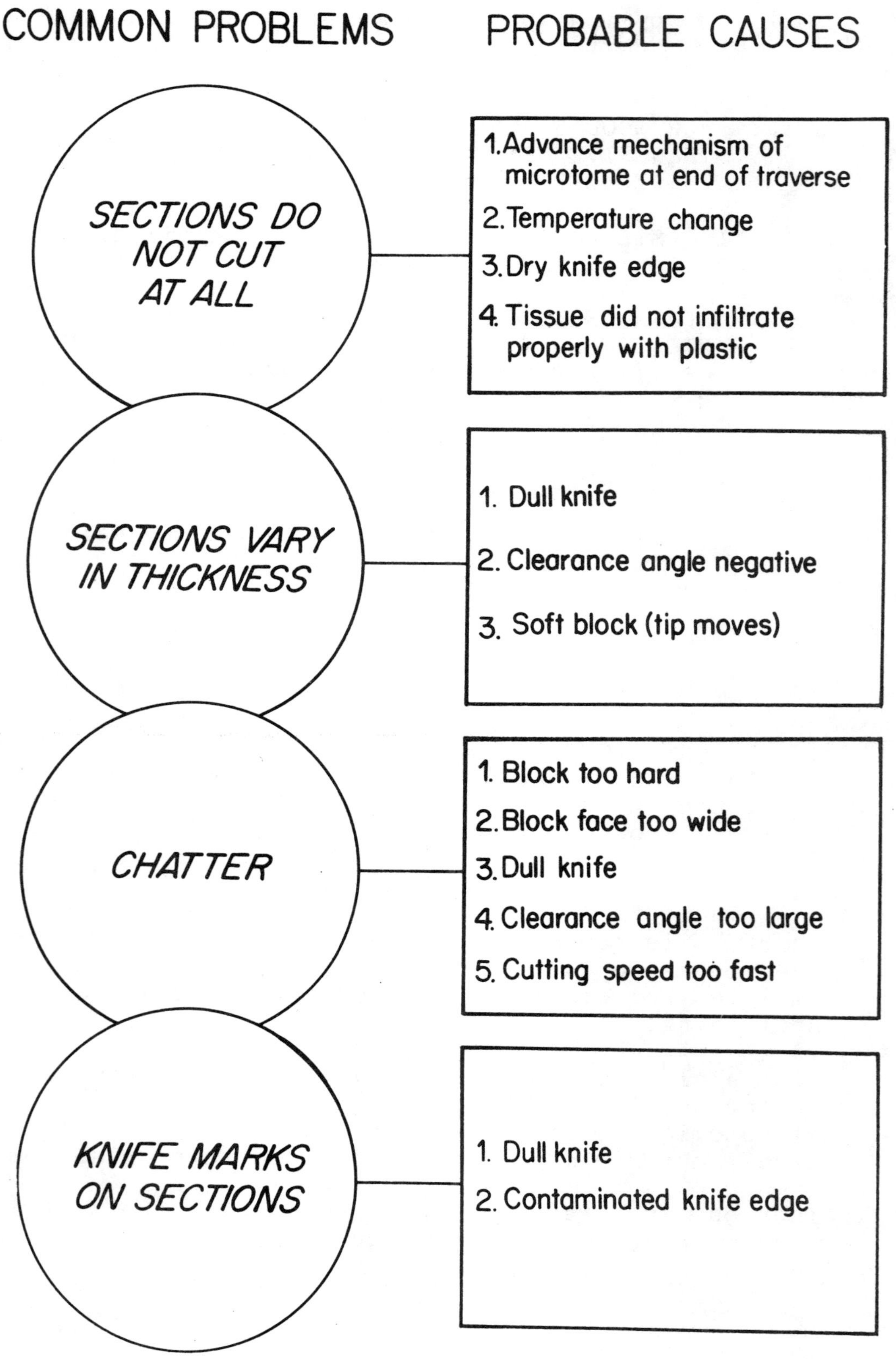

FIGURE 10-c. Common problems and probable causes during sectioning.

WETTING OF BLOCK FACE

1. Meniscus level of floatation fluid in reservoir too high
2. Incorrect knife angle

SECTIONS CONTAIN LINES

1. Small nick in knife edge

SECTIONS CRUMBLE OR STICK TO KNIFE EDGE

1. Low meniscus level of floatation fluid
2. Contaminated knife edge
3. Blunt knife edge
4. Knife angle too small

SECTIONS CONTAIN WRINKLES

1. Embedding not good
2. Dull knife edge
3. Knife edge not consistent in sharpness
4. Block face too large
5. Specimen and knife loose in holders

Problem	Causes
RIBBONS NOT STRAIGHT OR ONLY INDIVIDUAL SECTIONS OBTAINED	1. Sides of block not trimmed parallel 2. Compression on one side of block only, due to dull knife or poor impregnation
SECTIONS PULLED OVER KNIFE EDGE BY BLOCK	1. Too much water in the reservoir 2. Block too soft 3. Clearance angle too small 4. Block face too large 5. Section too thick
CANNOT SEE INTERFERENCE COLORS IN SECTIONS	1. Light not properly adjusted 2. Level of floatation fluid in knife reservoir too high or too low
SKIPPING, SECTIONS BEING CUT ONLY DURING EVERY OTHER CUTTING STROKE	1. Incorrect knife angle 2. Specimen and knife controls loose 3. Section thickness control too low 4. Poor embedding

HOLES IN SECTION

1. Contaminated knife edge
2. Air bubbles in block face

SPECIMEN FALLING OUT OF SECTION

1. Embedding process needs improving

SECTION HAS A MUGGY APPEARANCE

1. Dirt
2. Contaminated knife edge, floatation fluid and grid
3. Stain precipitate
4. Improper use of electron microscope

FATIGUE OF THE OPERATOR

1. Eyes tired
2. Patience and confidence lost
3. Overanxiousness and nervousness

I. QUESTIONS:

1. Why is it necessary to cut ultrathin sections for viewing in the electron microscope?

2. List several primary factors that determine the quality of ultrathin sections.

3. How is the thickness of a section assessed and how reliable is this method?

4. What is the thickness range in angstroms for ultrathin sections? What interference colors are observed in this range?

5. Briefly discuss the importance of ribbon formation in cutting ultrathin sections.

6. Briefly discuss a few methods of collecting sections mentioning the advantages and disadvantages of each method.

7. Discuss the advantages and disadvantages of serial sectioning in electron microscopy.

8. Briefly describe several types of grids available for section collection.

9. What are some ways of flattening sections to eliminate compression?

10. What would you consider to be four of the most common problems encountered in obtaining ultrathin sections? Why?

EXERCISE NUMBER 11

DOUBLE FIXATION OF SPECIMENS WITH GLUTARALDEHYDE AND OSMIUM TETROXIDE

Betty R. Jones, Ph.D.

A. OBJECTIVES:

1. To kill, immobilize and preserve the structure of cells with the minimum alteration from the living state

2. To protect the structure of cells from disruption during embedding and sectioning procedures

3. To prepare the structure of cells for staining and exposure to the electron beam

4. To understand the principles of procedures commonly used in the fixation of materials

5. To evaluate the quality of a tissue fixed using double fixation:glutaraldehyde and osmium tetroxide

B. MATERIALS:

1. 3% glutaraldehyde in phosphate buffer

2. Freshly prepared 2% buffered osmium tetroxide

3. A laboratory animal (rat or mouse)

4. 4 wood applicator sticks

5. 4 clean pasteur pipettes with a rubber bulb

6. 4 glass vials with caps

7. 4 disposable plastic petri dishes

8. Cold phosphate buffer for washing tissues

9. A covered tin container for storing osmium tetroxide bottled solution

10. Dissecting kit

11. New single edged razor blades cleaned with absolute alcohol

NOTE: OSMIUM TETROXIDE IS POISONOUS. USE ONLY UNDER A FUME HOOD. AVOID HAVING FUMES COME IN CONTACT WITH THE EYES OR NOSE. WEAR DISPOSABLE GLOVES.

C. RATIONALE AND DISCUSSION:

Fixation of tissues is the most critical step in preparation for structural analysis. Proper fixation leaves the components of tissues in a close to normal or in situ physiological state as is possible and allows for standardization of technique. Reproducibility of condition and comparison of replicate tissues should be possible after fixation.

By definition, a properly fixed tissue is one in which the architecture of the cells is maintained on a supramolecular level; enzymatic activity of proteins is demonstrable; ions and charged radicals are located in situ; biological membranes are intact; nucleoproteins are heterodispersed and no artifacts of space are introduced into the tissues.

The fixation process is at least a two-phase exposure when conducted in solution, as is most commonly the case. However, fixation may be accomplished through physical treatment, which is less disruptive to tissue chemistry. Also, unfixed material may be prepared for electron microscopy.

Glutaraldehyde, the dialdehyde of glutaric acid, is a superb fixative due to cross-linking of the terminal carbonyl groups with the free amino groups of cell proteins. It preserves structure without adding significant contrast over and above that of the native density of cellular constituents.

Glutaraldehyde-fixed tissues are of relatively low contrast due to retention of large amounts of cellular ground substance and because of lack of reactivity with certain lipids. All membranous differentiations of the cell are not apparent in the image obtained with the electron microscope.

Therefore, for excellent morphological detail, glutaraldehyde-fixed tissues are post-fixed with osmium tetroxide. Osmium tetroxide (OsO_4) is an excellent fixative for the general morphological study of the fine structure of cells. Carbohydrates and lipids are especially well preserved.

Osmium penetrates tissues very slowly and glutaraldehyde penetrates tissues more rapidly. Therefore, glutaraldehyde is used as the primary fixative and osmium tetroxide is used as the secondary fixative in conventional fixation procedures.

D. PROCEDURE:

1. See end of exercise for preparation of fixatives and buffers.

2. Assemble all materials listed under a fume hood.

3. Sacrifice a small laboratory rat or mouse by decapitation using an animal guillotine.

   It is important that death be quick and that the animal not be anesthetized with ether.

4. Expose internal viscera of animal and dissect out the organ of choice very quickly.

   Excise 1-3 cm of tissue from this organ. Rinse off any free blood by flushing with cold phosphate buffer.

   It has been the author's experience that kidney, spleen, liver and heart make excellent practice tissues.

5. Place the 1-3 cm of excised tissue in a petri dish and cover with several drops of 3% buffered glutaraldehyde using a pasteur pipette.

   This keeps the tissues from drying out during the mincing process.

6. Using a new, alcohol-washed razor blade mince the tissues into 1-2 mm^2. The final size of the tissue should not exceed 2 mm^2.

7. Transfer tissues to clean glass vials using applicator sticks.

Each vial should be properly labeled with the type of tissue, date and student's last name.

8. Add 5 ml of 3% glutaraldehyde to glass vials using a pipette. Cover the vials with cap.

9. Fix tissues for 2-4 hours in 3% buffered glutaraldehyde at 4°C.

NOTE: TISSUES MAY BE LEFT IN GLUTARALDEHYDE OVERNIGHT AT 4°C.

10. Following glutaraldehyde fixation, rinse tissues in four 15 minute changes of cold phosphate buffer for 1 hour at 4°C.

It is important to wash out all excess glutaralde-hyde from the tissues before post-fixing with osmium.

Failure to wash out excess glutaraldehyde may result in interference with osmium fixation.

NOTE: TISSUES MAY BE WASHED OVERNIGHT IN BUFFER AT 4°C.

11. Pour off last buffer wash and cover tissues with 3 ml of osmium tetroxide.

Cover vial tightly and post-fix under hood for 2 hours.

12. Discard used osmium in tin can under hood.

Cover tissues with phosphate buffer and proceed to the next exercise.

PREPARATION OF BUFFERS AND FIXATIVES

SODIUM PHOSPHATE BUFFER (0.2 M)

STOCK SOLUTIONS

SOLUTION A

1.	Monobasic sodium phosphate (NaH_2PO_4)	27.80 gm
2.	Distilled water	1000 ml

SOLUTION B

1.	Dibasic sodium phosphate ($Na_2HPO_4 \cdot 7H_2O$)	53.65 gm
	or	
	($Na_2HPO_4 \cdot 12H_2O$)	71.70 gm
2.	Distilled water	1000 ml

LABEL, DATE AND STORE IN REFRIGERATOR IN TIGHTLY SEALED BOTTLE.

WORKING SOLUTION (FOR pH OF 7.2-7.4)

SOLUTION A	23 ml
SOLUTION B	77 ml

Mix solution A and B. Add 1 ml of 1% calcium chloride and dilute to a total volume of 200 ml.

Adjust pH to 7.3 using HCl. Label, date and store in refrigerator in a tightly sealed bottle.

GLUTARALDEHYDE FIXATIVE

3% glutaraldehyde in phosphate buffer final pH 7.3

OSMIUM TETROXIDE FIXATIVE

Stock solution (2%)

Crystalline OSO_4	2 gm
Distilled water	100 ml

Stock solution in 0.1 N HCl

Concentrated HCl (36%)	8.6 ml
Distilled water	1000 ml

WORKING SOLUTION OF OSMIUM TETROXIDE

Stock buffer solution	5 ml
0.1 molar HCl	5 ml
Distilled water	2.5 ml
2% osmium tetroxide	12.5 ml

NOTE: EXERT EXTREME PRECAUTIONS IN HANDLING AND MAKING UP THESE SOLUTIONS.

E. QUESTIONS:

1. What is the most preferred method for sacrificing a laboratory animal and what are the advantages of this method?

2. Discuss the characteristics of a properly fixed tissue.

3. List and briefly discuss several factors that affect the quality of fixation.

4. From a molecular basis discuss how glutaraldehyde fixes and preserves the fine structure of tissues.

5. What is the rationale for post-fixing glutaraldehyde-fixed tissues in osmium tetroxide?

6. Briefly discuss the various methods of fixation and their advantages.

7. What are the advantages and disadvantages of osmium tetroxide as a fixative?

8. Briefly discuss the advantages and disadvantages of several other fixatives for electron microscopy.

EXERCISE NUMBER 12

DEHYDRATING AND CAPSULE EMBEDDING FIXED SPECIMENS

Betty R. Jones, Ph.D.

A. OBJECTIVES:

1. To remove all water tissues
2. To replace the water with a suitable organic solvent such as alcohol or acetone
3. To gradually replace the dehydration agent with a suitable embedding medium
4. To completely impregnate the tissues with embedding medium
5. To polymerize the infiltrated and embedded tissues

B. MATERIALS:

1. Bottles of ethyl alcohol in the following percentages: 30%, 50%, 70%, 83%, 95% and 100%
2. Propylene oxide
3. Several clean pasteur pipettes with rubber bulbs
4. Epon 812 embedding medium (See Table 2)
5. 5-10 beem capsules
6. 50 ml disposable plastic syringes without needle
7. Covered container for waste
8. Beem capsule holder
9. Pencil
10. Small paper strips for labeling

C. RATIONALE AND DISCUSSION:

The removal of all free water from tissues is important if tissues are to be embedded in water-insoluble resins. The water is replaced by an organic solvent such as acetone or alcohol.

When alcohol is the dehydrating agent a transition solvent such as propylene oxide must be used. This is because most embedding resins are not miscible with alcohol.

The duration of dehydration should be as short as possible provided each step is long enough to bring about a gradual replacement of water.

This will minimize any extraction of cellular materials due to prolonged exposure to the solvent.

Complete and uniform penetration of tissues by a suitable embedding medium is an essential requirement for good sectioning. This is carried out through processes of infiltration and embedding.

Infiltration involves a gradual decrease in the concentration of the solvent and an increase in the concentration of the embedding medium.

The embedding medium impregnates and supports the tissue internally and also provides a block strong enough to be handled during sectioning.

TABLE I

COMPONENTS FOR EPON 812 EMBEDDING MIXTURE

Mixture A	
Epon 812	5 ml
DDSA (Dodecenyl-succinic anhydride)	8 ml
Mixture B	
Epon 812	8 ml
NMA (Nadic Methyl Anhydride)	7 ml

Final Embedding Mixture

Mixture A	13 ml
Mixture B	15 ml
DMP - 30 (2,3,6-Tris-[dimethyl-aminomethyl] phenol)	16 drops

D. PROCEDURE A: PREPARATION OF EPON 812 EMBEDDING MIXTURE

1. Line a tray with aluminum foil and assemble the following supplies under a fume hood:

   a. 2 (20 ml) glass cylinders
   b. 50 or 100 ml disposable glass beaker
   c. 1 medicine dropper
   d. 3 glass stirring rods

2. Pour 5 ml of Epon 812 directly from bottle into a 20 ml glass cylinder.

NOTE: POUR THE RESIN DOWN THE MIDDLE OF THE CYLINDER TO PREVENT IT FROM STICKING ON THE SIDE.

3. Add 8 ml of DDSA. Carefully mix using a stirring rod. Stir slowly to avoid air bubbles. Label this as Mixture A.

4. Pour 8 ml of Epon 812 into a second glass cylinder and add 7 ml of NMA. Mix carefully. This is Mixture B.

5. Pour Mixture A and Mixture B into a beaker. Carefully mix using a stirring rod. Add 16 drops of DMP-30.

   This is the final embedding mixture. Mix thoroughly and cover with cardboard beaker top or parafilm.

NOTE: USE PRECAUTION IN HANDLING ALL CHEMICALS. DISPOSABLE GLOVES SHOULD BE WORN.

CLEAN STOCK BOTTLES BY SOAKING IN ACETONE AND WASHING LATER IN HOT SOAPY WATER. DISCARD ALUMINUM FOIL IN WASTE CONTAINER.

E. PROCEDURE B: DEHYDRATING AND INFILTRATING FIXED TISSUES

1. Assemble the following percentages of alcohol: 30%, 50%, 70%, 83%, 75% and 100%.

2. Set up a time schedule as in Table 3 and cross out each step as it is completed.

TABLE II

DEHYDRATION AND INFILTRATION SCHEDULE.

SOLUTION	DURATION	TIME STARTED
30% Alcohol	10 min.	
50% Alcohol	10 min.	
70% Alcohol	10 min.	
83% Alcohol	10 min.	
95% Alcohol	10 min.	
100% Alcohol	10 min.	
100% Alcohol	10 min.	
100% Alcohol	10 min.	
Propylene oxide (P.O.)	15 min.	
Propylene oxide	15 min.	
P.O/Epon 812 mixture 2:1	1 hr.	
P.O/Epon 812 mixture 1:2	1 hr. or preferably overnight	

3. Assembly vial containing tissues. Pour off buffer.

4. Cover fixed tissues with 4-5 ml of 30% alcohol.

   Agitate carefully and let dehydrate for 10 minutes.

5. Pour or pipette off used alcohol and repeat procedure number 4 by adding the next alcohol solution.

   Continue to follow the schedule in (Table II).

NOTE: THE TISSUES SHOULD NOT BE ALLOWED TO DRY OUT IN BETWEEN CHANGES.

FOLLOW THE SCHEDULE CAREFULLY MAKING SURE YOU CROSS OUT EACH STEP AS IT IS COMPLETED.

6. The ratios of propylene oxide/epon 812 mixture (i.e. 2:1 and 1:2) may be prepared by eye estimation in a glass vial, then added to the tissues. Gently agitate.

7. After adding the 1:2 propylene oxide/Epon 812 mixture to the tissues place vial under fume hood with cap off.

   Leaving the cap off will allow some of the propylene oxide to evaporate thus resulting in better infiltration of tissues with the remaining embedding medium.

PROCEDURE C: EMBEDDING

NOTE: SEVERAL HOURS BEFORE EMBEDDING TISSUE TURN OVEN ON AND MONITER TEMPERATURE.

FOR EPON 812 MIXTURE THE TEMPERATURE SHOULD BE AT 60°C.

FOR MEDIUM SUCH AS SPURR LOW VISCOSITY THE TEMPERATURE SHOULD BE AT 70°C.

PLACE BEEM CAPSULES IN CAPSULE HOLDER 2 HOURS BEFORE EMBEDDING.

THEN PLACE CAPSULE HOLDER CONTAINING BEEM CAPSULES IN OVEN FOR DRYING.

1. Remove capsule holder with beem capsule from oven and assembly on aluminun foil.

   In addition assembly the following other items:

   Labels, pencil, 50 ml disposable syringe, epon mixture, sheet of dental wax and applicator sticks.

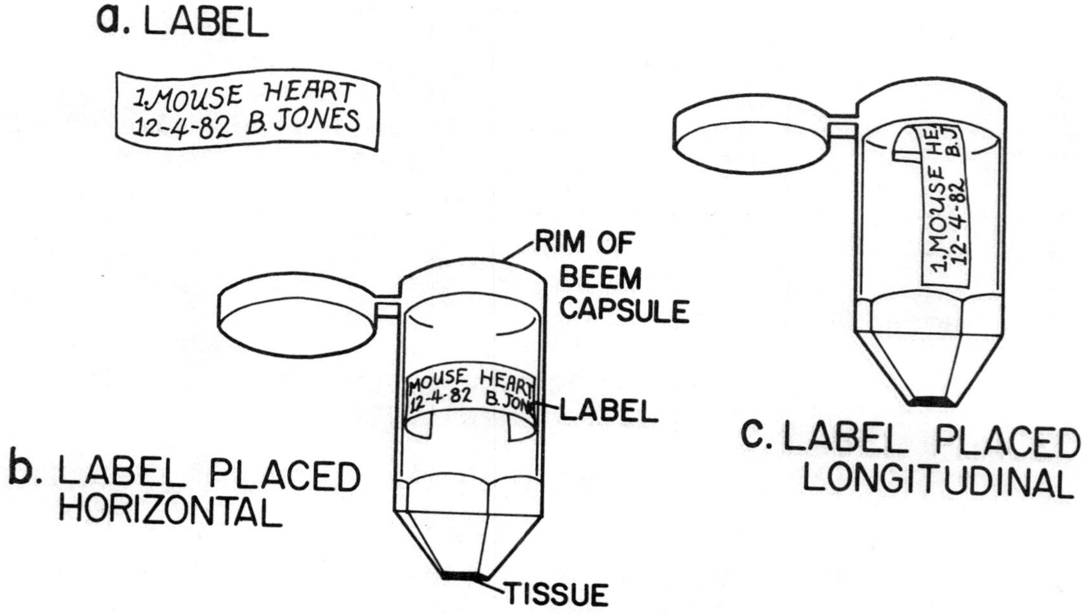

FIGURE 12-a-c. Appearance of label in beem capsule.

2. Remove vial containing tissues from hood and pour off the final propylene oxide/epon mixture.

   Carefully discard mixture in a covered container for waste.

3. Add 3 ml of fresh epon mixture to tissues. Turn vial upside down on sheet of dental wax. The tissues will fall onto dental sheet with epon mixture.

4. Use an applicator stick to select tissue specimens that are intact, solid, not showing fragmented edges and less than 1 mm in size. Select up to 10 specimens.

5. Using a 50 ml disposable syringe draw up some of the epon 812 mixture and place 2-3 drops in the bottom of each beem capsule in the capsule holder.

6. Gently pick up one piece of tissue with applicator stick.

   Tissue will adhere to applicator stick due to the sticky resin.

   Place one piece of tissue in the bottom of each beem capsule.

NOTE: NEVER PLACE MORE THAN ONE TISSUE IN THE SAME BEEM CAPSULE.

7. Place additional Epon 812 mixture in the beem capsule until the capsule is almost full.

   The epon resin should be 1-2 mm below the rim of the capsule.

8. Insert label in beem capsule by rolling it up on a pencil and lowering it into resin or using forceps to place it longitudinal along the side of the beem capsule (See Figure 12-a-c).

9. Check to make sure the tissue in capsule is oriented directly at the tip.

   If not at tip use an applicator stick to place it in position.

NOTE: DO NOT REMOVE THE BEEM CAPSULE FROM CAPSULE HOLDER.

F. QUESTIONS:

1. Why is it essential to wash tissues following fixation?

2. What is the purpose of dehydration?

3. Discuss the advantages and disadvantages of the two most commonly used dehydrating agents.

4. Why is it necessary to use a transition solvent when dehydrating tissues with alcohol?

5. Briefly discuss two transition solvents that can be used following alcohol dehydration.

6. Define infiltration and embedding.

7. What are some factors that determine a good embedding medium?

8. What are the roles of DDSA and DMP 30 in the embedding mixture?

9. Discuss the advantages and disadvantages of two common methods of polymerization.

10. Briefly describe the components and discuss the advantages and disadvantages of the following embedding resins: Epon 812, Araldite 502, Spurr low viscosity, methacylate, vestopal W, Rigolac, polyethylene glycol, maraglas, and glycol methacrylate.

NOTES

EXERCISE NUMBER 13

DOUBLE STAINING FOR ELECTRON MICROSCOPY: URANYL ACETATE AND LEAD CITRATE

Betty R. Jones, Ph.D.

A. OBJECTIVES:

1. To obtain better contrast and resolution in ultrathin sections by the deposition of a heavy metal
2. To facilitate better focusing
3. To provide a means of obtaining ultracytochemical data from structural components
4. To protect sections from damage caused by the electron beam by increasing their resistance
5. To provide clarity in the fine structure of cells

B. MATERIALS:

1. 3% solution of uranyl acetate
2. 0.2% lead citrate stain
3. 4 glass petri dishes
4. Boiled, millipore filtered distilled water
5. Epon 812 embedding mixture (40 ml)
6. Fresh sodium hydroxide pellets
7. Tweezers
8. Grids containing section
9. Time clock
10. Clean pasteur pipettes with rubber bulbs
11. 2 clean 50 ml beakers

12. 2 wooden spatula's

13. Wax pencil

14. Whatman's filter paper

NOTE: SEE END OF EXERCISE FOR PREPARATION OF STAINS.

C. RATIONALE AND DISCUSSION:

Staining in light microscopy differs from that of electron microscopy in that a large number of various differential staining techniques can be used to identify various tissue components.

In electron microscopy the structures of the cell can only be identified by their morphology and electron opacity.

Various areas of an electron micrograph are seen because of a difference in contrast.

This contrast is due to differential electron opacity based on a scattering or deflection of electrons in the section.

The deposition of a heavy metal such as uranyl acetate or lead citrate increases the mass of certain selective sites inherent in the tissue.

In other words, most atoms in plant and animal tissues have a low atomic number and the electron scattering ability of these light elements is small.

Therefore, the incorporation of heavy metal atoms throughout the tissue provides a sufficient range of densities in structures of the tissue thereby providing contrast in the section.

NOTE: HANDLE URANYL AND LEAD SALTS WITH EXTREME CARE! LEAD SALTS ARE POISONOUS!

OPEN BOTTLES CAREFULLY AND AVOID CONTACT WITH SKIN!

MAKE SURE GLASSWARE USED FOR STORING STAINS IS CHEMICALLY CLEAN BEFORE RETURNING TO THE GENERAL STOCKROOM.

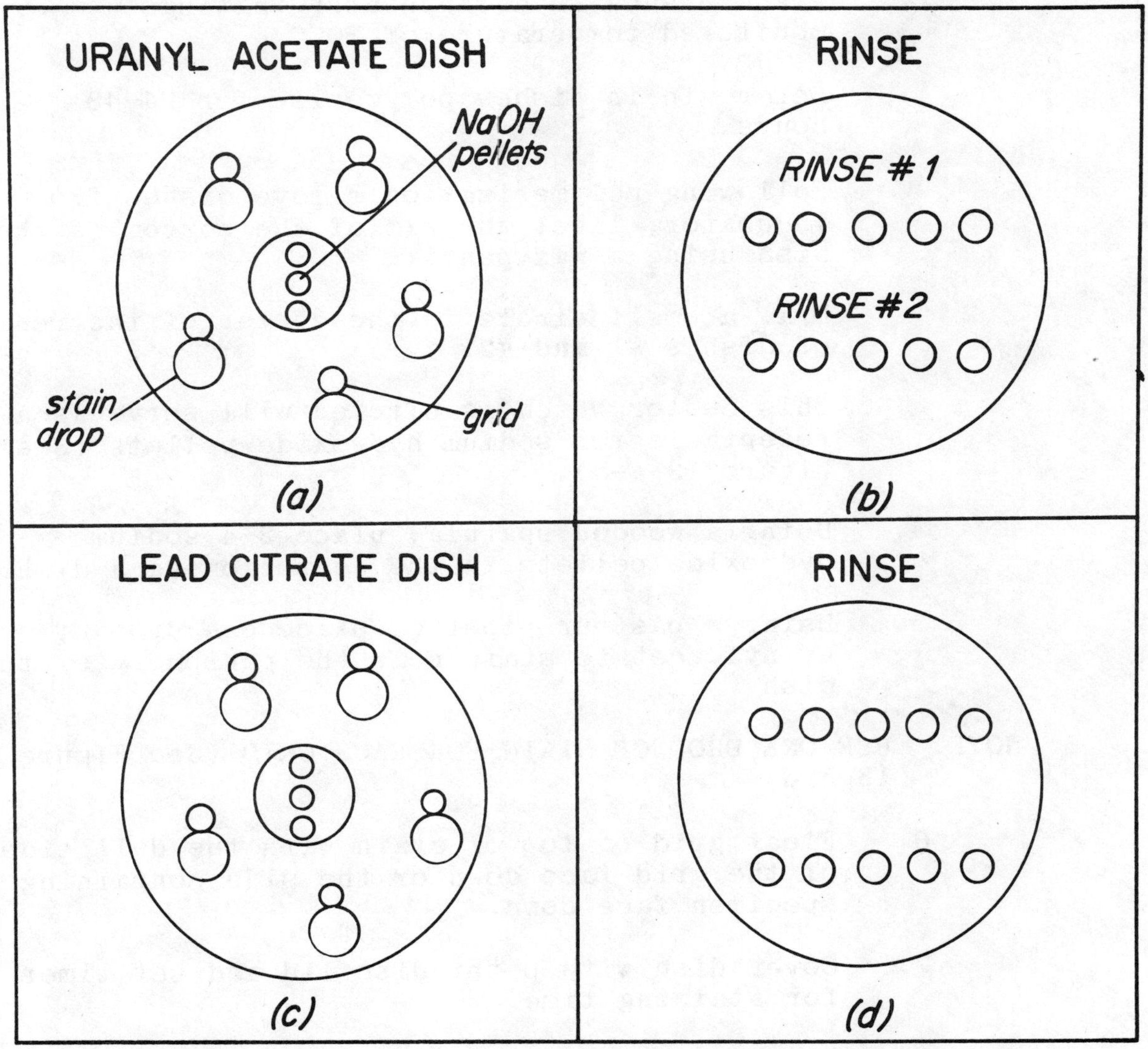

FIGURE 13-a-d. Uranyl acetate and lead citrate staining procedure.

D. PROCEDURE:

1. Prepare smooth, clean staining surfaces in petri dishes by pouring 10 ml of Epon 812 embedding resin in each dish.

NOTE: PETRI DISHES WILL BE PREPARED 2 DAYS IN ADVANCE.

2. Place dishes in an incubator that has a monitored temperature of 60°C.

   Let resin in dishes polymerize for 24-48 hours.

3. Following polymerization remove dishes from incubator, label the rim of the bottom of the dish using a wax pencil.

   Draw a small circle in the center of the resin of dishes #1 and #2.

   This center which is circled will serve as a receptacle for sodium hydroxide pellets (See Figure 13-a-d).

4. Using a wooden spatula, place 2-4 sodium hydroxide pellets in the center of each dish.

5. Using a pasteur pipette, place one drop of uranyl acetate stain near the periphery of the dish.

NOTE: USE ONE DROP OF STAIN FOR EACH GRID (See Figure 13-a-d).

6. Float grid on top of stain with the dull side of the grid face down or the side containing specimen face down.

7. Cover dish with petri dish lid and set timer for staining time.

8. Allow grids to stain for 10 minutes.

9. Using another clean pipette, place drops of distilled water in petri dish #2 according to the number of grids (1 drop for each grid).

   Float grids on water surface (specimen side down) for 2 seconds in first rinse and for 2 seconds in second rinse.

10. Using a clean pipette, place drops of lead

citrate stain in petri dish #3.

Float grids on surface of the stain.

11. Allow grids to stain for 2 minutes.

12. Rinse grids in petri dish #4 (2 rinses) for 2 seconds each.

13. Place stained grids back into clean dry dish on filter paper with specimen side facing up.

NOTE: GRIDS WILL DRY WITHIN SECONDS.

14. You are now ready to examine stained grids in the electron microscope.

If contrast is poor, grids can be restained for a longer period of time.

NOTE: STAINING TIME VARIES WITH THE TYPE OF TISSUE.

However, for most tissues ten minutes in uranyl acetate and ten minutes in lead citrate should be sufficient if the lead citrate is not made up concentrated.

The lead citrate stain prepared in this exercise is concentrated, therefore staining time should be short (45 seconds to 2 minutes).

NOTE: FOLLOW INSTRUCTIONS CAREFULLY FOR PREPARING URANYL ACETATE AND LEAD CITRATE STAINS.

AT THIS STAGE, CONTAMINATION OF SECTIONS CAN EASILY OCCUR.

E. EVALUATION:

PRESENT 4 STAINED GRIDS TO THE INSTRUCTOR FOR EVALUATION IN THE ELECTRON MICROSCOPE.

F. PREPARATION OF URANYL ACETATE STAIN:

Stock Solution

3% uranyl acetate in water

Store in a dark bottle in the refrigerator

NOTE: THE STOCK SOLUTION IS SATURATED.

Working Solution

Filter before use. Centrifuge for 20 minutes the amount to be used for staining.

G. PREPARATION OF 0.2% LEAD CITRATE STAIN:

1. Place 20 mg of lead citrate into test tube with screw cap.

2. Add 9 ml of fresh boiled distilled water.

3. Add 1 ml of 1 N of sodium hydroxide.

4. Shake the solution until clear. The pH of the solution should be around 11.

NOTE: THE STAIN SHOULD BE CLEAR AND FREE OF PRECIPITATES.

IF AIR IS INTRODUCED INTO THE TUBE, THIS MAY LEAD TO PRECIPITATION OF LEAD CARBONATE AND A CLOUDY STAIN.

5. Centrifuge stain for 20 minutes.

6. Label tube and use immediately.

7. Grids can be stained from 45 seconds to about 2 minutes.

   However, this time is dependent on the type of tissue and embedding resin. Staining time also depends on the objectives of the study. The staining time will vary.

NOTE: USE THIS STAIN ONLY ONCE. IT SHOULD BE MADE UP FRESH EACH TIME.

8. To clean lead citrate from test tube, discard remaining stain and wash tube with a few sodium hydroxide pellets and water.

   Afterwards, wash in soap and water and rinse several times in warm distilled water.

   It is even good practice to let the tube soak for a few days then rinse after soaking.

F. QUESTIONS:

1. What method is most commonly used for staining grids and why?

2. Under what circumstances might the immersion method of staining grids be used?

3. Briefly discuss what is believed to be the mechanism of staining.

4. Name and briefly describe the action of four fixatives that act as stains.

5. Why are heavy metal salts used as stains?

6. What are some of the disadvantages of staining during fixation or dehydration?

7. What are some factors that effect the contrast of a section?

8. List eight characteristics that are essential in assessing the duration of staining.

9. Briefly discuss the specificity of lead and uranyl stain preparations in regard to the chemical composition of cell structures.

10. Briefly discuss lead citrate contamination in a section.

11. Under what conditions might one not stain a tissue section and why is there still some degree of contrast.

12. What is meant by en bloc staining?

13. Compare the advantages and disadvantages of en bloc staining to that of grid staining.

EXERCISE NUMBER 14

INTRODUCTION TO THE USE OF THE TRANSMISSION ELECTRON MICROSCOPE

Betty R. Jones, Ph.D.

A. OBJECTIVES:

1. To introduce the student to the parts and functions of the transmission electron microscope (TEM)

2. To demonstrate to the student operations and procedures for using the transmission electron microscope

B. MATERIALS:

1. Petri dish containing stained grids

   NOTE: GRID MODULES FOR STORING GRIDS MAY BE OBTAINED FROM EM SUPPLIERS.

2. A pair of fine tip forceps

3. Log book for record keeping

   NOTE: IN MOST EM FACILITIES A LOG BOOK IS KEPT NEAR THE MICROSCOPE, SEE SAMPLE LOG SHEET, TABLE II.

C. RATIONALE AND DISCUSSION:

The construction and use of the transmission electron microscope in the biological and physical sciences represents one of the most novel approaches and advances in these fields. Many of our present-day concepts on the cell structure of viruses, prokaryotic and eukaryotic organisms have developed at rapid momentum due to electron microscopy. Today electron microscopy is used in a variety of disciplines by students and investigators as a valuable teaching and research tool.

The transmission electron microscope is similar in function to the light microscope but methods of specimen preparation may vary. The TEM requires the use of thin sections (500-800 angstroms) that will allow electrons to penetrate whereas thick sections (0.5 - 1 micron) are adequate for light microscopy. The major advantages of the TEM over the light microscope lies in its resolution capabilities. The light microscope is capable of viewing particles as small as 0.2 microns whereas the TEM is capable of viewing molecules the size of 0.1 - 0.2 angstroms. This allows a great amount of structural detail to be visualized.

The first TEM with electromagnetic lenses was built by Knoll and Ruska in the 1930's. Since 1950 the TEM has been used for direct examination of cell structures down to the level of macromolecules. It has bridged the gap between examining cells by light microscopy and analyzing the molecular structure of cells by x-ray diffraction and birefringent microscopy.

In principle, the TEM uses an electron beam emitted from a tungsten wire filament as the source of illumination. The image of the specimen is formed by focusing the electron beam through a series of magnetic lenses. As the beam is transmitted through a thin specimen electron scattering occurs due to dense components in the specimen. As a result of this phenomenon shadows of scattered electrons are casted on the photographic emulsion plate or film.

Living or wet specimens cannot be viewed in the TEM. Therefore, special fixation, embedding, sectioning and staining procedures are required to prepare specimens for viewing in the microscope. Procedures on the use and operation of the microscope have been omitted due to the different types of models available.

The use and operation of a TEM would depend upon the model type. Manufacturers of TEMs are helpful in providing detailed instruction manuals and service people to train one on how to use the instrument. The use of the TEM today and its future possibilities in many disciplines especially the physical and biological sciences are immeasurable.

D. PROCEDURE:

NOTE: THE INSTRUCTOR WILL EXPLAIN AND DEMONSTRATE TO THE STUDENT THE PARTS AND FUNCTIONS AND HOWS TO USE THE

ELECTRON MICROSCOPE IN A STEPWISE SEQUENCE.

A special handout on the operation procedures of the microscope based on the supplier's instruction manual will be given to the student. This will include a schematic drawing of the microscope type and parts.

TABLE I

General Basic Parts and Functions of the Transmission Electron Microscope. Study and Become Familiar With the Following Parts and Their Functions Before Attempting To Use the Microscope.

PARTS	FUNCTIONS
A. Illumination System	This system produces the source of radiation and focuses it onto the specimen. It is composed of two unit parts: the electron gun and the condenser lens.
1. Electron Gun	The gun contains the source of electrons and consists of three components: the filament, cathode shield, and anode plate aperture.
a. Filament	The filament which may also be referred to as the cathode or emitter is a small, symmetrical, v-shaped piece of tungsten wire 0.1 mm in diameter.
	When the filament is heated by an electrical current it becomes energized and emits electrons. A cloud of electrons will be emitted and concentrate at the tip of the filament.
B. Cathode Shield	The shield houses the filament and contains an aperture located below the tip of the filament.
	It acts as a weak lens accumulating the electrons into a dense cloud. The

	shield is also referred to as the focusing electrode or grid cap.
C. Anode Plate Aperture	The anode is a circular plate that contains a central aperture through which the electron beam can pass. It is aligned in such a way that it lies on an axis running through the filament tip and cathode shield aperture. The anode may also be called the positive electrode. It has a positive potential with respect to the cathode which is held at a negative potential. The electrons will be accelerated down the column by the voltage difference between the filament and anode plate.
2. Condenser Lens	The condenser lens focuses the electron beam emerging from the filament of the electron gun onto the specimen. An intense spot of illumination will be produced on a small part of the specimen and then will be spreaded over a larger area at reduced illumination for visualization and recording.
B. Imaging System	This system is responsible for magnifying the final image of the specimen. It is composed of two lenses: the objective lens and the projector lens.
1. Objective Lens	The objective lens allows the electron beam that has passed through the specimen to come into focus again forming a magnified intermediate image.

	The image produced by these lenses will determine the quality of the image that is finally recorded through photography. The objective lens thus determines the resolution of the microscope.
2. Projector Lens	The projector lens may also be referred to as the ocular lens. These lenses permit magnification of a part of the intermediate image to form the final image.
C. Image Recording System	This system transforms the electron beam into a permanent image that can be viewed directly. The primary components are a fluorescent screen and photographic emulsions.
1. Fluorescent Screen	The fluorescent screen allows direction of orientation and location of the desired field of view. It enables one to study the characteristics of the image produced by the specimen and also to focus correctly the lenses and align the microscope.
2. Photographic Emulsions	The photographic emulsions are electron image plates or films that provide the final permanent recording of the image.

E. RESULTS:

As a result of this exercise students should be able to do the following:

1. Turn on water circulator controls.

2. Turn on and pump down microscope.

3. Turn on the high voltage.

4. Turn on the electron beam.

5. Adjust and align the illumination of the beam on the fluorescent screen.

6. Place the grid in the specimen holder.

7. Insert specimen holder into the path of the beam.

8. Focus the specimen image on the screen.

9. Select the area of interest for photographic recording.

10. Select desired magnification.

11. Focus the image again.

12. Expose electron microscope plates or films.

13. Record information in log book.

14. Remove exposed plates or films from the microscope.

15. Reload microscope cassettes in magazine with unexposed plates or films.

16. Pump microscope plates or films.

17. Shut off microscope.

18. Develop plates or films (See Exercise Number 18A).

TABLE II

SAMPLE LOG SHEET

Name and Model of
Transmission Electron Microscope: Zeiss 9S-2
Institution: Morehouse College
Department: Biology
City & State: Atlanta, Georgia

A. Name of Developer: D-19

B. Number of Times Used: 2

C. Name of Fixer: Kodak Fixer

D. Number of Times Used: 2

E. Name of Clearing Agent: Hypo (Optional)

F. Number of Times Used: 2

G. High Voltage: 60 Kv

H. KV Scale Reading when taking Photographs: 6

I. Plate or Film Emulsion Setting: 6

J. Log Sheet Number: 402

DATE	Plate or Film Number	Magnification	Specimen	Scale Reading & Emulsion	Operator
1/10/83	4,448	14,000	Liver	50 Scale 6 emulsion	John Doe

TABLE III

COMMON PROBLEMS THAT OCCUR WITH ELECTRON MICROSCOPES.

The basic and most likely problems to occur with microscopes and their possible causes are listed in the Table below. This list is by no means exhaustive (courtesy of Dr. Brenda S. Weakley, 1972, A Beginner's Handbook in Biological Electron Microscopy pp. 192-198, Churchill Livingstone, London.

PROBLEMS	POSSIBLE CAUSES
No beam when filament is energized	1. High voltage off
	2. Filament blown
	3. Fluorescent screen up
	4. Specimen chamber closed off from column
	5. Something else obstructing beam
	6. Condenser 2 defocused
	7. Magnification not at lowest setting
	8. Specimen traverse rods at end of runs
	9. Column misaligned
Beam disappears during operation	1. High voltage cut off due to low water pressure on vacuum leak
	2. Filament blown
	3. Specimen traverse rods at end of run
	4. Some alignment control inadvertently moved
	5. Objective aperture has shifted

Beam cross section is elliptical or distorted	1. Condenser astigmatism needs correction 2. Contamination of condenser too great for astigmatism corrector to compensate
Flickering of Beam	1. Filament about to blow 2. Contamination associated with gun or gun chamber 3. Electrical instability
Reduced Illumination	1. Gun misaligned 2. Filament too far from cathode shield aperture
Beam sweeps across screen when condenser, current is varied	1. Condenser aperture off-center
Image moves across screen when objective focus is varied	1. Condenser lens not properly aligned with respect to objective
Image moves across screen when magnification is varied	1. Projector lens not properly aligned with respect to rest of column
Image will not focus	1. Course focus control out of focus 2. Astigmatism of objective lens 3. Specimen holder not properly placed 4. Electrical fault

Blurring of diffraction fringe in one direction around a hole in a carbon film	1.	Mechanical vibration near microscope or an unidirectional varying magnetic field
Blurring of diffraction fringe all around hole in carbon film	1.	Instability in lens current on accelerating voltage
Image drift	1.	Thermal expansion of specimen or grid
	2.	Torn specimen
	3.	Dirt on specimen
	4.	Dirty specimen holder
	5.	Defective specimen stage
Short Filament Life	1.	Beam current too high
	2.	Insufficient vacuum
	3.	Filament position incorrect with respect to cathode shield aperture
Discharges in gun	1.	Insufficient vacuum
	2.	Contamination in gun chamber
High contamination rate	1.	Dirty specimen holders
	2.	Insufficient vacuum
	3.	"Cracked oil"
Layer of contamination visible on fluorescent screen and/or	1.	Oil in column due to failure of value between diffusion pump and

viewing port	column; electricity failure; or failure of pump cooling system

F. EVALUATION

The instructor will evaluate plates or films following Exercise Number 18A.

G. QUESTIONS:

1. Give a brief synopsis of the history of electron microscopy mentioning the contribution made by pioneers in the field.

2. By diagram show the sequential steps that occur as the current of the filament is increased.

3. Which is a better electron source, a pointed filament or a non-pointed filament? Explain why.

4. What is the advantage of a microscope having a double condenser lens?

EXERCISE NUMBER 15

ELECTRON OPTICS

Raymond L. Hayes, Ph.D.

A. OBJECTIVES:

1. To relate the light microscope to the electron microscope in terms of resolving capacity and magnification

2. To understand the lens system of the electron microscope

3. To appreciate the significance of components of the electron image

4. To appreciate the quality of the specimen for electron microscopy

B. MATERIALS:

1. Electron microscope

2. Column diagrams of the microscope

C. RATIONALE AND DISCUSSION:

To understand the design and operation of an electron microscope requires application of physical principles of visible light optics to a system using a beam of electron particles.

Principles

The generator of electrons in the electron microscope is the "gun" consisting of a tungsten wire cathode (-) which is heated to incandescence. A cloud of electrons is generated by thermionic emission and is accelerated toward an anode (+) by a differential of 50+ kilovolts. Under these circumstances the electrons travel in wave form with a = 0.05 nm. Hence the electron beam is 100,000 times shorter in wave length than visible light.

The lens system of the electron microscope consists of a stacked series of electromagnetic coils, each of which is toroidal in shape. These lenses are able to deflect the beam of electrons and as such magnify images produced after passing through a specimen. The condenser lenses condense the electron beam, while the objective and projector lenses magnify the image.

The column of the electron microscope is under vacuum to prevent gaseous particles from deflecting the beam of electrons. This vacuum is developed by a rotary pump and maintained by an oil diffusion pump.

Magnification in the microscope is adjusted by changing the amount of current through the objective lens. The projector lens is a fixed magnification and from there the image of the specimen is projected onto a fluorescent screen or photographic film. Preservation of resolving quality depends upon grain size of the film emulsion and focusing of the image on the fluorescent screen.

Contrast is introduced into the microscope by using small apertures (25-50 microns) in the lenses. The smaller the aperture, the greater the contrast. Resolution is increased at higher electron accelerating voltages but at the sacrifice of contrast.

D. PROCEDURE:

1. With instructor at microscopic controls, observe the influence of altering the objective lens current upon magnification.

2. By changing apertures to different sizes, show how contrast within biological specimens are altered.

3. By comparing unstained and stained sections, show how contrast is introduced within the specimen to facilitate electron microscopic examination.

E. RESULTS:

As the objective lens current increases, magnification increases. Contrast increases with smaller apertures. Unstained sections of biological material reveal no structural detail, since the electron density of organic tissue is very low. Stains introduce contrast and permit structural detail to be defined within thin sections.

F. QUESTIONS:

1. Define the following:

   Resolving power, diffraction, electron cloud, accelerating voltage, refractive index, magnification, magnetic coils, magnetic field.

2. Describe the basic components and functions of an electron microscope system.

3. Diagrammatically compare the light microscope with the electron microscope.

NOTES

EXERCISE NUMBER 16

RESOLUTION

Raymond L. Hayes, Ph.D.

A. OBJECTIVES:

1. To explain the optical features of the human eye

2. To apply principles of visible light optics to the design of a microscopic system

3. To define resolution as influenced by inter-particle distance and distance from the eye

B. MATERIALS:

1. Series of sized particles

2. Series of poster board backgrounds

3. Measuring tape

4. Graph paper for recording data

C. RATIONALE AND DISCUSSION:

Since the objective in microscopic examination of any structure is to reveal structural detail, the factors limiting or determining elucidation of detail must be appreciated.

Principles

Resolution is defined as the ability to distinguish the individuality of two particles or lines when juxtaposed next to one another. To understand resolution in the human visual system, it is important to recognize that the eye contains photosensitive cells in the retina. These receptor cells, known as rods and cones, constitute an epithelial sheet in which each cell is roughly 4 mm in diameter.

To resolve two particles in the environment at least two receptor cells in the retina must be stimulated giving information which must be preserved with spatial separation through integrative relations in the visual cortex.

Clinical testing has established that optimal resolution of the human visual system occurs under conditions where two particles are separated by 0.1-0.2 mm and are held 10 inches (254 mm) from the surface of the cornea.

The capacity of the eye to resolve declines at closer or further distances provided the size of the particles remains constant. The size of the particles has no influence upon resolution until they are so small as not to be perceived.

To change resolution, the distance of the particles from the eye must change or the particles must be magnified through the addition of an external lens system.

D. PROCEDURE:

1. Place two particles 1 mm apart and 10" (254 mm) from the cornea. Vary the distance of the particles from the eye, recording the interval at which resolution is lost.

2. Vary the interparticle distance, keeping the eye-to-particle distance at a fixed interval of 20" (508 mm), recording the interval at which resolution is lost.

3. Repeat #1 and #2 with particles of larger or smaller size to evaluate the influence of this variable upon resolution.

4. With particles used in #1 and #2, vary intensity of illumination or grayness of background to evaluate the influence of contrast upon resolution.

E. RESULTS:

Resolution of visible particles by the unassisted human eye depends upon two distances: interparticle distance and eye-to-particle distance. Differentiation of the particles from the background improves clarity of resolution. Greater contrast is introduced with more intense illumination and distinction in shade or color.

F. QUESTIONS:

1. Discuss the importance of resolution with respect to fine structural details.

2. Why is it essential to understand the optical features of the human eye to appreciate resolution?

3. Briefly discuss resolution in relation to wave length.

NOTES

EXERCISE 17

MAGNIFICATION

Raymond L. Hayes, Ph.D.

A. OBJECTIVES:

1. To explain how objects are magnified in physical terms

2. To determine the theoretical resolving power of a microscope

B. MATERIALS:

1. Light microscope

2. Immersion oil

3. Sample grid or stage micrometer

C. RATIONALE AND DISCUSSION:

In order to increase resolving power of the human eye, scientists apply system which magnify particles. How magnification improves resolution is critical for the understanding of how a microscope functions.

D. PRINCIPLES:

When visible light passes from air to another medium such as oil or water, light waves are refracted or bent. Magnification of a visible object is essentially the consequence of an orderly change in the path of light from that object to produce an enlarged image. According to the Abbe formula,

$$D = \frac{0.61\ x}{n\ \sin}$$

Where D = resolution
n = refractive index of medium between particles and eye
x = half angle (See Figure 17-a).

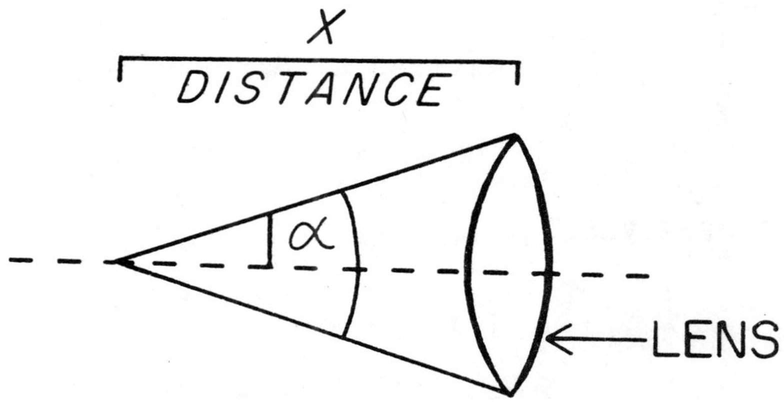

λ = WAVELENGTH (NANOMETERS)

$n \sin \alpha$ = N.A. (NUMERICAL APERTURE OF LENS)

FIGURE 17-a. Semi-aperture angle (α) limited by lens diameter.

As X (distance from particle or object to lens) decreases, approaches 90° and sin approaches 1.

In light microscopy where the wavelength of visible light is about 5,000 nm and where n of inversion oils is about 1.5,

$$D = \frac{0.61 \times 5{,}000}{1.5 \sin} = 0.2 \text{ m}$$

Theoretically, as sin approaches 1, D reaches a theoretical limit of 0.2 m.

If the optical system magnifies the object 1000 X the interval of 0.2 m is magnified to 0.2 mm. If the interparticle distance is less than 0.2 m no amount of magnification will effectively resolve the object.

If the wavelength of visible light used decreases such as ultraviolet light (2500 nm) then resolution decreases accordingly. If quartz lenses are used and not glass, resolving power increases.

D. PROCEDURE:

1. N.A. of a lens is engraved by the manufacturer upon the lens mount. Introduce this factor as the denominator of the Abbe equation.

2. Under oil immersion, examine the resolution of a sample in white light.

3. Introduce colored filters to change and determine the influence of light upon resolution with the micrometer.

4. Record the effect of wavelength upon resolving capacity of the microscope.

5. If oils of different refractive indices are available substitute and repeat evaluation of resolution to determine effect of n upon the optical capacity of the instrument.

E. RESULTS:

As wavelength of visible light shortens, resolution decreases. As refractive index of oil increases, resolution increases.

F. QUESTIONS:

1. Briefly discuss magnification in relation to Abbe's equation.

2. How does one determine the final magnification of an electron micrograph?

EXERCISE NUMBER 18A

PHOTOGRAPHIC TECHNIQUES

DEVELOPMENT OF ELECTRON MICROSCOPE PLATES OR FILMS

Betty R. Jones, Ph.D.

A. OBJECTIVES:

1. To record images viewed in the electron microscope
2. To process plates or film to permit effective printing and storage of information
3. To eliminate artifacts of photographic origin

B. MATERIALS:

1. Electron image plates or films
2. D-19 Developer (1 part developer: 2 parts tap water)
3. Kodak fixer (full strength)
4. Kodak Hypo clearing agent (1 part hypo: 4 parts tap water)
5. Distilled water
6. Equipped darkroom
7. Timer
8. Safelight, #1 wratten filter

NOTE: PREPARE SOLUTIONS BY CAREFULLY FOLLOWING DIRECTIONS ON PACKAGE.

C. RATIONALE AND DISCUSSION:

The utilization of electron microscopy techniques efficiently requires obtaining artifact-free data from photographic electron image plates or films. The information is recorded on very fine grain sensitive emulsion plates or films that are selected for their sensitivity to the electron beam rather than to visible light. The emulsion overcoating is gelatin and it reduces the susceptibility of the plates or films to abrasion.

Under the illumination of the safelight in the darkroom, the emulsion-coated side of the plate or film appears lighter and glossier than the red-dyed surface. Electron image plates or films are designed specifically for electron micrography and electron microscopes.

Photographic procedures for standard photography utilizes light waves reflected from the specimen. The light waves pass through an aperture and strike the light sensitive emulsion. The emulsion is a suspension of silver halide aggregates.

When light impinges the silver halide grains, energy from light is lost to the grains. The result of this is the formation of small silver specks, which must be large enough so that the grain can be developed. Therefore, when the exposed plate or film is placed in a developing solution in the dark, the silver halide grains which contain large specks are activated.

The developer is essentially a reducing agent. A pattern appears on the plate or film, giving a negative image of the specimen or person which was photographed. The emulsion is washed free of developer and placed in a fixative.

The fixative removes the color or opalesce and converts all the underdeveloped silver halide in the emulsion to a water-soluble product. This water-soluble product is removed by washing and the dried negative can be stored until a positive print is obtained.

The same principles are used in electron photomicrography, only an electron beam is the source of energy instead of a beam of light. The reaction of electrons with silver halide grains is similar to that of light. Therefore, the emulsions used for light photography can be used for electron photography techniques.

NOTE: THE EXPOSED PLATE OR FILM IS YOUR PERMANENT RECORD OF THE ULTRATHIN STAINED TISSUE SECTION. ONCE YOU HAVE LOCATED A SECTION OF INTEREST IN THE ELECTRON MICROSCOPE, EXPOSE SEVERAL PLATES OF THE IMAGE IN A SERIES. THIS IS IMPORTANT IN CASE THE SPECIMEN IS DESTROYED.

GRIDS MAY BE STORED FOR LONG PERIODS OF TIME, BUT THEY ARE FRAGILE AND YOU MAY NEVER BE ABLE TO FIND THAT SAME SECTION AGAIN.

BEFORE DEVELOPING PLATES, MAKE SURE DARKROOM IS CLEAN AS POSSIBLE! PREPARE ALL SOLUTIONS IN ADVANCE, CHECK TEMPERATURE OF ROOM AND SOLUTIONS. MAKE SURE PANS ARE LABELLED PROPERLY.

D. PROCEDURE:

1. Remove plate receiver from the microscope and take into darkroom or store temporarily pending further treatment.

2. In the darkroom transfer exposed plates or films to plate or film holders or racks.

3. Immerse plates or films in D-19 developer and agitate at 30 second intervals.

4. Develop plates or films in D-19 for 4 minutes at 68° F.

5. Rinse plates or films for 90 seconds in running tap water at 65 to 70° F.

6. Fix plates on films with frequent agitation at 65 to 70° F for 8 to 10 minutes in kodak fixer.

7. Wash plates or films for 30 seconds in running tap water (65 to 70°F).

8. Rinse plates or films in kodak hypo clearing agent for 2 minutes with moderate agitation.

9. Rinse plates or films in running tap water for 10 minutes.

10. Soak plates or films in distilled water for 20 seconds.

11. Allow to dry in clean dust-free area.

12. Store in glassine envelopes.

13. Properly label envelope by placing the date the photograph was taken, magnification, plate number, specimen, operator and etc.)

14. Inspect plate or film with illuminator or ocular lens for evaluation.

FACTORS ESSENTIAL IN PLATE OR FILM EVALUATION

For each plate or film indicate:

1. Focus i.e., refocusing after each shot

2. Contrast - low, medium, etc.

3. Grain

4. Illumination

5. Information content

6. Artifact

E. RESULTS:

Obviously the content of the plate for informational purposes is most important. This content however, is affected by focus or out-of-focus image, by too little or too much contrast, by graininess of the plate, by irregularity of illumination or artifacts such as stain precipitates, dirt, folds or tears in the section.

F. EVALUATION:

PRESENT 4 PLATES OR FILMS OF HIGH QUALITY TO THE INSTRUCTOR FOR EVALUATION.

G. QUESTIONS:

1. Briefly discuss the advantages and disadvantages of using electron microscope plates verses electron microscope film.

2. What do you consider to be the four most essential characteristics in the development process of electron microscope plates or films?

3. Briefly discuss several factors essential in plate or film evaluation.

4. How does one assess what is a good produced plate or film?

5. Define the following terms: extrinsic artifacts, intrinsic artifacts, photographic emulsion, silver bromide grains, shading the print.

6. With reference to photographic paper, briefly discuss grade number series in conjunction with contrast.

7. What grade paper is considered normal for photographic reproduction?

8. Briefly discuss short, "normal", and long development of the images produced on photographic paper in relation to exposure time and lightness or darkness of the plate or film.

9. What is the significance of having study prints?

10. Differentiate between the action of electrons and light on photographic emulsions.

NOTES

EXERCISE NUMBER 18B

MAKING AND PRINTING ENLARGEMENTS OF ELECTRON MICROSCOPE PLATES OR FILMS

Betty R. Jones, Ph.D.

A. OBJECTIVE:

1. To obtain high quality electron micrographs

B. MATERIALS:

1. Enlarger (with correct type lens for plate or film size)

2. Easel to make 8 x 10 inch prints in addition to other desired sizes

3. Photographic paper (8 x 10 inches or preferred choice) in several ranges (i.e. 1 through 5 for Brovira; or F1 through F5 for Kodabrome)

4. Focusing instrument

5. 5 plastic trays (size 11 x 14 inches)

6. Plastic tongs for handling prints

7. Timers (at least 3)

8. Safety light (yellow filter)

9. Solutions:

   a. DEKTOL DEVELOPER (stock solution): Prepare according to instructions on package.

      Working solution of Dektol:
      1 part Dektol stock
      2 parts water

   b. STOP BATH SOLUTION: 16 cc of Acetic Acid/1 liter of water

   c. FIXER STOCK SOLUTION: Use kodak fixer

full strength. Prepare according to instructions on package.

d. STOCK HYPO CLEARING AGENT: Prepare according to instructions on package.

   Working solution: 1 part Hypo stock solution, 4 parts water

e. Plastic tray with running water

10. Print washer

11. Thermometer

12. Electron microscope plates or films (from Exercise Number 18A)

NOTE: RECHECK DARKROOM FOR CLEANLINESS AND MAKE SURE ALL EQUIPMENT IS WORKING PROPERLY.

C. RATIONALE AND DISCUSSION:

Preparation of specimens and recording images of the specimen on photographic emulsion plates or films are indeed key steps in carrying out electron microscopy work. But what is perhaps even more enjoyable and rewarding is enlarging and printing an EM plate or film on photographic paper. The electron micrograph and its interpretation represents the final end-product of the entire electron microscopy procedure.

It may be necessary to produce and assemble one to two thousand electron micrographs from a specimen before forming a final conclusion based on the data. The principles and techniques used for enlarging and printing electron microscope plates or films are the same as those used in any photographic laboratory.

A positive print is produced by projecting light through the plate or film onto paper which contains a coating of photographic emulsion. Light passes through the plate or film in inverse proportion to the varying densities that make up the negative image.

A latent image is not induced in the silver bromide grains in the paper below the opaque areas in the plate, whereas the lighter areas allow light to pass so that aggregates in the paper beneath them are energized and will darken upon development. After exposure, the photographic paper is developed, fixed, washed and dried.

TABLE I

General Major Parts and Functions for a Darkroom Enlarger

NOTE: BECOME FAMILIAR WITH THE PARTS AND FUNCTIONS OF THE ENLARGER.

	PARTS	FUNCTIONS
A.	Lamp	Contains the light source which may be a coiled filament, V-shaped filament or several fluorescent tubes depending on the type of enlarger.
B.	Condenser lens	Bends rays of light from the lamp source.
C.	Bellows	Permits movement of the focusing lens up and down.
D.	Focusing lens	Focuses the image and regulates the amount of light passing through the lens. The following F stop openings (diaphragm) are present. F/5.6, F/8, F/11, F/16, F/22, and F32.
E.	Focusing knob	Allows focusing lens to be raised up and down to bring image into focus.
F.	Magnification knob	Raises up and down the entire lens system to changing the degree of magnification of the negative (plate or film).

TABLE I (continued)

	PARTS	FUNCTIONS
G.	Negative carrier	Supports and holds the negative in place. It is placed between the condenser lens and the bellows.
H.	Automatic timer and light control	The timer and light control is placed between the power source and the enlarger. The power source will come from a socket in the wall. There are two major controls: 1) Toggle switch - turns the light source on from the enlarger and permits focusing. NOTE: DO NOT REMOVE PHOTOGRAPHIC PAPER FROM CONTAINER WHEN THIS LIGHT IS ON. 2) Button - allows the enlarger's light source to become active for the amount of time that has been pre-set on the timer. The light source from the enlarger will go off automatically at the end of the pre-set time. This means the exposure time is completed. Use this button when the photographic paper has been positioned in the easel for exposure.

TABLE II

Key Words and Important Points for
Enlarging and Printing

	KEY WORDS	IMPORTANT POINTS
1.	Negative (EM plate or film)	Always place in enlarger with emulsion side (shiny side) up.
2.	Paper	Place in easel with emulsion side (shiny side) up.
3.	F stop or diaphragm	Focus image on easel with F stop open to the largest amount of light (smallest F stop number F/5.6)
4.	Paper grade	Choose the appropriate grade of paper according to the appearance of the negative. Become familiar with the following points. a) Flat or soft negatives do not show that much difference between light and dark areas. b) Contrasty or hard negatives reveal the sharpest difference between light and dark areas. c) For highly contrasty or hard negatives use soft paper (For example grades F1 or F2 kodabrome or BN1-1 or BN1-2 brovira). d) For flat or soft negatives use harder paper grades. F3, F4 or F5, or BN1-3, BN1-4 or BN1-5). NOTE: SOFT PAPER IS MORE SENSITIVE AND QUICK THAN HARD PAPER. THUS, WHEN MAKING AN EXPOSURE ON HARD PAPER, THE EXPOSURE

		TIME OR AMOUNT OF LIGHT MUST BE INCREASED.
5.	Exposure time	a) The denser or darker a negative, the longer exposure time needed. b) The more a negative is enlarged, the longer exposure time needed. c) The higher the F stop number, the less light transmitted. NOTE: THE DIAPHRAGM OPENINGS GENERALLY RANGE FROM F/5.6, F/8, F/11, F/16, F/22 and F/32. THIS MAY VARY DEPENDING ON THE TYPE OF ENLARGER.
6.	Print exposed correctly	a) Faint images should begin to appear 30-40 seconds after being in the developer. b) Print should be fully developed in 1.5 - 2 minutes NOTE: USE TIMER FOR ACCURACY THROUGHOUT ENTIRE PRINTING PROCESS. c) When print develops quickly and turns black in 2 minutes, it is over exposed. Reduce amount of light or cut down on the exposure time by 1/2 for the next print. d) When print does not develop or develops faintly in 2-4 minutes, it is under exposed. Increase the amount of light or double the exposure time for the next print.

D. PROCEDURE:

1. Check work area for cleanliness and set up the following trays with solutions:

a. Tray #1 - Dektol developer working solution
b. Tray #2 - Stop Bath
c. Tray #3 - Fixer stock solution
d. Tray #4 - Hypo clearing agent working solution
e. Tray #5 - Running water

NOTE: MAKE SURE FINAL SOLUTION HAS PROPER TEMPERATURE (20°C or 68°F). ICE CUBES OR WARM WATER MAY BE ADDED TO SOLUTIONS IN THEIR PREPARATION TO OBTAIN THE CORRECT TEMPERATURE.

2. Shut off room lights and work under safety lights during the entire printing process. Take a small amount of paper from box and store in paper safe or drawer for printing.

3. Insert negative, shiny side up, into negative carrier and place carrier into the enlarger.

4. Focus with open diaphragm on easel or white paper.

5. Close diaphragm to desired F stop (preferably F/11) and shut off enlarger light.

6. Place a sheet of (8" x 10") of photographic paper (i.e. BN1-2 brovira) into easel with emulsion side (shiny side) up.

7. Cover 4/5ths of this paper with a piece of cardboard or black construction paper.

8. Push automatic timer and expose 1/5th of the paper for 10 seconds.

9. Move cardboard over another 1/5th of the area and expose again for 10 seconds.

10. Repeat this process the entire sheet has been exposed.

NOTE: THIS WILL RESULT IN 5 SECTIONS OF THE SHEET, EXPOSED FROM 10 SECONDS IN THE FIRST AREA DOWN TO 50 SECONDS IN THE LAST AREA.

11. Developer paper for 1.5 - 2 minutes in tray #1. Agitate at intervals using tongs.

12. Drain print off and proceed to tray #2. Leave in stop bath for 10 seconds.

13. Drain print and transfer to tray #3. Fix print for 5-8 minutes. After 2 minutes, the room lights can be turned on to inspect the print.

14. Select the area that provides the best exposure time and repeat on the next 8" x 10" print.

NOTE: THE ABOVE STRIP IS CALLED A TEST STRIP AND IS USEFUL IN DETERMINING THE PROPER EXPOSURE TIME FOR A NEGATIVE.

IT PREVENTS WASTING ENTIRE SHEETS OF PHOTOGRAPHIC PAPER.

15. Clear prints in Hypo clearing agent for 2 minutes. The use of Hypo solution cuts down on print washing time.

16. Wash in running tap water for 5 minutes.

17. Wash in print washer for 5 minutes.

18. Dry in print dryer.

NOTE: PRINTS CAN BE SOAKED OVERNIGHT IN DISTILLED WATER IF NOT DRIED THE SAME DAY.

AFTER DARKROOM WORK IS COMPLETED, MAKE SURE ALL TRAYS, BENCHES, ENTIRE WORK AREA AND ETC. ARE CLEANED AND EQUIPMENT COVERED.

E. QUESTIONS:

1. How is a positive print produced from an electron microscope plate or film?

2. What is the length of time the image on photographic paper should be developed and why?

3. How can one assess the proper exposure time for a plate or film and what advantages does this procedure offer?

4. Before drawing a conclusion about the ultra-structural data from a specimen, what range of micrographs are usually printed?

5. During development of plates or film, why is agitation essential?

6. What problems should one look for if plates or films are overdeveloped?

7. What problems may occur after prolonged exposure of plates or films in the fixer?

8. What essential factors should be considered in choosing an emulsion for electron microscopy?

9. Briefly describe two types of enlargers for electron microscopy work with reference to the principles of how light is produced.

10. Discuss the advantages and disadvantages of a print processing machine.

NOTES

EXERCISE NUMBER 19

TRANSMISSION ELECTRON MICROGRAPH INTERPRETATION

Betty R. Jones, Ph.D.

A. OBJECTIVES:

1. To identify, describe and interpret data from transmission electron micrographs
2. To correlate fine structure of organelles with possible functions within a given system
3. To compare fine structure with other published or known cell structures
4. To utilize histology and EM atlases as a guide in interpreting data
5. To produce thousands and thousands of micrographs of similar and dissimilar sections and to draw conclusive results from this large poole
6. To be able to show sections cut in different angles and planes
7. To be able to construct a montage of a given tissue or organ
8. To write a detailed caption or description for each electron micrograph of interest
9. To store micrographs properly

B. MATERIALS:

1. Electron micrographs
2. Magnifying lens
3. Pencil and record book
4. Sheet protectors
5. Binder

C. RATIONALE AND DISCUSSION:

Interpretation of transmission electron micrographs can be considered the final phase of the electron microscopic procedure. In this phase of the work, thousands and thousands of micrographs may be produced before a conclusive and definitive answer can be given for a particular problem investigated.

Carefully determine if the data gathered provide descriptive evidence with clarity for the objectives that were initiated. Write a detailed description for each electron micrograph. One of the rules that I generally follow is to interpret everything thats visible to the naked eye in the micrograph.

Use magnifying lens to check focusing and to better visualize structures. Once a detailed caption has been written for each micrograph of interest it is important to store micrographs properly. Place your micrographs in sheet protectors and store in a binder with captions.

Proper storage will eliminate grease and dirt from hands during handling. The correct magnification should be included as part of the caption for each micrograph. The final magnification should be written at the end of the caption.

TABLE I

SUMMARY OF MAJOR STEPS FOR PREPARATION OF SPECIMENS FOR TEM

1	2	3
FIX IN GLUTAR-ALDEHYDE	RINSE IN BUFFER	POSTFIX IN OSMIUM TETROXIDE
4	5	6
RINSE IN BUFFER	DEHYDRATE IN ALCOHOL SERIES	INFILTRATE IN PROPYLENE OXIDE AND EPON 812
7	8	9
EMBED IN BEEM CAPSULES IN FULL STRENGTH EPON 812	POLYMERIZE IN OVEN	TRIM BLOCKS
10	11	12
CUT THICK SEC-TIONS FOR LIGHT MICROSCOPY	CUT ULTRATHIN SEC-TIONS FOR TEM	PICK UP ON 200 or 300 COPPER MESH GRIDS

13	14	15
STAIN IN URANYL ACETATE AND LEAD CITRATE SOLUTIONS	VIEW GRIDS IN TEM	TAKE PHOTOGRAPHS OF SECTIONS AT DESIRED MAGNIFICATIONS
16	**17**	**18**
DEVELOP PLATES OR FILMS	MAKE PRINT FROM EM PLATES OR FILMS	DRY ELECTRON MICROGRAPHS
19	**20**	**21**
INTERPRET ELECTRON MICROGRAPHS	LABEL AND STORE MICROGRAPHS IN SHEET PROTECTORS	PLACE SHEET PROTECTORS IN BINDER

D. SELECTED SAMPLE RESULTS:

NOTE: SEVERAL SAMPLE TRANSMISSION ELECTRON MICROGRAPHS ARE INCLUDED IN THIS EXERCISE (SEE FIGURES 1-9)

The micrographs are on the five structure of the miracidium of _schistosoma mansoni_ (schistosomiasis). Tissues were processed for EM by B.R. Jones following the published procedures of Dr. Steve Chia Tung Pan, Department of Tropical Public Health, Harvard School of Public Health, Boston, Massachusetts.

Users of this manual that are interested should refer to the reference below for an excellent and detailed description of the five structure of the miracidium of _S. mansoni_.

Reference:

Pan, Steve Chia-Tung (1980). The fine structure of the Miracidium of _Schistosoma mansoni_. J. Invertebr. Pathol., 36:307-372.

E. EVALUATION:

EXAMINE EACH MICROGRAPH CAREFULLY AND ANSWER THE QUESTIONS AT THE END OF THIS EXERCISE. A DISCUSSION OF ANSWERS TO QUESTIONS WILL FOLLOW.

NOTE: STUDENTS MAY WISH TO REFER TO EM ATLASES AND CELL BIOLOGY TEXTS FOR ANSWERING SOME OF THE QUESTIONS IN THIS EXERCISE.

(SEE APPENDIX B FOR LIST OF EM REFERENCE TEXTS). ONE OF THE EM TEXTS I FIND VERY USEFUL IS LISTED BELOW.

Fawcett, Don Wayne (1981). _The Cell_. W.B. Saunders Company, Philadelphia, Pennsylvania.

FIGURE 1. Miracidium of Schistosoma mansoni; Epidermal System: (ep), epidermal plate displaying cilia, electron-dense bodies and mitochondria; (b), basal lamina; (n), nucleus of ridge cyton; cytoplasm of cyton contains golgi-complex, granular er, mitochondria, glycogen particles, ribosomes, (ic), distinct processes of interstitial cells containing glycogen; (arrows), note: membrane-bound vesicles. The vesicles are round to oval, contain a dense central zone and measure 500 nm at the greatest diameter. Cytoplasm of the ridge and its cytons is moderately electron-dense and finely granular in texture. A small number of microvilli can be seen on the ridge surface (33,200X).

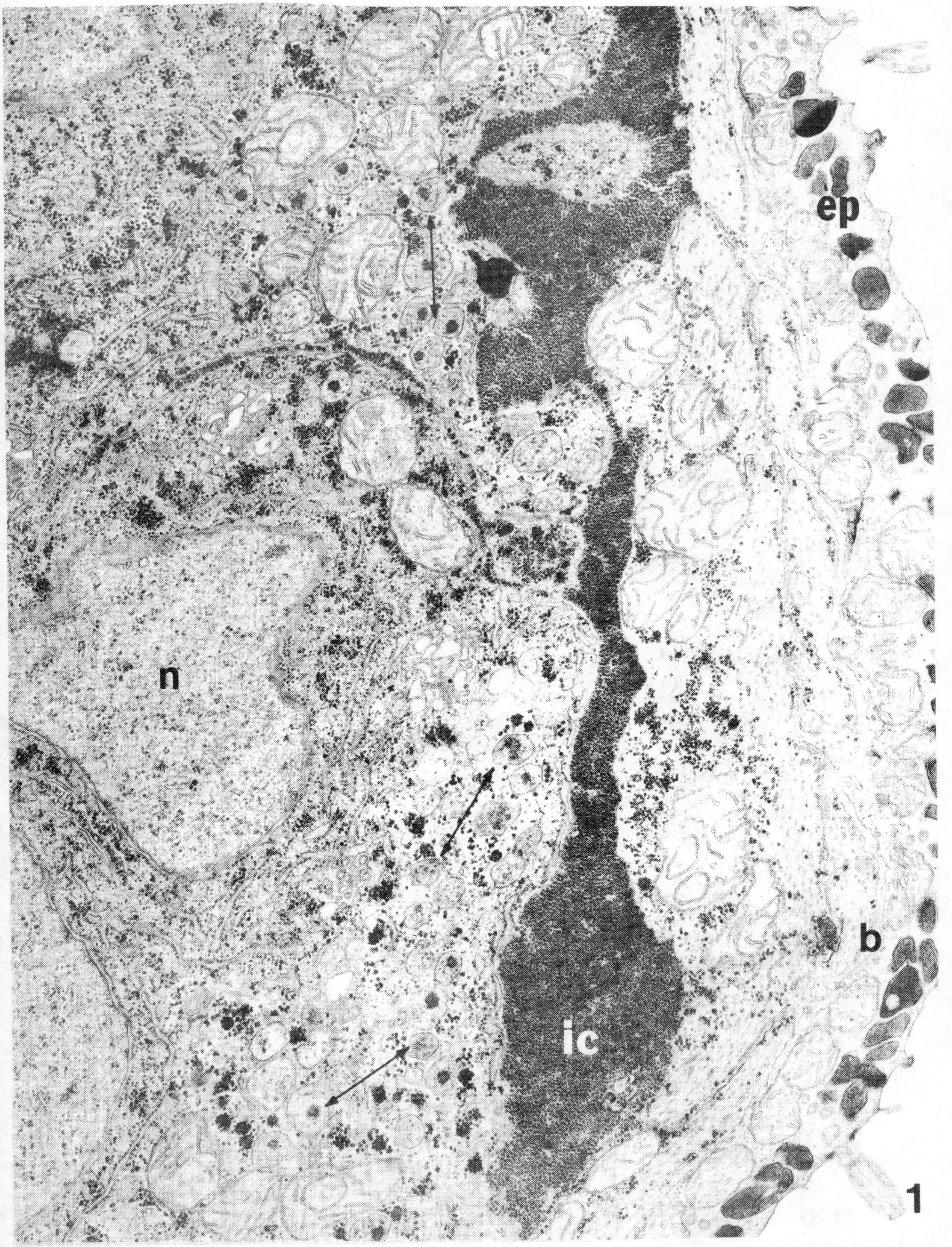

ep
n
b
ic
1

FIGURE 2. Miracidium of *Schistosoma mansoni*; Epidermal System: Part of epidermal ridge cyton (38,500X).

ERRATA

pp. XIV, XV, XVI are correctly printed, but out of sequence.

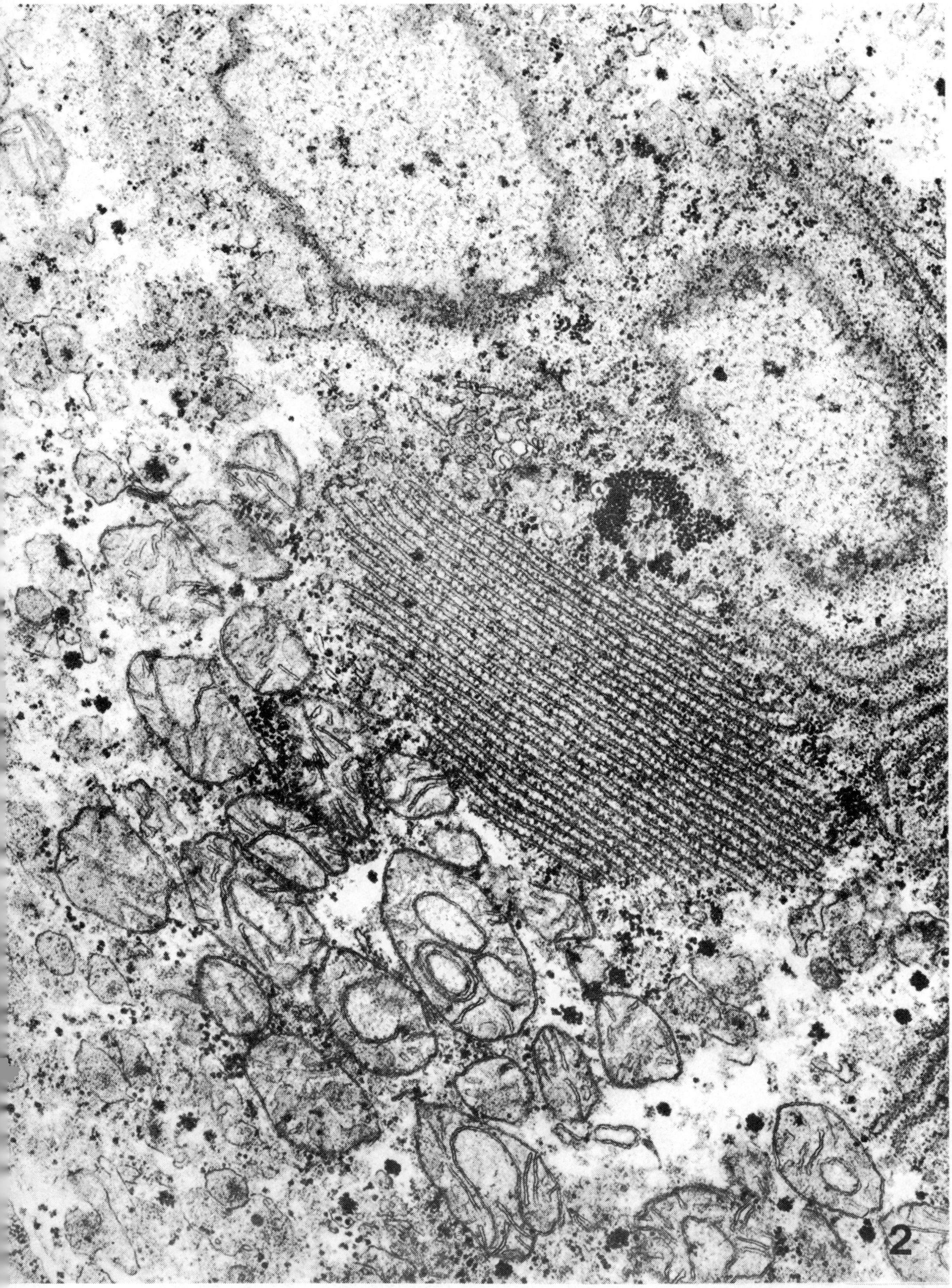

2

FIGURE 3. Miracidium of *Schistosoma mansoni*; Nervous System: (arrows), numerous "neurosecretory" membrane-bound vesicles in the region of the ridge cyton; (b), note: cytoplasmic bridge connecting epidermal ridge cyton to epidermal plate and (r) ridges separating plates in successive tiers; (ic), process of interstitial cell; (n), nucleus; (ger), granular endoplasmic reticuli; (m), mitochondria (33,200X).

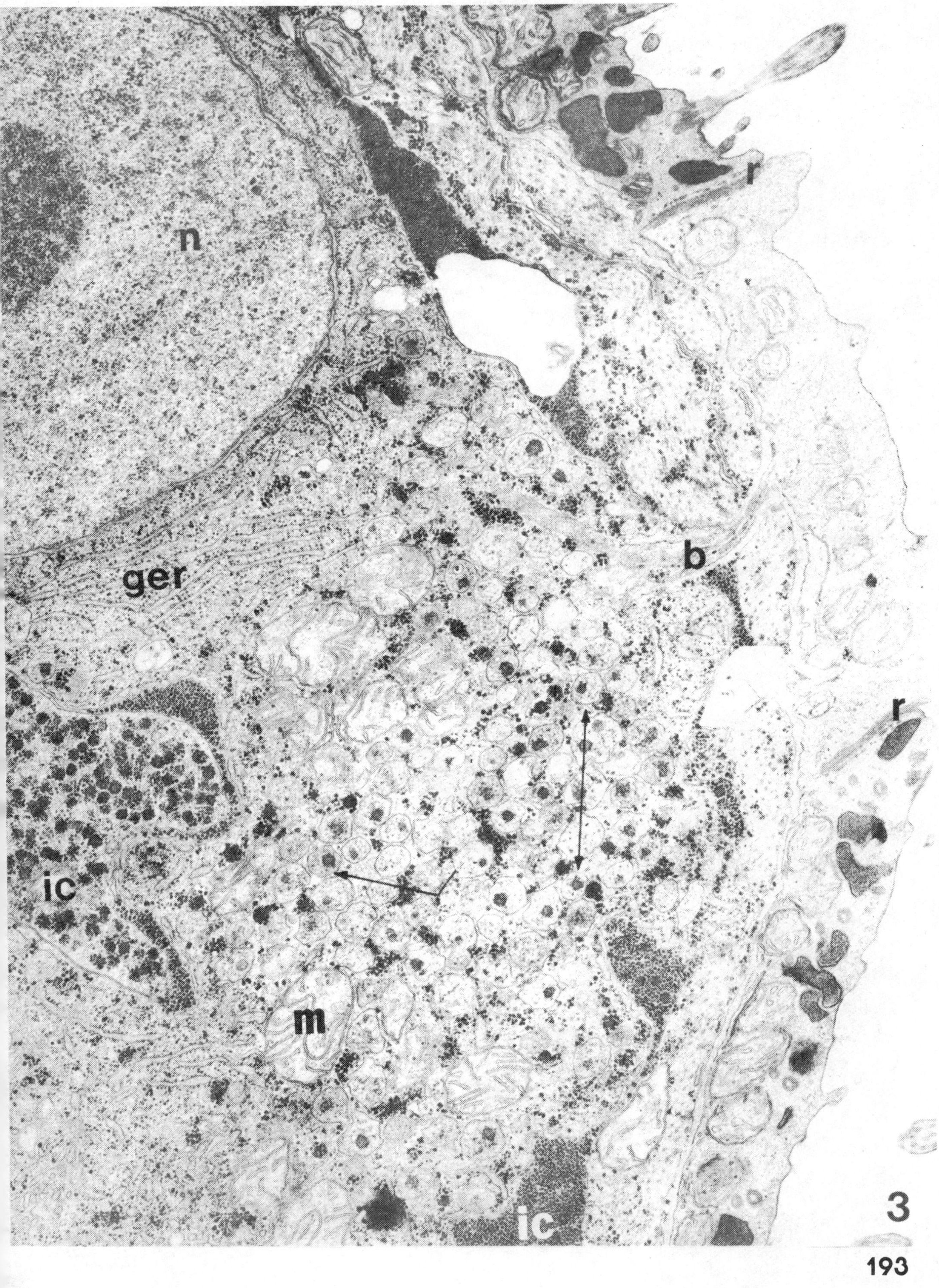
r
n
b
ger
r
ic
m
ic
3

FIGURE 4. Miracidium of _Schistosoma_ _mansoni_; Nervous System: Neurons in the neural mass; (n), nucleus of neuron; (arrows), "neurosecretory" vesicles; (g), glycogen particles; (m), mitochondria; (a), axon containing "neurosecretory" vesicles (49,500X).

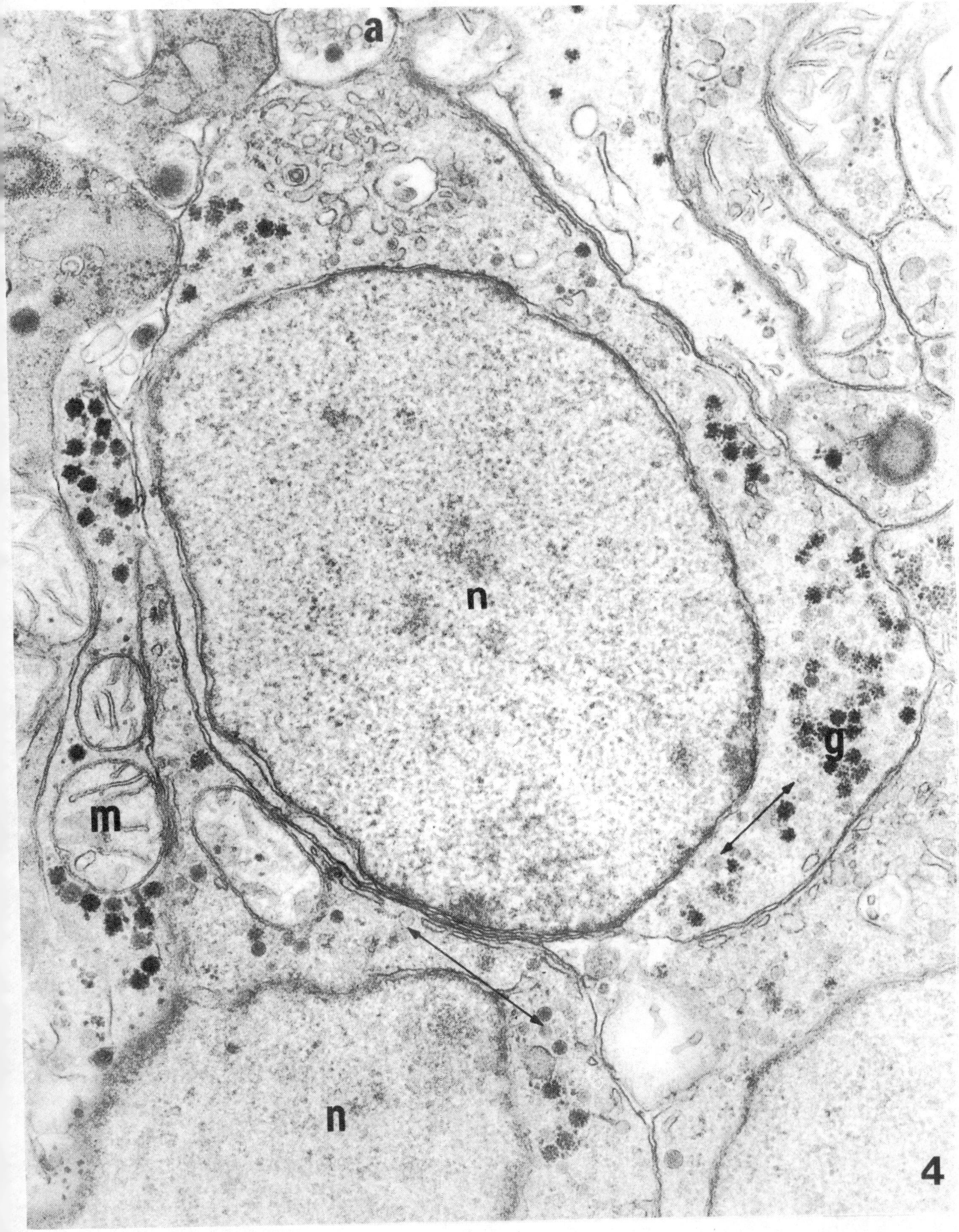
a
n
g
m
n
4

FIGURE 5. Miracidium of *Schistosoma mansoni*; Excretory System: (fc), flame cell consisting of cyton, cylinder or barrel and cilia; (l), short leptotriches in peri-flame cell space; (t), common excretory tubules containing lumen and mitochondria; (g), glycogen (26,900X).

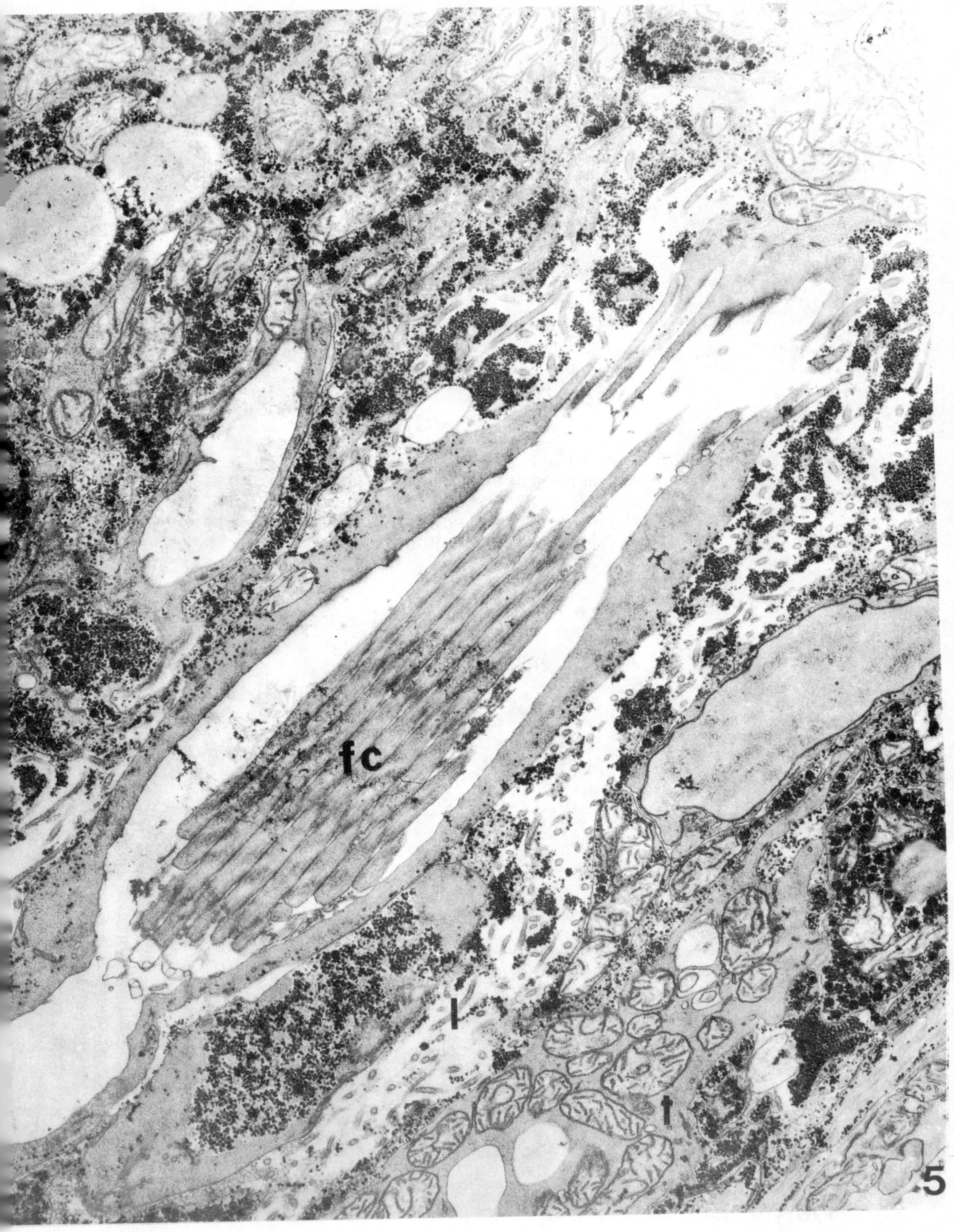
fc
g
l
t
5

FIGURE 6. Miracidium of *Schistosoma mansoni*; Excretory System: (arrows), long leptotriches; (fc), flame cell, note microtubules; (ic), processes of interstitial cells containing glycogen; (n), nucleus (26,900X).

n
fc
ic
n
6

FIGURE 7. Miracidium of *Schistosoma mansoni*; Excretory System: flame cell showing 9 + 2 pattern of cilia and septate desmosome joining margins of cylinder wall; (ld), lipid droplets; (n), nucleus; (ic), processes of interstitial cell containing glycogen (21,500X.

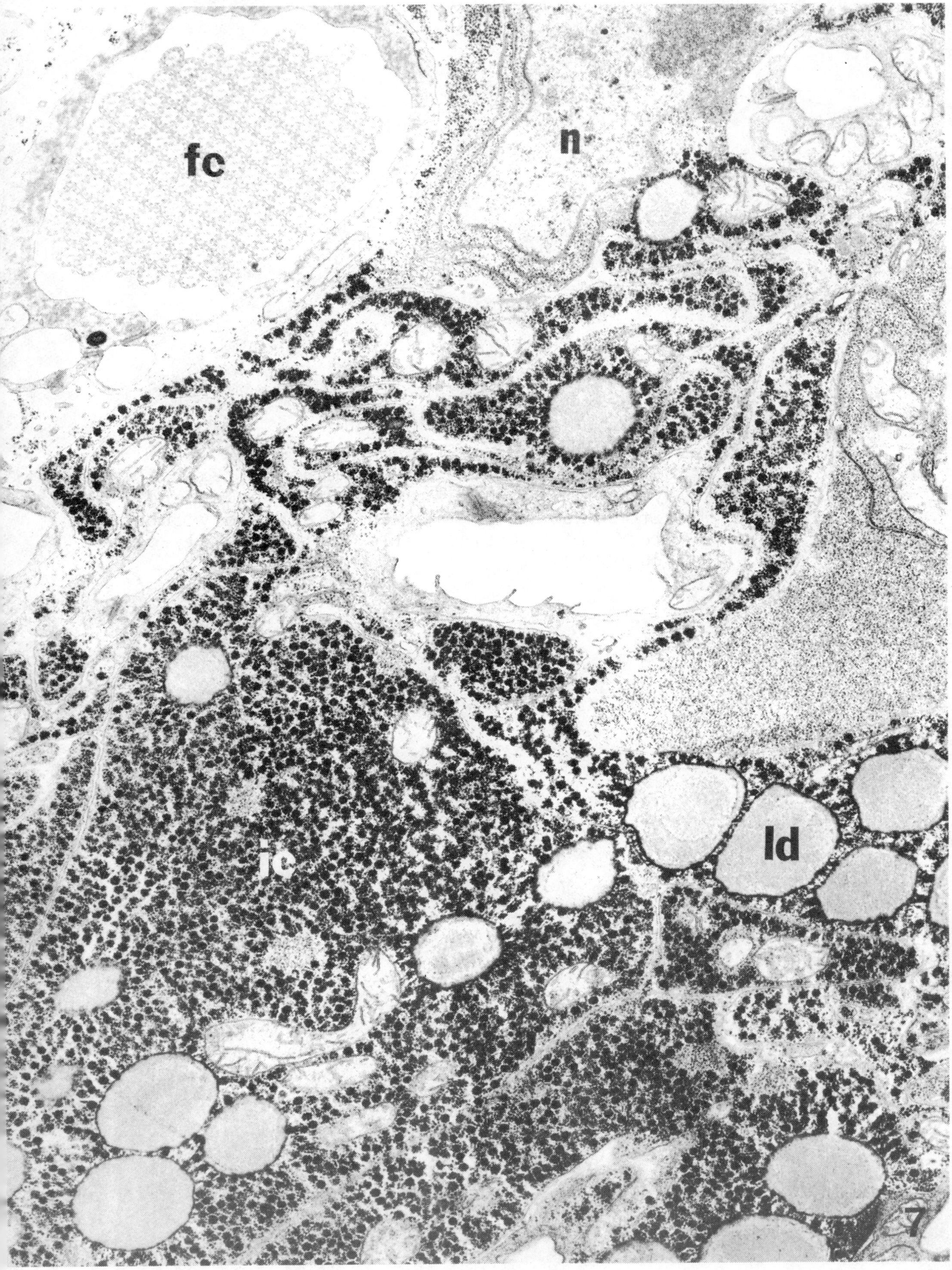
fc
n
jc
ld
7

FIGURE 8. Miracidium of *Schistosoma mansoni*; Excretory System: (P), perikaryons with nuclei and mitochondria; (l), lumen of excretory tubules; (ic), interstitial cells containing glycogen (26,900X).

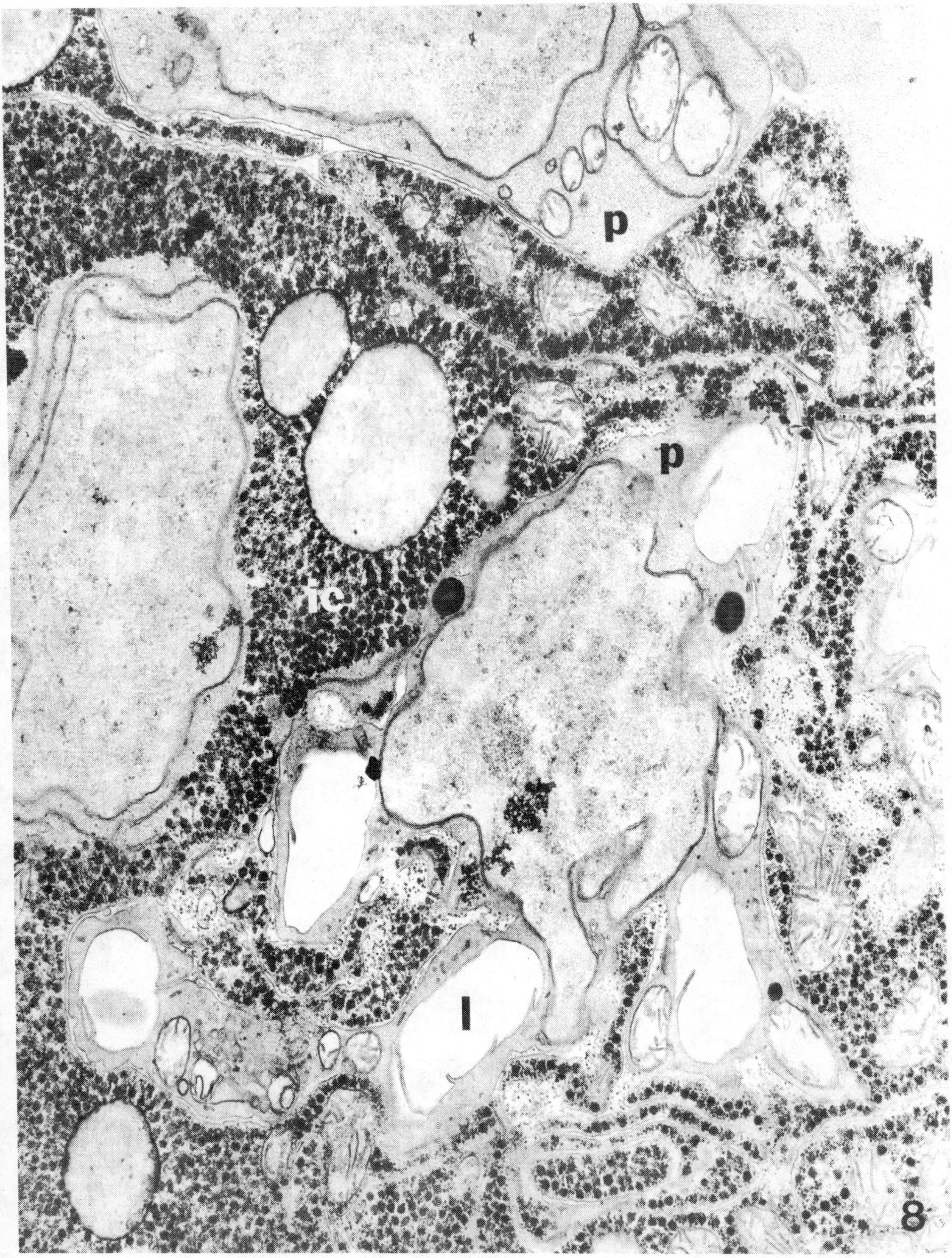
p
p
fc
l
8

FIGURE 9. Miracidium of *Schistosoma mansoni*; Pentration Glands: Part of the apical penetration gland containing (sd), secretory droplets and (g), glycogen; (n), nucleus; (m), mitochondria of nearby cyton (33,200X).

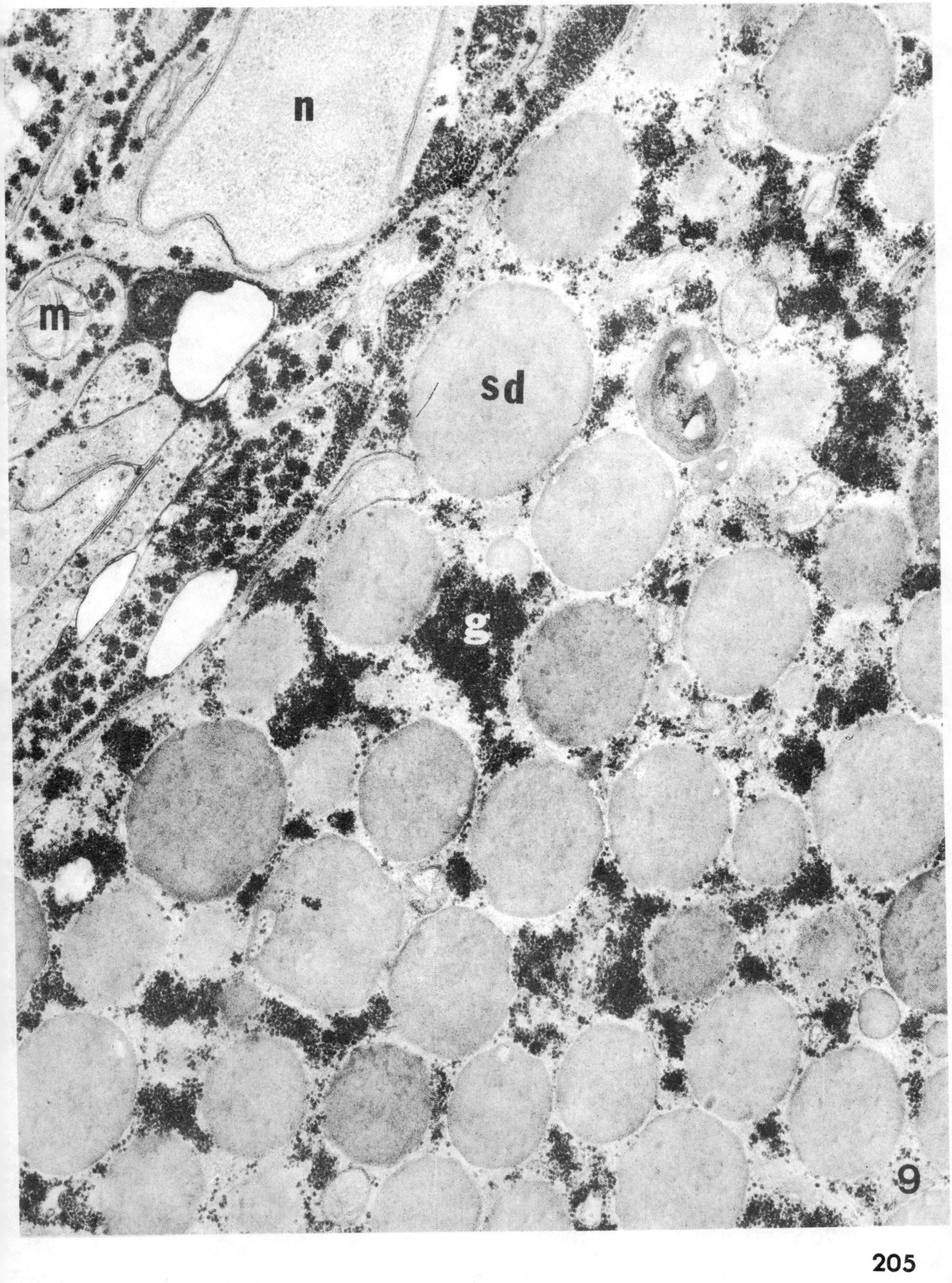
n
m
sd
g
9

F. QUESTIONS:

1. The interstitial cell in Figure 1 is filled with glycogen. Glycogen is found in many cell types.

   a. What is the structure and function of glycogen?

   b. How is it broken down to yield glucose?

   c. How in turn does glucose supply energy and provide short-chain carbon skeletons that are resused in the synthesis of essential constituents of the protoplasm?

   d. With reference to EM preparation what makes glycogen demonstrable in an electron micrograph?

e. Differentiate morphologically between alpha and beta glycogen particles.

f. Are the glycogen particles distributed throughout the epidermal system of Figure 1 alpha or beta or both?

g. In Figure 1 glycogen occurs as individual particles in the muscle fiber. Why are the muscle fibers in close association with mitochondria?

2. From Figure 2 list at least eight cellular constituents of the epidermal ridge cyton.

3. The basal lamina of the miracidium of *Schistosoma mansoni* is a thin layer (450-900 Å thick) that follows the basal contour of the epidermal plates and ridges. The basal surface of each plate is attached to muscle fibers (Pan, 1980).

a. In Figure 3 locate and draw an arrow to the basal lamina.

b. Locate and draw an arrow to a septate desmosome.

c. Where do you think the components of the basal lamina are synthesized and secreted?

d. What is the function of the basal lamina?

e. Are desmosomes morphologically similar from tissue to tissue and what is their function?

f. Define the following junctional specializations: Tight junctions (zonula occludens), adhering junctions (zonula adherens)/and Gap Junctions (Nexuses).

g. Morphologically describe the vesicles in Figure 3.

4. Figure 4 shows neurons of the central nervous system of the parasite.

   Describe ultrastructurally the appearance of axons in Figure 4. Compare these axons with those of vertebrates at the EM level.

5. The micrographs of the excretory system in Figures 5,6, and 7 show a flame cell of the miracidium consisting of cilia.

   a. With reference to sectioning techniques explain the differences in the appearance of the flame cells.

   b. What is the difference between cilia, flagella, and microvilli?

c. Briefly discuss the structure of cilia in the cells of protozoa and invertebrates compared to that of higher animals.

6. In Figure 8, the lumen of the excretory tubules contain a few cytoplasmic projections. What do you think could be a possible function for these projections?

7. Numerous electron dense secretory droplets of various sizes can be seen in Figure 9.

   a. What do you think these droplets are most likely comprised of?

   b. How might one explain the differences in their electron densities?

c. If one of your classmates presented you a micrograph of a section through the apical penetration gland using the same method of preparation as in Figure 9 and the droplets were non-osmiophilic or showed little electron density, what conclusions could be drawn from this?

EXERCISE NUMBER 20A

PREPARATION OF SPECIMENS FOR SCANNING ELECTRON MICROSCOPY

FIXATION OF SPECIMENS FOR SCANNING ELECTRON MICROSCOPY

Betty R. Jones, Ph.D.

A. OBJECTIVES:

1. To examine and describe the surfaces of biological and nonbiological specimens
2. To become familiar with three-dimensional images of the surfaces of specimens
3. To examine fixed or unfixed specimens

B. MATERIALS:

1. Laboratory rat or mouse
2. 3% glutaraldehyde in phosphate buffer
3. Cold phosphate buffer for washing in squirt bottle
4. Freshly prepared osmium tetroxide
5. 1 vial with cap
6. 2 wood applicator sticks
7. 2 precleaned single edged razor blades
8. Alcohol series (30, 50, 70, 83, 95 and 100%)
9. Amyl acetate solutions (70 and 100%)
10. Dissecting kit
11. 2 clean pasteur pipettes with a rubber bulb

C. RATIONALE AND DISCUSSION:

The scanning electron microscope permits viewing of the surface topography of specimens from a range of 5 to 10 nanometers. The fixation procedure for specimen preparation is similar to that of transmission electron microscopy. However specimens may be viewed unfixed and embedding and sectioning are not necessary.

For good correlation one should practice viewing the internal surface topography of tissues in addition to the external surface topography. Specimens prepared for SEM should be critical point dried and coated with a heavy metal. These topics are dealt with in Exercises 20B and 20C.

The SEM can be the instrument of choice when information is needed on the external and internal fine surface topography of specimens.

D. PROCEDURE:

NOTE: FOR PREPARATION OF FIXATIVES AND BUFFERS SEE EXERCISE NUMBER 11.

1. Sacrifice rate or mouse by decapitation.

2. For practice dissect and remove a section of the trachea. This organ is simple and easy to work with for demonstration purposes.

3. Cut and open several pieces of the trachea about 4 mm square. Opening of the trachea will expose the inner lining of the tissue.

4. Rinse the surface with cold phosphate buffer by flushing onto the tissues. Rinse 4 to 5 times.

5. Place tissues in 3% glutaraldehyde in phosphate buffer.

6. Fix tissues for 2 hours at $4^{o}C$.

NOTE: THE GLUTARALDEHYDE CAN BE CHANGED TO A FRESHER GLUTARALDEHYDE SOLUTION BEFORE THE FIXATION PERIOD IS COMPLETED.

7. Rinse with cold phosphate buffer (4 to 5 different rinses).

8. Postfix in 2% buffered osmium tetroxide for 2 hours.

NOTE: POSTFIXATION IS OPTIONAL. HOWEVER IT HAS BEEN THE AUTHOR'S EXPERIENCE THAT WHEN TISSUES ARE OSMICATED LESS COATING IS REQUIRED AND CONTRAST IS ENHANCED. IF THIS STEP IS OMITTED GO DIRECTLY TO STEP 10.

9. Rinse tissues in phosphate buffer 4 to 5 times.

10. Dehydrate tissues in a graded series of alcohol (30, 50, 70, 83, 95 and 100%) for 10 minutes each and two changes of 100%, 10 minutes each.

11. Immerse tissues in 70% amyl acetate solution for 15 minutes and then into 3 changes of 100% amyl acetate, 15 minutes each. Tissues may be left in amyl acetate up to 48 hours.

NOTE: AMYL ACETATE SHOULD BE MADE UP IN ALCOHOL. IT IS TOXIC AND SHOULD BE USED UNDER THE HOOD. THIS STEP IS ALSO OPTIONAL AND IF OMITTED GO DIRECTLY TO EXERCISE NUMBER 20B.

E. QUESTIONS:

1. What would you consider to be one disadvantage of fixing specimens for SEM?

2. Is there an advantage to postfixing specimens for SEM in osmium tetroxide?

NOTES

EXERCISE NUMBER 20B

CRITICAL POINT DRYING OF SPECIMENS FOR SCANNING ELECTRON MICROSCOPY

Betty R. Jones, Ph.D.

A. OBJECTIVES:

1. To place a specimen in a liquid such as carbon dioxide, nitrous oxide, or freon

2. To carry to a critical pressure and temperature in which the specimen is surrounded by gas both externally and internally without any surface tension

3. To remove the gas above the critical temperature leaving the specimen intact and completely dry

4. To eliminate forms of distortion such as shrinkage and wrinkles that ordinarily might occur if a specimen is air dried

B. MATERIALS:

1. Critical point dryer

2. Fixed specimen from Exercise Number 20A

3. Carbon dioxide

C. RATIONALE AND DISCUSSION:

Most biological tissues contain a relatively large amount of water. If tissues are air dried this water will evaporate but in doing so distortion may occur. The distortion is primarily caused by the surface tension of the water as it is withdrawn in and around the cells of the tissue.

The water in the tissue is first replaced by a dehydrating solvent such as alcohol then amyl acetate. The use of amyl acetate is optional (See Exercise Number 20A). Then finally a transitional fluid such as carbon dioxide, nitrous oxide or freon is used inside a tightly sealed chamber with the tissue for the critical point drying.

When the tissue is placed in this environment the first phase which is called the vapor phase causes the fluid in the cells to pass into a vapor state. When this occurs the surface tension of the liquid is zero.

When the critical point drying chamber is heated enough to bring the temperature of the transitional fluid such as CO_2 to its critical point the vapor phase of fluid is bled out of the chamber slowly.

This leaves the tissue dry and intact and the specimen at this point has been critical point dried. By definition critical point means that point at which the fluid phase and the vapor phase have the same density.

TABLE I

LIST OF INTERMEDIATE CHEMICAL AGENTS AND TRANSITIONAL FLUIDS MOST COMMONLY USED AT VARIOUS CRITICAL POINT TEMPERATURES AND PRESSURES. (Dawes, C.J. 1979. Biological Techniques for Transmission and Scanning Electron Microscopy. Ladd Research Industries, Burlington, Vermont).

Intermediate Chemical Agent	Transitional Fluid	Critical Temperature	Critical Pressure
Ethanol, Amyl Acetate, Acetone	Liquid carbon dioxide	31.0	1,073
Acetone, Freon 113	Freon 13	29.0	561
Acetone, Freon 113	Freon 116	19.2	432
Ethanol, Freon 113	Freon 23	26.0	702
	Nitrous oxide	36.0	1,048

D. PROCEDURE:

NOTE: STUDENTS SHOULD NOT USE THE CRITICAL POINT DRYER WITHOUT THE APPROVAL OF THE INSTRUCTOR. THE EXACT PROCEDURE WILL DEPEND ON THE TYPE OF INSTRUMENT AND TRANSITIONAL FLUID USED.

THE INSTRUCTOR WILL EXPLAIN AND DEMONSTRATE TO THE STUDENT THE PARTS AND FUNCTIONS AND HOW TO USE THE CRITICAL POINT DRYER.

A SPECIAL HANDOUT LISTING A SEQUENCE OF STEPWISE PROCEDURES WILL BE GIVEN OUT.

E. EVALUATION:

PRESENT ONE CRITICAL POINT DRIED SPECIMEN TO THE INSTRUCTOR FOR EVALUATION.

F. QUESTIONS:

1. Why is it necessary to critical point dry some specimens for preparation of SEM?

2. Under what circumstances might critical point drying not be necessary?

NOTES

EXERCISE NUMBER 20C

COATING SPECIMENS FOR SCANNING ELECTRON MICROSCOPY

Betty R. Jones, Ph.D.

A. OBJECTIVES:

1. To first coat the specimen with a thin carbon base
2. To secondly coat the specimen with a heavy metal such as gold or a combination of gold and palladium
3. To increase the contrast of the specimen

B. MATERIALS:

1. Critically point dried specimens
2. Sputter coater or vacuum evaporator
3. Gold and palladium wires

C. RATIONALE AND DISCUSSION:

The specimen should first be coated with carbon. The carbon will allow the heavy metal to adhere to the specimen and prevent charging of electrons by causing the electron current to flow off the specimen. The second coating which is that of a heavy metal such as gold or a combination of gold-palladium serves as a source of secondary electrons.

All coatings should be casted on as evenly as possible over the entire specimen in order to avoid artifacts and to form an image identical to the original specimen. Avoid coating the specimen too heavily. This may obscure surface structural details.

D. PROCEDURE:

NOTE: COATING SPECIMENS FOR SEM CAN BE CARRIED OUT BY USING A SPUTTER COATER MACHINE OR BY EVAPORATING THE METALS ONTO THE SPECIMEN FROM A ROTARY STAGE. IN THE LATTER CASE THE INSTRUMENT USED IS A VACUUM EVAPORATOR.

THE INSTRUCTOR WILL DEMONSTRATE HOW TO USE THE SPUTTER COATER OR VACUUM EVAPORATOR (See Exercise Number 32).

HANDOUTS WILL BE PASSED OUT PRESENTING ADDITIONAL DETAILED STEPS TO FOLLOW BASED ON MODEL TYPE.

E. EVALUATION:

PRESENT ONE COATED SPECIMEN TO THE INSTRUCTOR FOR EVALUATION.

Q. QUESTIONS:

1. Briefly discuss the advantages for carbon and metal coating of specimens for SEM.

2. How does the choice of accelerating voltages in the scanning electron microscope influence coating of the specimen?

EXERCISE NUMBER 21

INTRODUCTION TO THE USE OF THE SCANNING ELECTRON MICROSCOPE

Betty R. Jones, Ph.D.

A. OBJECTIVES:

1. To introduce the student to the parts and functions of the scanning electron microscope (SEM)

2. To demonstrate to the student operations and procedures for using the scanning electron microscope

B. MATERIALS:

1. Specimen

NOTE: SPECIMEN SHOULD ALREADY BE CRITICAL POINT DRIED, COATED WITH A METAL AND MOUNTED ON SPECIMEN STUB (SEE EXERCISE NUMBER 20B AND 20C).

2. Polaroid film (Type 52 or Type 55)

NOTE: IF FILM TYPE 55 IS USED A NEGATIVE CAN BE REPRODUCED.

3. Sodium sulfate solution

NOTE: SODIUM SULFATE SOLUTION IS USED FOR PROCESSING NEGATIVES FROM TYPE 55 FILM

4. Felt-tipped pen for labeling scanning electron micrographs

C. RATIONALE AND DISCUSSION:

The development and use of the scanning electron microscope occurred during the latter part of the 1960's. One might think of an EM facility being complete when a SEM and TEM are both available with accompanying supplies and equipment for specimen preparation. The SEM is capable of providing a three-dimensional view of the fine external and internal surfaces of specimens whereas the TEM permits only a two-dimensional view of thin sectioned materials.

Thin sectioning is not necessary for SEM work. If an x-ray microanalyzer is attached to the SEM information on the chemical composition of specimens can be obtained. This relationship between surface structure and chemical analysis of components of the specimen is a unique feature of the scanning microscope. The SEM is capable of producing three-dimensional images because of its large depth of field. The depth of field is 3-500 times that of the light microscope.

In principle the electron beam of the SEM is organized into an electron probe that goes across the surface of the specimen. The focused beam will strike the surface of the specimen and cause secondary electrons to be emitted in addition to backscattered electrons, x-rays and cathodeluminescence or light. These different forms of released energy can be used in studying and obtaining information from specimens in the SEM. For this reason the scanning microscope has multiple uses in biology, engineering, and physics.

The methods of specimen preparation for scanning electron microscopy are similar to that of transmission electron microscopy except that thin sectioning, embedding and conventional EM staining with heavy metals are not required. In some cases fixation may be optional. However in general specimens are fixed, dehydrated, critical point dried coated with a heavy metal and viewed in the SEM.

The SEM is used in a broad capacity in biology. This can perhaps be attributed to the rapid progress of procedures for specimen preparation and to the design of the microscope itself. Today the capabilities of the SEM are many.

There are scanning microscopes designed to study the distribution of enzymes, antigens, elemental composition of specimens, frozen and unfixed specimens and even the localization of reaction sites in cells. The scanning scope has become a valuable research tool and its possibilities in many fields in the future are like the TEM immeasurable.

D. PROCEDURE:

NOTE: THE INSTRUCTOR WILL EXPLAIN AND DEMONSTRATE TO THE STUDENT, PARTS AND FUNCTIONS AND HOW TO USE THE SCANNING ELECTRON MICROSCOPE.

A special handout put together based on instructions from the microscope manual will be given out. This handout will consist of schematic drawings of the microscope type and parts.

E. RESULTS:

As a result of this exercise students should be able to do the following:

1. Turn on the high voltage.

2. Turn on the electron beam.

3. Adjust and align the illumination of the beam on the visual cathode ray tube.

4. Adjust the brightness and constrast of the illumination of the beam on the cathode ray tube.

5. Insert a mounted specimen into the column of the SEM.

6. Focus the image of the specimen on the cathode ray tube.

7. Take photographs of the specimen.

8. Process negatives if type 55 PN film is used.

9. Turn off SEM.

F. EVALUATION:

PRESENT 4 PHOTOGRAPHS TO THE INSTRUCTOR FOR EVALUATION.

G. QUESTIONS:

1. How does preparative techniques for scanning electron microscopy differ from those of transmission electron microscopy?

2. How does the final image produced in the scanning electron microscope differ from that produced in the transmission electron microscope?

3. How does the scanning electron microscope differ from the light microscope with respect to department of field?

4. Briefly discuss how the interest for scanning electron microscopy developed?

13	14	15
WASH NEGATIVES IN RUNNING TAP WATER	HANG TO DRY	INTERPRET POSITIVE PRINTS
16	17	18
LABEL AND STORE POLAROID POSITIVES IN SHEET PROTECTORS	USE POLAROID POSITIVES FOR STUDY PURPOSES	PRINT NEGATIVES USING STANDARD DARKROOM PROCEDURES
19	20	21
DRY MICROGRAPHS	LABEL AND STORE MICROGRAPHS IN SHEET PROTECTORS	PLACE SHEET PROTECTORS IN BINDER

D. SELECTED SAMPLE RESULTS:

NOTE: SEVERAL SAMPLE SCANNING ELECTRON MICROGRAPHS ARE INCLUDED IN THIS EXERCISE (SEE FIGURES 1-10).

The first series of micrographs (Figures 1-3) are on the fine surface topography of *Cysticercus fasciolaris*, the larval form of the adult cat taperworm, *Hydatigera taeniaeformis* (published micrographs of B.R. Jones).

The second series of micrographs (Figures 4-10) consist of scans of the fine surface topography of various snail vectors involved in the transmission of schistosomiasis (unpublished micrographs of B.R. Jones).

Snails for carrying out SEM work were provided by Dr. Edward H. Michelson, Department of Tropical Public Health, Harvard School of Public Health, Boston, Massachusetts.

E. EXAMINE EACH MICROGRAPH CAREFULLY AND ANSWER THE QUESTIONS AT THE END OF THIS EXERCISE.

A DISCUSSION OF ANSWERS TO QUESTIONS WILL FOLLOW. STUDENTS MAY WISH TO CONSULT REFERENCE TEXTS FOR ANSWERING SOME OF THE QUESTIONS. (SEE APPENDIX B FOR LIST OF REFERENCE TEXTS).

NOTES:

FIGURE 1. Scanning electron micrograph of the scolex of the larval tapeworm, *Cysticercus fasciolaris*: (r), rostellum; (s), rostellar sac; (arrows), hooks; (su), suckers. Pit-like indentions are regularly distributed at the apex of the scolex (350X).

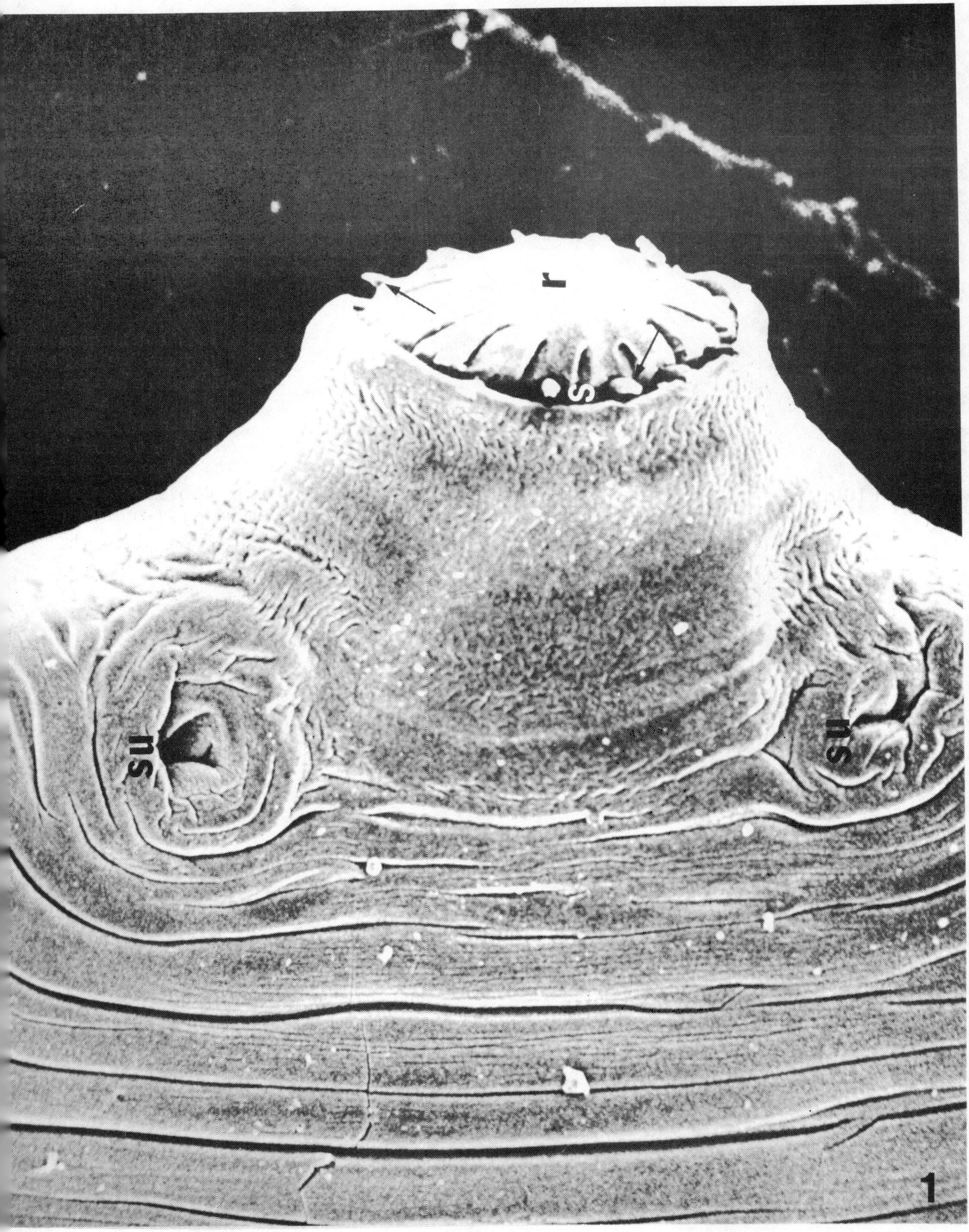
r
s
ns
ns
1

FIGURE 2. Scanning electron micrograph of the scolex of Cysticercus *fasciolaris* taken from a different angle (390X).

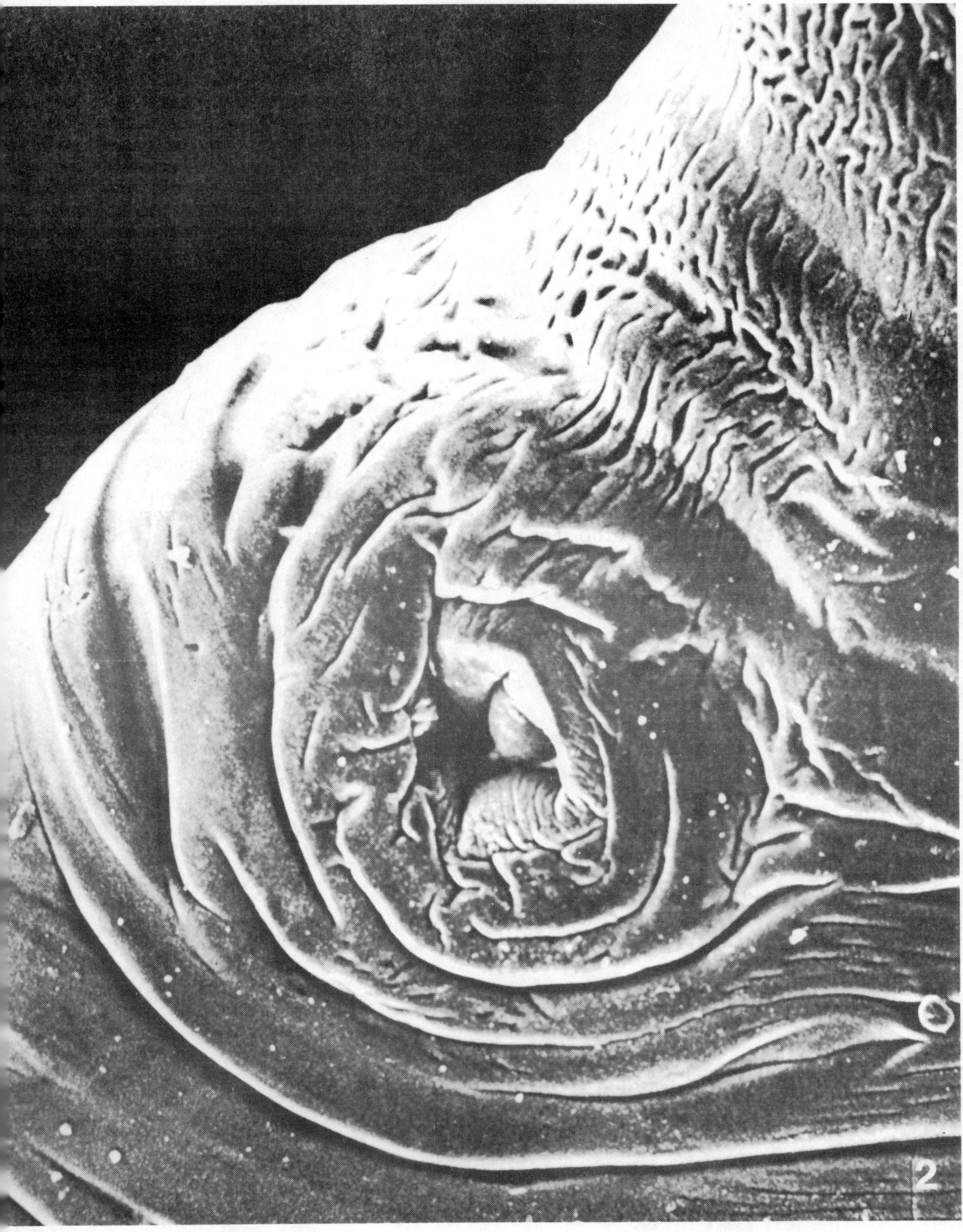
2

FIGURE 3. Scanning electron micrograph of the scolex of *Cysticercus fasciolaris* showing the inner surface features of a sucker (700X).

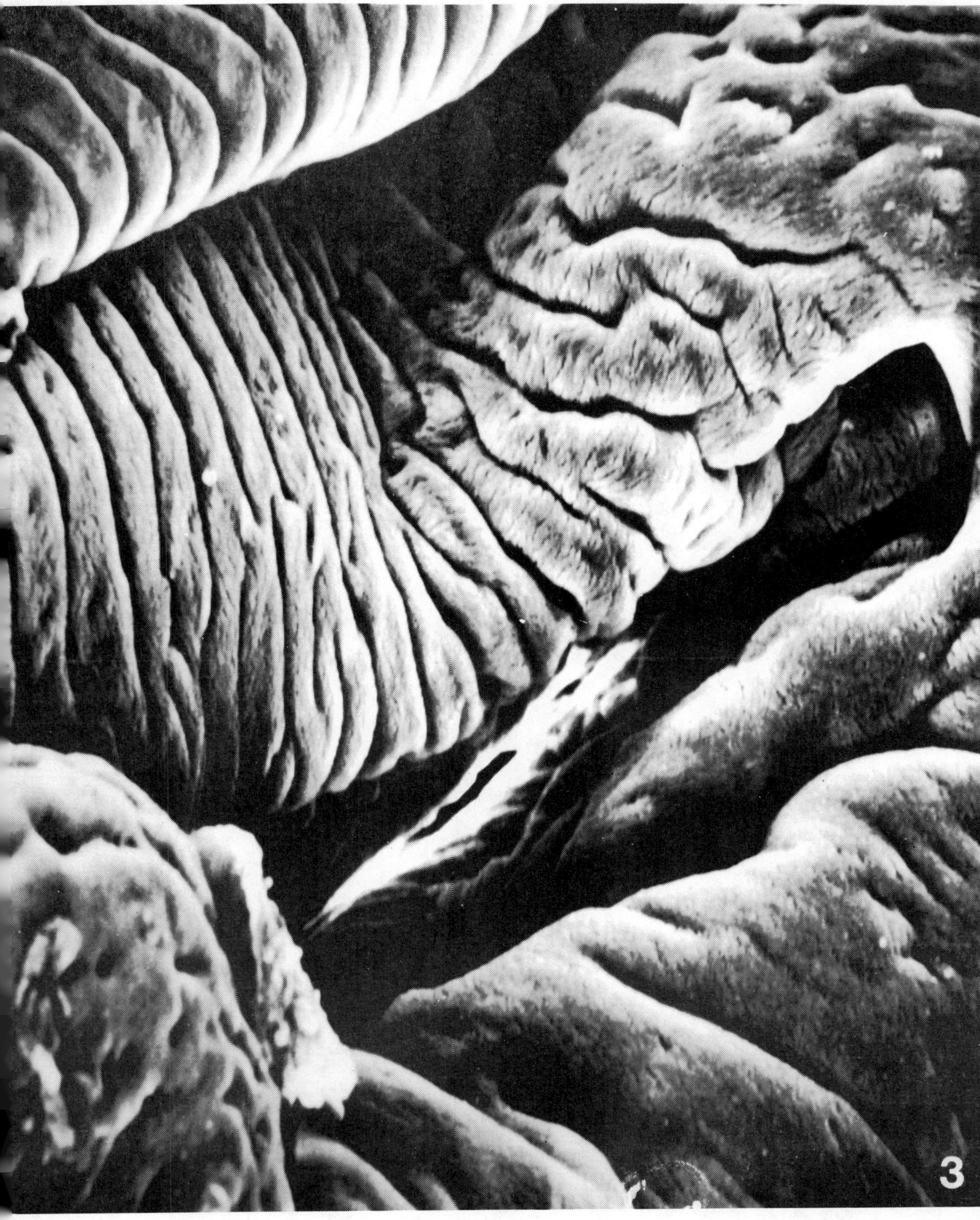

FIGURE 4. Scanning electron micrograph of the head-foot region of the snail, *Biomphalaria glabrata*: (h), head; (t), tentacles; (mc), mantle collar; (f), foot (35X).

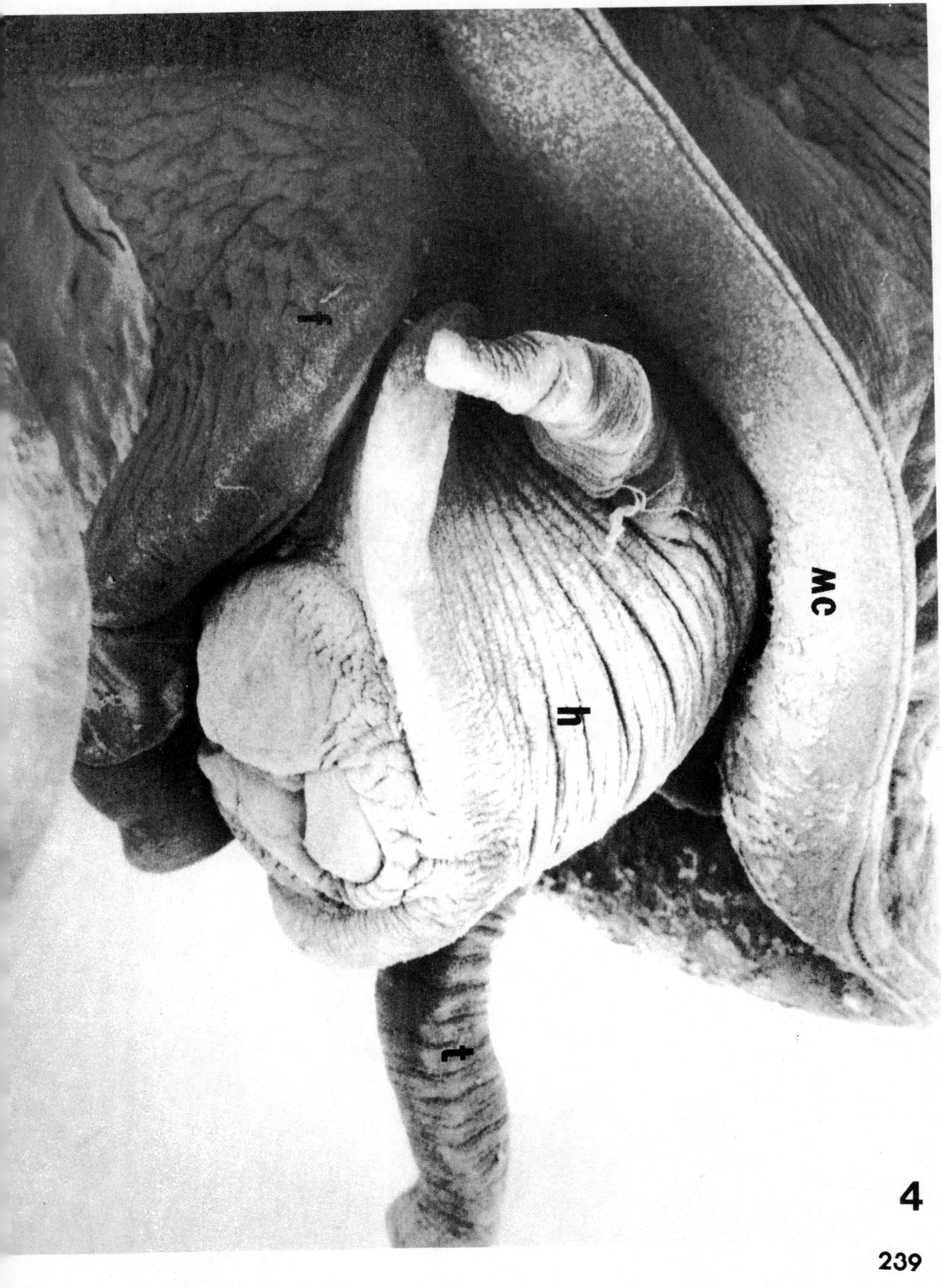

4

FIGURE 5. Scanning electron micrograph of the inner surface of the mantle collar of *Biomphalaria glabrata* showing tufts of cilia (2,500X).

FIGURE 6. Scanning electron micrograph of the mouth region of *Biomphalaria glabrata* (100X).

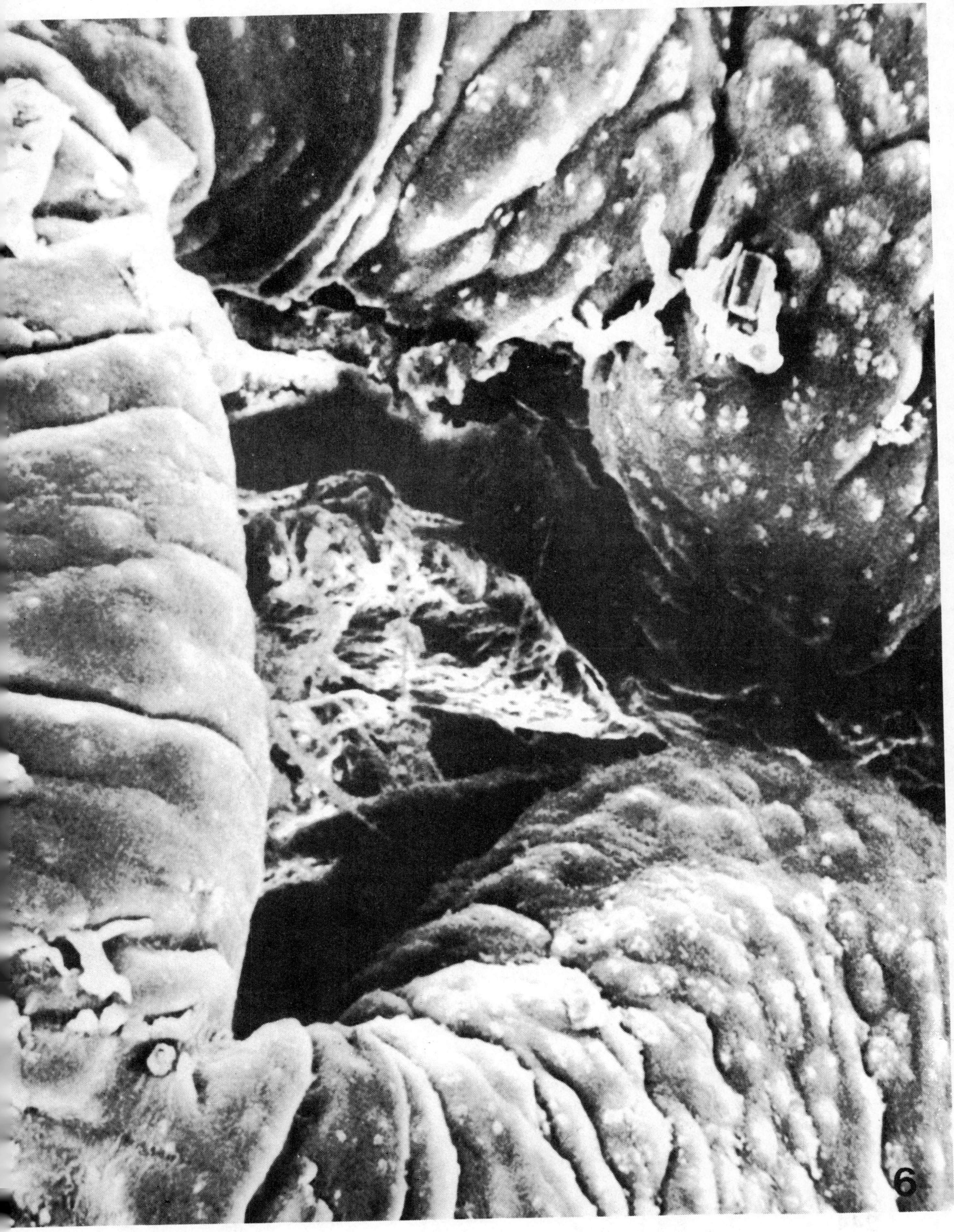
6

FIGURE 7. Scanning electron micrograph of the mantle collar of Biomphalaria straminea showing numerous cilia (6,400X).

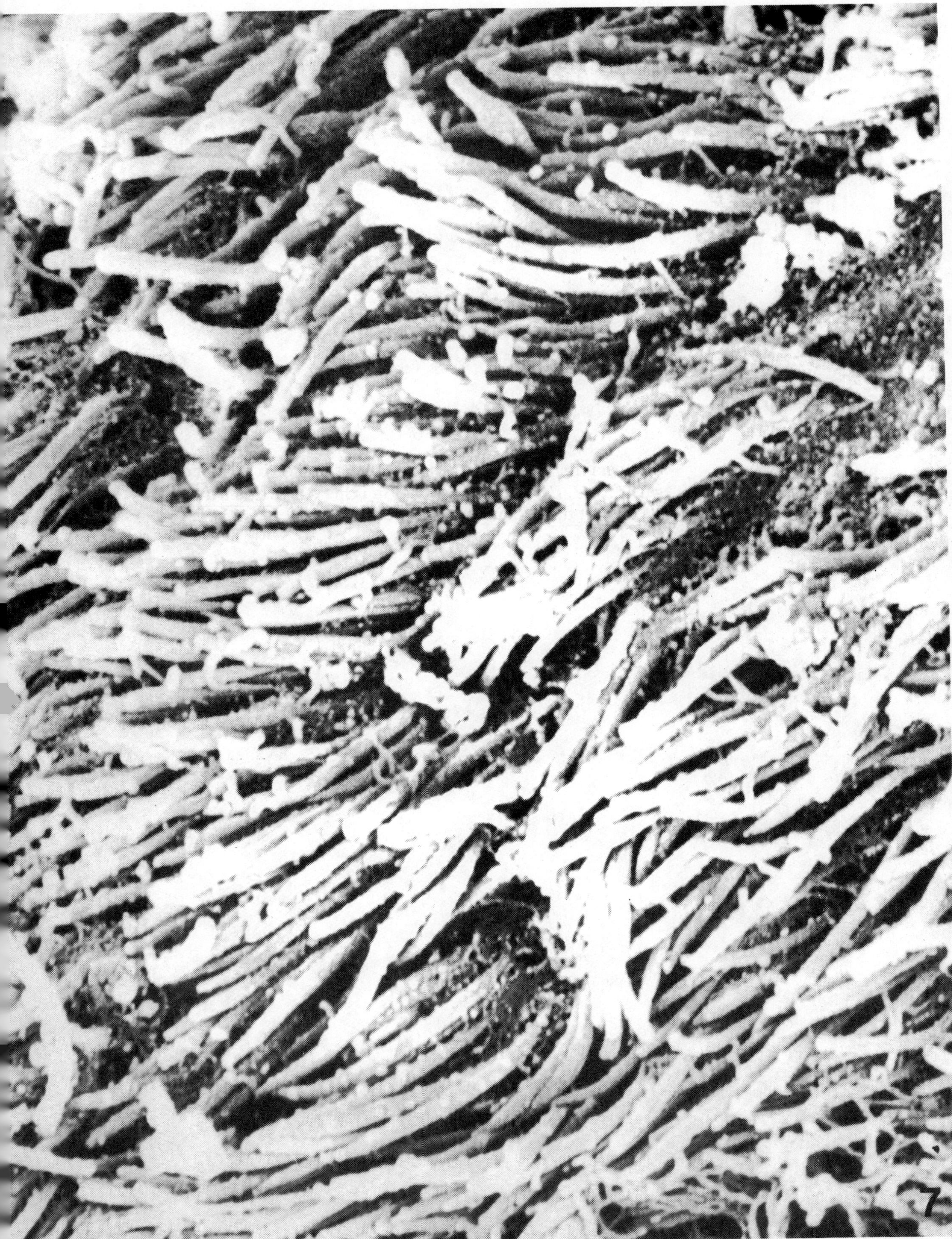

7

FIGURE 8. Scanning electron micrograph of the top surface of the head region of *Bulinus truncatus* showing tufts of cilia (100X).

8

FIGURE 9. Scanning electron micrograph of *Bulinus truncatus* showing at higher magnification of cilia from Figure 8 (15,000X).

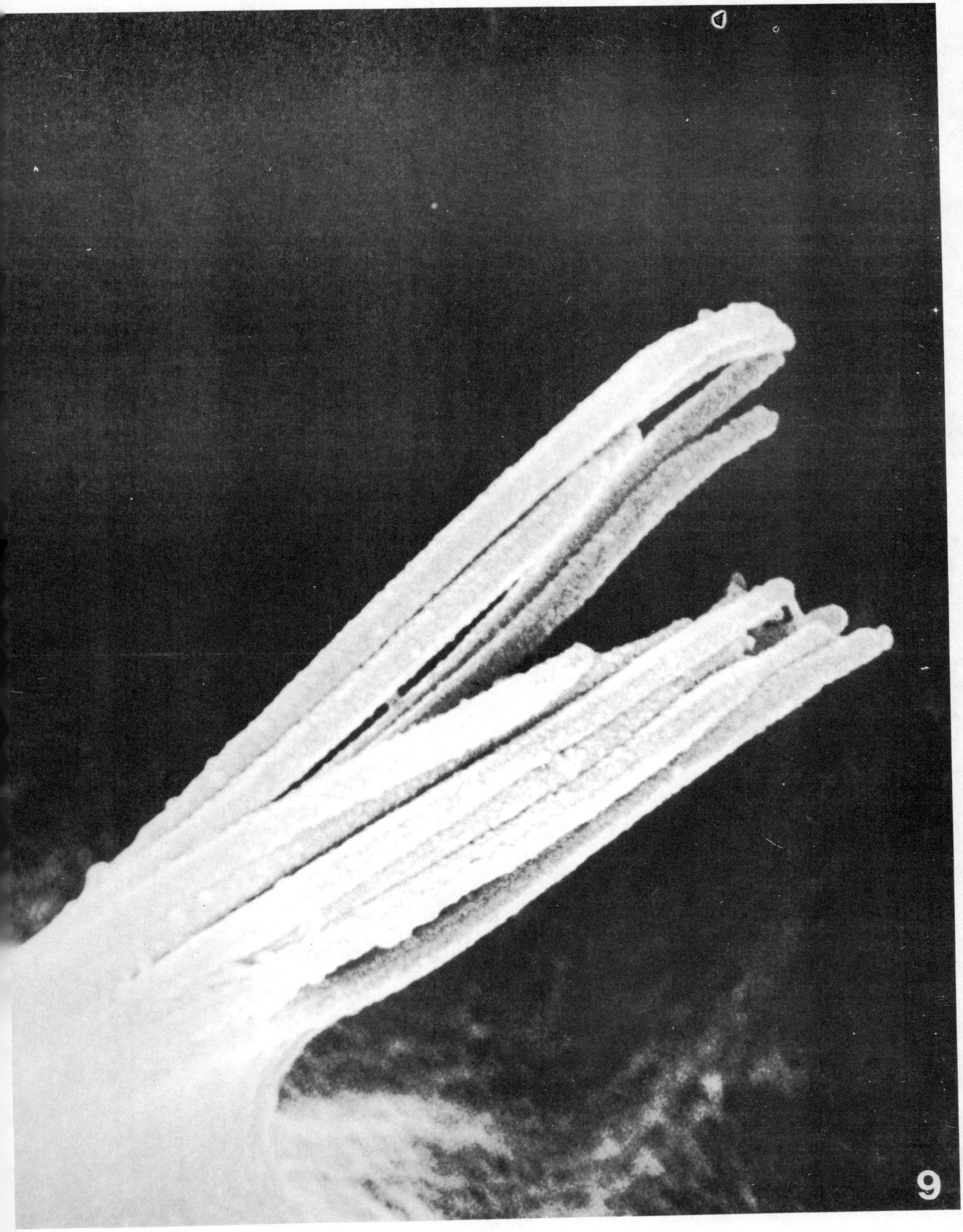

9

FIGURE 10. Scanning electron micrograph of the mantle collar of *Biomphalaria tenagophila* showing tufts of cilia (20,000X).

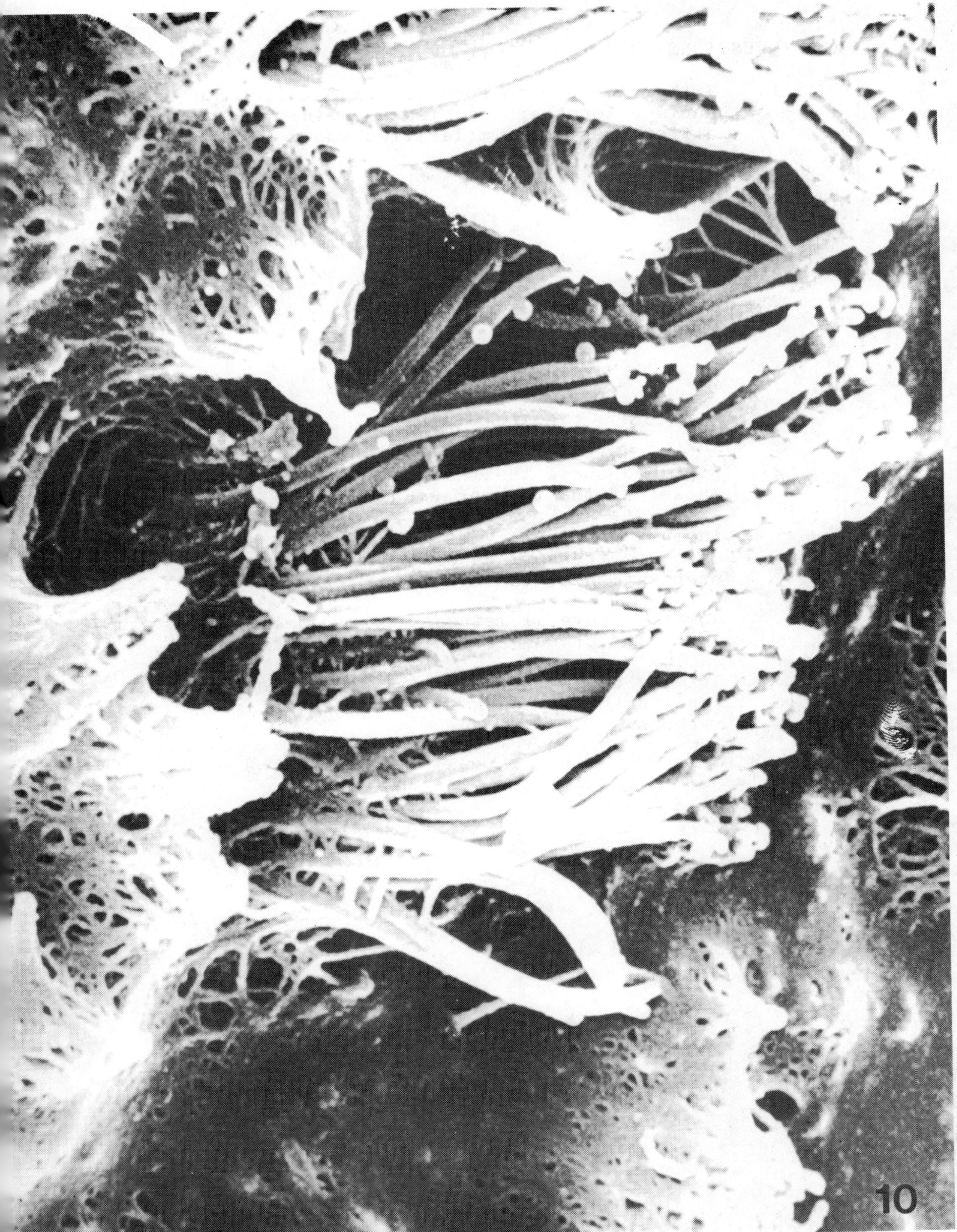

10

E. QUESTIONS:

1. With reference to specimen orientation and fine surface topographical features what are the differences and similarities among Figures 1, 2, and 3?

2. Figure 3 shows the inner surface topography of a sucker of the tapeworm.

   In this particular instance it was not necessary to exposure the surface of the sucker during preparation of the specimen.

   However, several methods are available for exposing the interior surfaces of specimens for SEM.

   Briefly discuss at least two of these methods. Mention the approach and major advantages of the methods.

EXERCISE NUMBER 23

NEGATIVE STAINING OF SPECIMENS

Betty R. Jones, Ph.D.

NOTE: FOR DETAILED AND SPECIFIC INFORMATION ON NEGATIVE STAINING SUCH AS VARIOUS METHODS OF APPLICATION SEE EXERCISE NUMBER 30 OF THIS MANUAL.

A. OBJECTIVES:

1. To provide direct visualization of macromolecules and particulate specimens without the use of conventional procedures and techniques

2. To reveal structural details at the subparticle level in unfixed specimens

3. To provide a simple method for showing structural detail and appropriate contrast for observing specimens

4. To study and measure the dimensions of particles and subunits from specimens

5. To determine the molecular weight of particles from specimens

6. To use as a "negative stain" a solution that has high electron scattering ability, high solubility, high density, high melting point and nonreactivity with the specimen such as uranyl acetate and phosphotungstic acid

7. To apply the drop, spray or float method for staining specimens

8. To obtain micrographs which reveal contrast and can be used to garn structural information from the specimen which corresponds to its appearance in solution

B. MATERIALS:

1. Specimen (bacterial or viral suspension)

2. 2% uranyl acetate (100 ml)

NOTE: THE URANYL ACETATE SHOULD BE FRESHLY PREPARED AND FILTERED.

3. Cold distilled water (200 ml)

4. Forceps with a fine tip

5. Several pasteur pipettes with rubber bulbs

6. Filter paper (Whatman's #3)

7. Formvar-carbon coated grids

NOTE: COATED GRIDS SHOULD BE PREPARED WITHIN ONE WEEK. SEE EXERCISE NUMBERS 31 and 32 FOR COATING OF GRIDS.

8. 3 petri dishes

C. RATIONALE AND DISCUSSION:

The basic principles of negative staining are not difficult to understand. A stain with high electron scattering properties such as uranyl acetate or phosphotungstic acid is deposited on a suspension of the specimen on a formvar-carbon coated grid.

In addition to high electron scattering properties the stain selected should be highly soluble, nonreactive with the specimen, have a high density and a high melting point.

The specimen suspension and stain are allowed to dry on the coated grid. The resulting product is electron micrographs in which the specimen appears white outlined by a dark background.

Negative staining may be the method of choice when structural detail information is needed at the subparticle level for particulate specimens.

Due to the technique, various subparticle details such as the subunits of viruses and proteins have been revealed. The technique is also popular because it is simple to carry out and the possibility of obtaining negative results are very slim.

D. PROCEDURE:

1. Using a clean pasteur pipette place a drop of specimen suspension on a formvar-carbon coated grid.

2. Remove any excess fluid by touching the side of droplet with filter paper.

3. Place stain droplet over specimen suspension (See Figure 23-a-c).

4. Let mixture dry in closed petri dish lined with filter paper.

5. Observe in the transmission electron microscope.

6. Take pictures at desired magnifications.

7. Develop and print EM plates or films.

NOTE: REFER BACK TO EXERCISE NUMBER 18 ON PHOTOGRAPHIC TECHNIQUES.

E. RESULTS:

The specimen will appear white outlined by a dark background.

F. EVALUATION:

PRESENT ONE ELECTRON MICROGRAPH TO THE INSTRUCTOR FOR EVALUATION.

G. QUESTIONS:

1. Discuss the principles of negative staining versus positive staining.

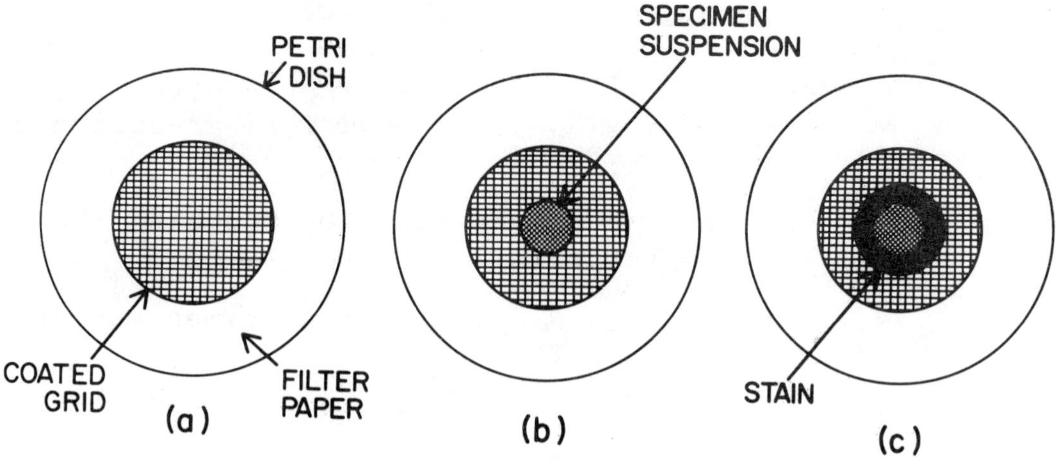

FIGURE 23-a-c. Negative staining of specimens.

(a) Formvar-carbon coated grid on filter paper in petri dish.

(b) Drop of specimen suspension placed on grid.

(c) Drop of stain placed on specimen suspension.

---

2. Under what circumstances might one employ the negative staining technique?

3. What investigators were responsible for first working out a routine procedure for negative staining? Describe the essential features of their first experiment.

4. Briefly discuss how artifacts may creep into specimens using this technique.

5. Discuss two common chemicals that are used as negative stains and factors that may render them a negative stain instead of a positive stain.

SECTION II

EXERCISE NUMBER 24

ENZYME CYTOCHEMISTRY LIGHT MICROSCOPY AND ELECTRON MICROSCOPY

John M. Robinson, Ph.D.

INTRODUCTION

Enzyme cytochemistry, at both the light and electron microscope levels, has provided valuable information about cell structure and tissue organization. The purpose of these exercises is to illustrate principles of enzyme cytochemistry by localizing selected enzymes.

Enzymes are special proteins which catalyze reactions in biological samples. In order to study an enzyme one must be able to follow some part of the reaction (e.g. loss of substrate, change in a co-factor, product formation, molecular rearrangements or other physical properties).

The biochemist quantifies an enzyme reaction using instruments such as spectrophotometers, fluorometers, scintillation counters, and respirometers. The cytochemist localizes the activity of an enzyme spatially within a cell or tissue using light or electron microscopes.

The cytochemist views structure and function as an integral part of tissue study, and therefore attempts to correlate the chemical and metabolic features of a tissue with its morphology.

In all types of cytochemistry (enzyme cytochemistry, immunocytochemistry, and the cytochemistry of structural components such as carbohydrates, lipids, proteins, nucleic acids) a reaction product which is visible or can be rendered visible at the light or electron microscope level must be formed.

It is of primary importance to understand what reaction products are formed and how they are formed. In this introduction the discussion will be limited to the formation of a reaction product by a) metal precipitation methods, b) coupling azo dye methods, and c) oxidation-reduction methods.

The principle which is most frequently employed in the formation of reaction product is known as the simultaneous capture reaction. The general form for simultaneous capture reactions is defined in the reaction scheme of Figure 24-a.

FIGURE 24-a. General scheme for simultaneous capture reactions.

An example of a simultaneous capture reaction as applied to metal precipitation methods is shown in Figure 24-b.

Figure 24-b shows the use of metal precipitation reaction for localization of acid phosphatase.

The "final reaction product", the insoluble metal phosphate, can be visualized directly at the electron microscope level (a);

Or can be further modified so that it can be visualized at the light microscope level (b).

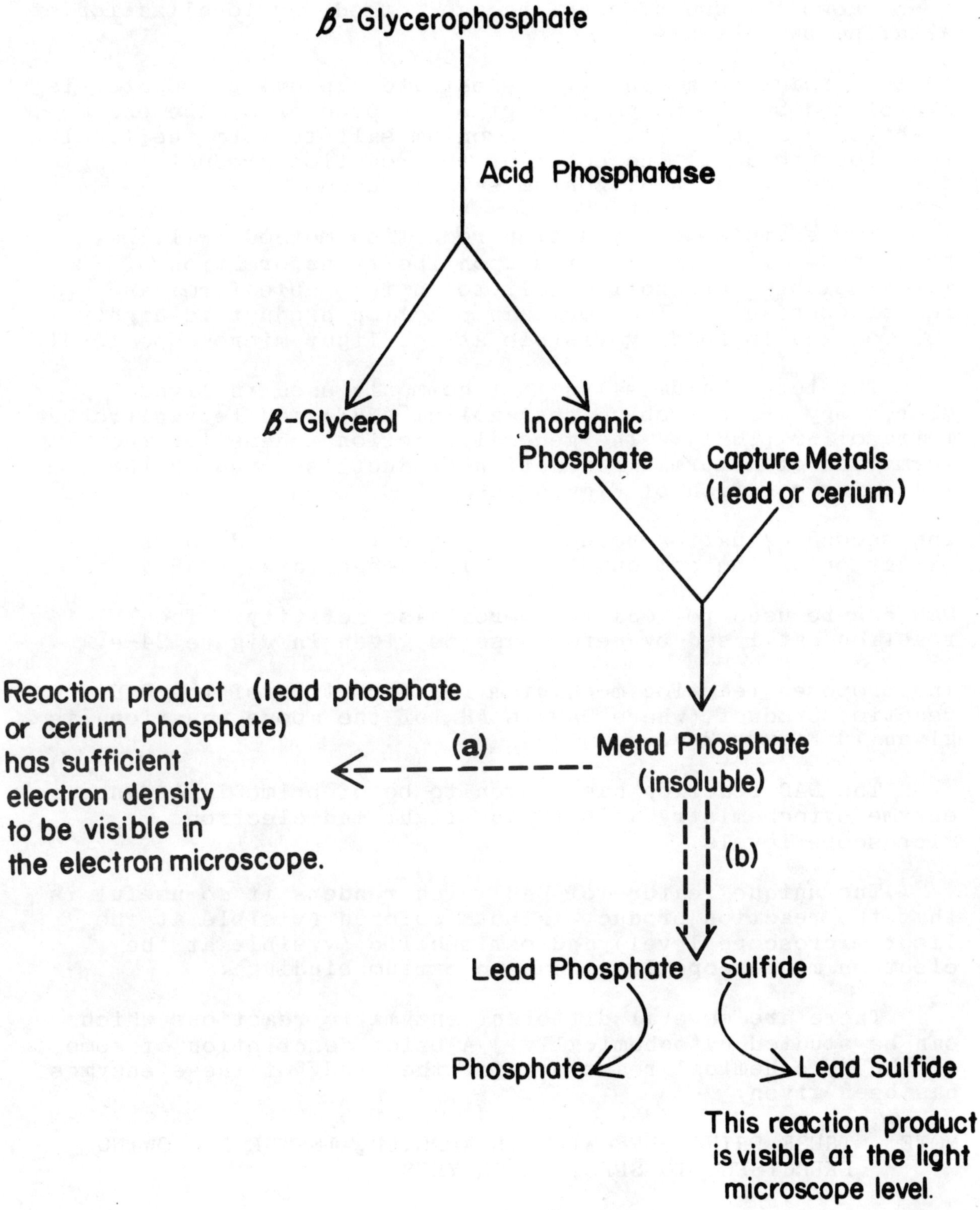
β-Glycerophosphate
Acid Phosphatase
β-Glycerol
Inorganic Phosphate
Capture Metals (lead or cerium)
Reaction product (lead phosphate or cerium phosphate) has sufficient electron density to be visible in the electron microscope.
(a)
Metal Phosphate (insoluble)
(b)
Lead Phosphate
Sulfide
Phosphate
Lead Sulfide
This reaction product is visible at the light microscope level.

FIGURE 24-b.

An example of a simultaneous capture reaction as applied to coupling azo dye methods is shown in Figure 24-c. Figure 24-c shows the use of coupling azo method for localization of alkaline phosphatase.

In this reaction a synthetic substrate napthyl phosphate, is hydrolyzed by the enzyme. Napthol, a product of the enzyme reaction combines with the diazonium salt to form the final reaction product. The color of the reaction product is determined by which diazonium salt is used.

Two examples of oxidation-reduction methods will be presented. The first relies upon the transformation of a water-soluble tetrazolium salt to an insoluble formazan reaction product. The formazan reaction product is highly colored and is readily visible at the light microscope level.

The tetrazonium salts most commonly used in light microscopy are nitroblue tetrazolium (NBT) and tetranitroblue tetrazolium (TNBT). The general reaction scheme for the formation of a formazan reaction product is given in the following equation of Figure 24-d.

The second oxidation-reduction method to be studied is the oxidation of the compound 3,3'-diaminobenzidine (DAB).

DAB can be used to localize peroxidase activity. The reaction catalysed by peroxidase is given in Figure 24-e.

The proposed reaction mechanism for formation of the DAB reaction product (where DAB is AH_2 of the above equation) is given in Figure 24-f.

The DAB reaction has proven to be of prime importance in enzyme cytochemistry at both the light and electron microscope levels.

The unique feature of DAB which renders it so useful is that the reaction product is both colored (visible at the light microscope level) and osmiophilic (visible at the electron microscope level due to osmium binding).

There are several different enzymatic reactions which can be studied cytochemically. A brief description of some of the cytochemical reactions for the study of these enzymes has been given.

NOTE: THESE PRINCIPLES WILL BE APPLIED, IN THE FOLLOWING EXERCISES, TO SPECIFIC ENZYMES.

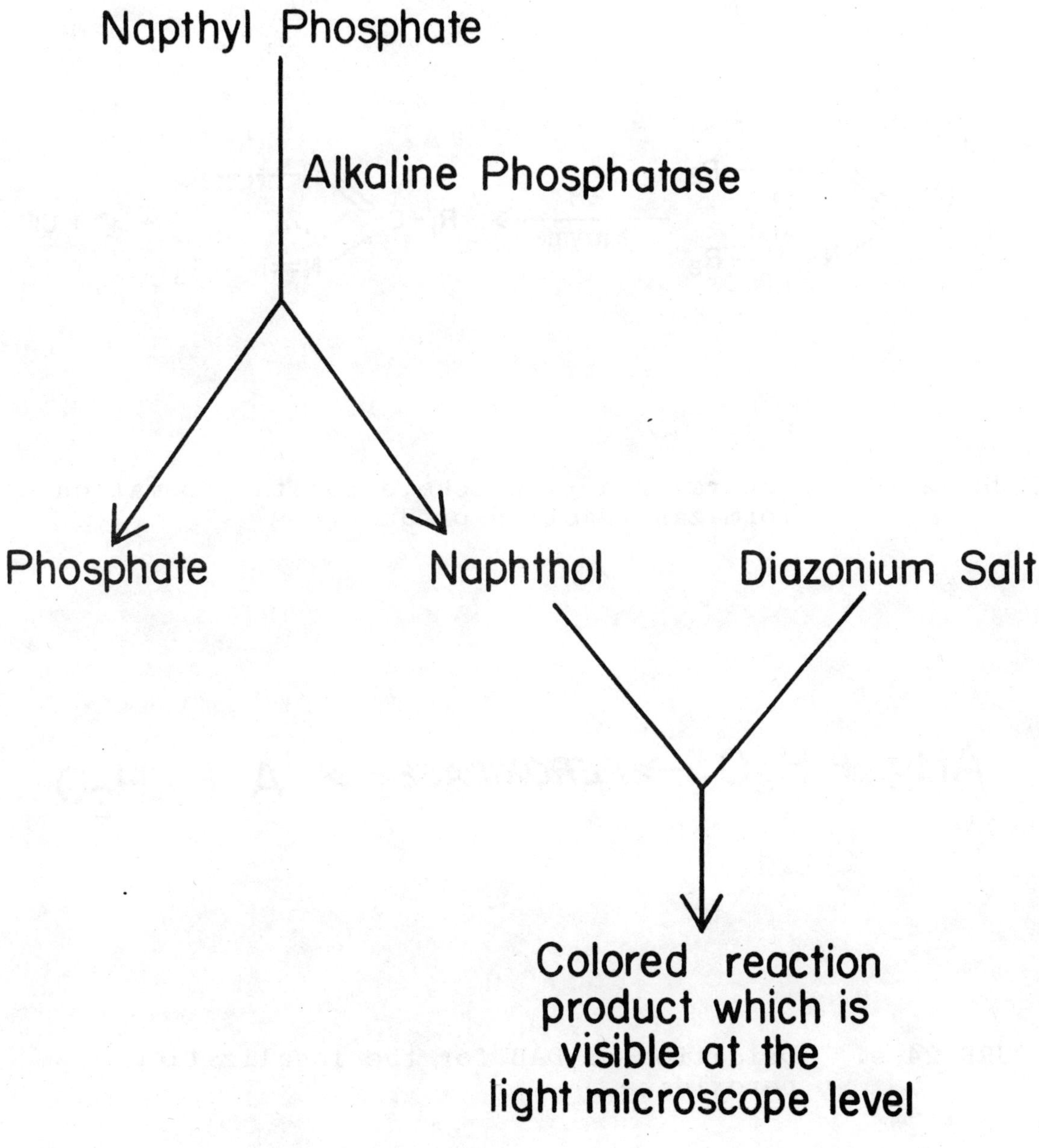
Napthyl Phosphate
Alkaline Phosphatase
Phosphate
Naphthol
Diazonium Salt
Colored reaction
product which is
visible at the
light microscope level

FIGURE 24-c.

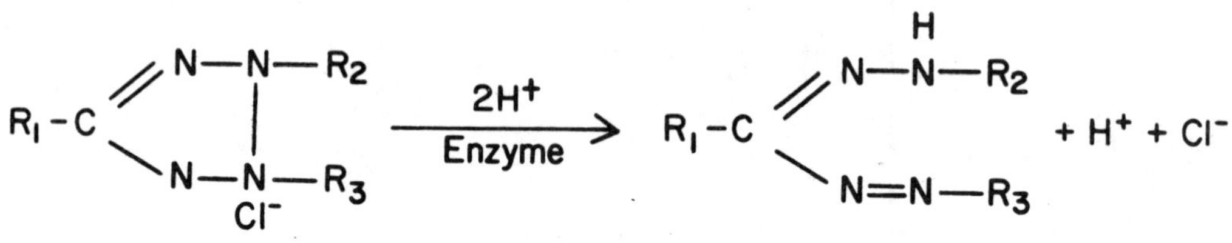

FIGURE 24-d. General reaction scheme for the formation of a formazan reaction product.

$$AH_2 + H_2O \rightarrow PEROXIDASE \rightarrow A + 2H_2O$$

FIGURE 24-e. Oxidation of DAB for the localization of peroxidase.

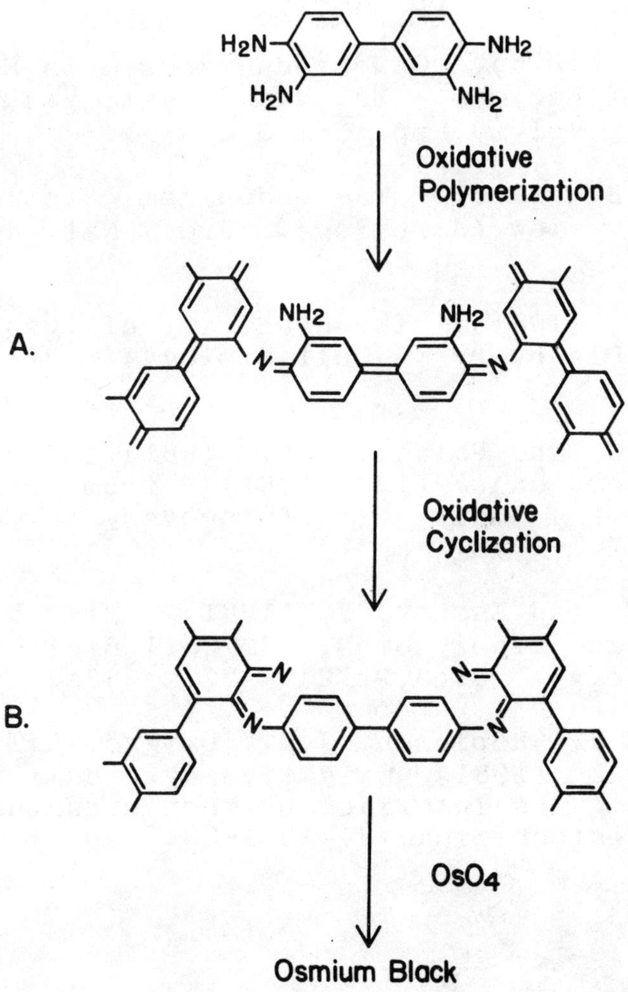

FIGURE 24-f. Proposed reaction mechanism for the formation of DAB reaction product (courtesy of A.M. Seligman, M.J. Karnovsky, H.L. Wasserkrug, and J.S. Hanker (1968). J. Cell Biol., 38:1-14. (Copyright permission, Rockefeller University Press).

Selected general references for outside reading:

1. Barka, T. and Anderson, P.J. (1963). Histochemistry. Theory, practice and bibliography. Harper and Row, New York.

2. Essner, E. (1973). Phosphatases. In Electron Microscopy of Enzymes. Ed. M.A. Hayat, Van Nostrand Reinhold Co., Vol 1, pp. 44-76.

3. Hanker, J.S. (1975). Oxidoreductases. In Electron Microscopy of Enzymes. Ed. M.A. Hayat, Van Nostrand Reinhold Co., Vol. 4, pp. 1-139.

4. Novikoff, A.B. (1976). The endoplasmic reticulum: A cytochemist's view (A review). Proc. Nat. Acad. Sci. 73:2781-2787.

5. Bainton, D.F. (1981). The discovery of lysosomes. J. Cell Biol. (Discovery in Cell Biology). 91(3, Pt. 2):66s-76s.

6. Farquhar, M.G. and Palade, G.F. (1981). The Golgi Apparatus (Complex) - (1954-1981) - from artifact to center stage. J. Cell Biol.(Discovery in Cell Biology) 91(3,Pt.2): 77s-103s.

7. Tolbert, N.E. and Essner, E. (1981). Microbodies: Peroxisomes and glyoxysomes. J. Cell Biol. (Discovery in Cell Biology). 91(3,Pt.2): 271s-283s.

8. Karnovsky, M.J., Robinson, J.M., Briggs, R.T., Karnovsky, M.L. (1981) Oxidative cytochemistry in phagocytosis: the interface between structure and function. Histochemical J. 13:1-22.

EXERCISE NUMBER 24A

LOCALIZATION OF ACID PHOSPHATASE AT THE LIGHT MICROSCOPE AND ELECTRON MICROSCOPE LEVEL BY A METAL PRECIPITATION METHOD

John M. Robinson, Ph.D.

A. OBJECTIVES:

1. To obtain precise localization of the enzyme acid phosphatase at both the light and electron microscope levels

2. To determine in which subcellular components the acid phosphatase activity resides

3. To compare the cytochemical localization of acid phosphatase in two different tissues

4. To determine the specificity of the cytochemical reaction by the use of control experiments

B. MATERIALS:

1. A laboratory animal (mouse or rat)

2. 2% glutaraldehyde in 0.1 M sodium cacodylate buffer (pH 7.2)

3. 2% osmium tetroxide in 0.1 M sodium cacodylate buffer (pH 7.2)

4. 0.2 M sodium cacodylate buffer (pH 7.2)(and proper dilutions of the buffer as the experiment requires)

5. 0.2 M acetate buffer (pH 5.0)(and proper dilutions of the buffer as the experiment requires)

6. Cytidine monophosphate, disodium salt or β-glycerophosphate, disodium salt; either prepared as 20 mM stock solutions

7. Lead nitrate (12% aqueous)

8. Filter paper and funnel

9. Petri dishes

10. Dilute ammonium sulfide (2% aqueous)

11. Cryostat

12. Water-soluble tissue mounting medium for cryostat

13. Tissue chopper (Smith-Farquhar type)

14. 10% agar solution

15. Clean razor blades

16. Glass tissue vials

17. Clean glass cover slips (22 mm square)

18. Clean glass microscope slides

19. Glycerol

20. A graded series of ethanol solutions for dehydration (See Exercise Number 12)

21. Embedding resin for specimen preparation for electron microscopy (See Exercise Number 12)

C. RATIONALE AND DISCUSSION:

The goal of all cytochemical experiments is the localization of enzyme activity which reflects the true localization of the enzyme within the cell or tissue. Precise localization requires that the capture reaction be very rapid so that precipitation of the reaction product occurs at the site of the enzyme and that the reactant compounds do not diffuse away. The lead capture reaction which is employed in this exercise does give an accurate representation of the subcellular localization of acid phosphatase activity.

It is important to maintain the integrity of the cells or tissue as much as possible without destroying the activity of the enzyme. Whenever possible it is best to chemically fix the tissue. Some enzymes, however, cannot survive even the mildest fixation conditions. In enzyme cytochemistry there is always a compromise between maintaining good morphological detail and detectable enzyme activity.

Acid phosphatase, the enzyme under consideration in this exercise, retains much of its activity after mild fixation.

All of the components of a cytochemical reaction mixture must have access to the cells and subcellular compartments. This is more of a problem with solid pieces of tissue than with suspensions of single cells. In this exercise two methods are used to overcome this problem. The two methods are, however, similar; they both enable one to cut the tissue into pieces much thinner than could be reproducibly done free-hand. The first of these methods utilizes a cryostat, which is a refrigerated compartment containing a rotary microtome. The cryostat can cut frozen sections thin enough to permit direct viewing with a light microscope (routinely 5-20 µm sections are cut). The second method uses an instrument called a tissue chopper; the most popular chopper is the Smith-Farquhar type. With this instrument non-frozen sections can be cut (30-100 µm) which are thin enough for most enzyme cytochemical incubations.

In this exercise acid phosphatase will be localized at both the light and electron microscope levels. Results obtained at the light microscope level will illustrate the overall distribution of the enzyme activity within the tissue. Results obtained at the electron microscope level will show which subcellular compartment(s) contain this enzyme activity. Control experiments will illustrate that deposition of reaction product is enzyme mediated.

D. PROCEDURE:

1. Prepare the primary fixative (2% glutaraldehyde in 0.1 M sodium cacodylate buffer, pH 7.2) and chill on ice.

   Prepare the glutaraldehyde solution as described in (Exercise Number 11).

2. Sacrifice the animal and quickly remove pieces of liver and a kidney.

3. Transfer the tissue into separate petri dishes containing cold fixative

4. Quickly trim the tissue into small pieces approximately 1 mm wide and 3-5 mm long.

   Keep an accurate record of the time the tissue is in the fixative.

5. Keep the tissue in ice cold fixative for 1 hr

NOTE: THIS MODE OF FIXATION IS TERMED <u>IMMERSION FIXATION</u>.

IT IS BETTER TO FIX TISSUE SUCH AS LIVER AND KIDNEY BY <u>PERFUSION</u>, THAT IS BY INFUSING THE FIXATIVE DIRECTLY INTO THE CIRCULATORY SYSTEM.

HOWEVER, FOR THE PURPOSE OF THIS EXERCISE IMMERSION FIXATION WILL BE SUFFICIENT.

6. Following fixation, wash the tissue in ice cold 0.1 M cacodylate buffer several times (4 changes of buffer over a 1 hr period).

7a. For light microscope localization of acid phosphatase embed pieces of tissue in cryostat mounting medium and freeze in the cryostat.

   Use specimen mounts and freezing procedure for the particular cryostat which is available to you.

7b. For electron microscope localization of acid phosphatase embed pieces of tissue in molten 10% agar (42-45°C) on mounts designed for the the tissue chopper which is available to you.

   Transfer the mounts to a refrigerator to solidify the agar.

8a. Cut cryostat sections at 10 µm.

8b. Cut tissue chopper sections at 70 µm.

9a. Pick up cryostat sections on glass cover slips (try to collect 2-3 sections per cover slip).

9b. Transfer chopped sections to ice cold 0.1 M cacodylate buffer.

10a. Air dry cryostat sections for a few minutes (5-10 min) at room temperature.

10b. Transfer chopped sections to 0.1 M acetate buffer, pH 5.0. Change the solution 3 times over 30 minutes.

11. Prepare the cytochemical medium.

The final medium contains 0.1 M acetate buffer (pH 5.0), 2 mM cytidine monophosphate or β-glycerophosphate, and 1.2% lead nitrate.

The control medium is the same except that it lacks the substrate for the enzyme.

12a. Transfer the cover slips with cryostat sections to a small petri dish.

Have separate containers for medium with and without substrate.

12b. Transfer the chopped sections to vials containing medium (preferably several sections per vial).

Have separate vials for medium with and without substrate.

13. Incubate the specimens at room temperature or in a 37°C incubator if available for 30 min. for the cryostat sections and 60 min for the chopped sections.

During the incubation there should be gentle agitation to insure proper mixing of the reaction components and to prevent local exhaustion of these components.

14. Replace cytochemical reaction medium with 0.1 M acetate buffer.

Wash the samples 3 times over 30 minutes.

15a. Develop the cytochemical reaction in the cryostat sections by incubating for a few minutes (5-10 minutes is probably sufficient) with dilute ammonium sulfide.

See figure 24-a of the introduction for the rationale for this portion of the reaction.

NOTE: AMMONIUM SULFIDE IS POISONOUS AND GIVES OFF NOXIOUS VAPORS.

THIS COMPOUND MUST BE USED IN A FUME HOOD.

IT IS BEST TO WEAR VINYL GLOVES WHEN HANDLING AMMONIUM SULFIDE.

15b. Wash the chopped sections in 0.1 M sodium cacodylate buffer following the acetate buffer.

16a. Rinse cryostat sections in distilled water to remove all of the ammonium sulfide.

16b. Post fix the chopped sections in 2% OsO_4 in 0.1 M sodium cacodylate buffer for 1 hr at room temperature.

Prepare the osmium solution as previously described (See Exercise Number 11).

17a. Mount cover slips containing cryostat sections in a drop of glycerol and examine in the light microscope.

17b. Process the osmium treated sections for electron microscopy and examine thin sections as previously described (See Exercise Number 14).

E. PREPARATION OF SOLUTIONS:

1. Cacodylate buffer

Prepare Solution A. 0.2 M solution of sodium cacodylate (42.8 grams of $Na(CH_3)_2AsO_2 \cdot 3H_2O$ in 1,000 ml distilled H_2O).

Prepare Solution B. 0.2 M HCl

Adjust to desired pH (7.2) by mixing a portion of solution B with a portion of solution A and then dilute to give final desired concentration.

2. Acetate buffer

Prepare Solution A. 0.2 M solution of acetic acid (11.5 ml of glacial acetic acid diluted to 1,000 ml with distilled H_2O).

Prepare Solution B. 0.2 M solution of sodium acetate (16.4 grams of $C_2H_3O_2Na$ or 27.2 grams of $C_2H_3O_2Na\ 3H_2O$ in 1,000 ml of distilled H_2O) $C_2H_3O_2Na \cdot 3H_2O$

Adjust to desired pH (5.0) by mixing a portion of solution A with a portion of solution B.

Then dilute with distilled H_2O to give the desired concentration (0.12 M for the cytochemical medium and 0.10 M for the buffer washes).

3. Cytochemical Medium

To make Complete medium add:

8 ml of 0.12 M acetate buffer

1 ml of 20 mM cytidine monophosphate or β-glycerophosphate (prepared in distilled water).

1 ml of 12% lead nitrate (in distilled water)

Any volume of medium may be prepared as long as the proportion of components is maintained the same.

NOTE: ADD THE LEAD NITRATE SLOWLY (DROPWISE) WITH CONSTANT MIXING.

THE SOLUTION MAY BE SLIGHTLY CLOUDY, IF SO, FILTER BEFORE USE.

4. To make Control Medium add:

8 ml of 0.12 M acetate buffer

1 ml of distilled water

1 ml of 12% lead nitrate

5. Agar

To make 10% agar:

Dissolve 5 grams of agar in 50 ml of water.

Bring the water to a boil to get agar into solution.

Maintain in hot water bath to prevent agar from solidifying.

F. RESULTS:

Lead phosphate reaction product will be formed at the sites of enzyme activity (acid phosphatase). The lead phosphatase is electron dense and is directly visible in thin sections in the electron microscope.

Lead phosphate is not visible at the light microscope level; incubation with ammonium sulfide converts the reaction product into lead sulfide which will appear as dark deposits.

G. EVALUATION:

Examine slides with the light microscope and record whether reaction product is formed and if formed where it is located. Present slides of liver and kidney from control and experimental incubations to instructor for evaluation.

Examine grids with the electron microscope and record whether reaction product is formed and if formed where it is located. Present electron micrographs, from the ultrastructural portion of the experiment to the instructor.

H. QUESTIONS:

1. What is the distribution of reaction product at the light microscope level in liver and kidney?

2. What types of cells contain reaction product? Does the distribution of reaction product vary from cell type to cell type?

3. Which organelles contain reaction product?

EXERCISE NUMBER 24B

LOCALIZATION OF ALKALINE PHOSPHATASE AT THE LIGHT MICROSCOPE LEVEL BY A COUPLING AZO DYE METHOD

John M. Robinson, Ph.D.

A. OBJECTIVES:

1. To obtain precise localization of the enzyme alkaline phosphatase at the light microscope level with a coupling azo dye method
2. To compare the cytochemiical localization of alkaline phosphatase in several different tissues
3. To determine the specificity of the cytochemical reaction by the use of control and inhibitor experiments

B. MATERIALS:

1. A laboratory animal (mouse or rat)
2. Cold neutral buffered formaldehyde
3. Commercial albumin fixative, for mounting sections on slides
4. 0.05 M Tris-maleate buffer, pH 8.3
5. Napthol AS-MX (used as the enzyme substrate)
6. N,N-dimethyformamide
7. Diazonium salt such as Fast Blue RR salt or Fast Violet B salt
8. Levamisole
9. Petri dishes
10. Clean razor blades

11. Cryostat

12. Water-soluble tissue mounting medium for cryostat

13. Clean glass cover slips (22 mm square)

14. Clean glass microscope slides

15. 0.9% NaCl

16. Filter paper and funnel

17. Glycerin gelatin

C. RATIONALE AND DISCUSSION:

The rationale for various steps in enzyme cytochemical experiments given in Exercise Number 24A also apply to localization of alkaline phosphatase with a coupling azo dye method. In this exercise fixed tissue will be used to maintain cellular integrity as much as possible. The coupling azo dye methods have been shown to give reliable localization of phosphatases. Cryostat sections will be used so that components of the cytochemical reaction mixture will have access to cell compartments containing the enzyme. Lastly, control experiments and an inhibitor specific for most alkaline phosphatases will be used to assess the specificity of the cytochemical localization.

D. PROCEDURE:

1. Prepare the fixative (neutral buffered formaldehyde) and chill on ice.

2. Sacrifice the animal and quickly remove pieces of tissue. Remove a kidney, pieces of liver, and pieces of small intestine.

3. Transfer the tissue to a petri dish containing ice cold 0.9% NaCl and rinse.

   (Take special care to rinse the luminal contents from the small intestine).

4. Transfer the rinsed tissue to cold neutral buffered formaldehyde.

   Use a separate petri dish for each tissue.

5. Cut the tissue into small pieces (1X3 mm pieces for liver and kidney and 3-4 mm circles of small intestine).

6. Leave the tissue in cold fixative for 2-3 hours.

7. Wash the pieces of tissue several times in cold 0.9% NaCl. This step could be left overnight in a refrigerator. The tissue should also be washed several times in 0.05 M Tris-maleate, pH 8.3.

8. Embed pieces of tissue in cryostat mounting medium and freeze in the cryostat.

9. Cut cryostat sections at 10 μm.

10. Pick up sections on albuminized cover slips.

    To prepare albuminized cover slips a very small drop of the albumin fixative is placed on the cover slip.

    It is then smeared around with a finger tip until a very thin layer is formed.

11. Air dry the cryostat sections (30-60 min) at room temperature.

12. Prepare the <u>complete cytochemical medium</u>, the <u>control medium</u>, and the <u>complete medium with the inhibitor</u> levamisole.

NOTE: DIRECTIONS FOR PREPARATION OF THESE MEDIA ARE GIVEN AT THE END OF THIS SECTION.

13. Transfer the cover slips with cryostat sections to small petri dishes (or some other suitable staining dish).

    (Have separate dishes for each of the different media).

14. Incubate the specimens in the different media at room temperature or in a 37°C incubator for 15-60 min.

    Incubate until the reaction product forms.

    (The time required may vary from tissue to tissue).

15. Rinse cover slips in distilled water.

16. Mount cover slips in melted glycerin gelatin on glass microscope slides.

PREPARATION OF SOLUTIONS

1. Neutral buffered formaldehyde solution (pH 7.0)

   To make neutral buffered formaldehyde add:

   10 ml of 37% formaldehyde

   90 ml of distilled water

   0.4 grams of sodium phosphate, monobasic

   0.65 grams of sodium phosphate, dibasic

   (Check pH and adjust to pH 7.0 if necessary)

2. Tris-maleate buffer

   Prepare Solution A: 0.2 M solution of Tris (24.2 grams of tris (hydroxymethyl) aminomethane and 23.2 grams of maleic acid in 1,000 ml of distilled water

   Prepare Solution B: 0.2 M NaOH

   Adjust to desired pH (8.3) by mixing a portion of solution B with a portion of solution A and then dilute to give desired concentration

3. Cytochemical Medium

   To make complete medium add:

   5 mg of napthol AS-MX to 1 ml of N,N-dimethylformamide and dissolve

   50 ml of 0.05 M Tris-maleate (pH 8.3)

   40 mg of diazonium salt

NOTE: FILTER THE SOLUTION PRIOR TO USE

To make control medium add:

   1 ml of N,N-dimethyl formamide

   50 ml of 0.05 M Tris-maleate (pH 8.3)

   40 mg of diazonium salt

NOTE: FILTER THE SOLUTION PRIOR TO USE.

To make complete medium with inhibitor add:

5 mg of napthol AS-MX to 1 ml of N,N-dimethylformamide and dissolve

50 ml of 0.05 M Tris-maleate (pH 8.3)

40 mg of diazonium salt

2.4 mg of levamisole

NOTE: FILTER THE SOLUTION PRIOR TO USE.

4. Glycerin gelatin

To make glycerin gelatin dissolve 20 grams of gelatin in 100 ml of distilled water for 2-3 hours.

Add 125 ml of glycerol and a small amount of melted phenol (1-2 ml) as a preservative.

Heat until mixture is homogenous (in a fume hood).

Aliquot into small pyrex bottles and store in a refrigerator.

Melt over low heat when needed.

E. RESULTS:

The sites of alkaline phosphatase activity are indicated by azo dye deposits. The color of the azo dye depends on which diazonium salt is used.

In this exercise the color will be red to blue. The control sections should be devoid of reaction product.

Levamisole should inhibit the enzyme and thus formation of reaction product except in the intestine.

F. EVALUATION:

Examine the slides with the light microscope and record whether reaction product is formed and if formed where it is located.

PRESENT SLIDES TO INSTRUCTOR FOR FURTHER EVALUATION

G. QUESTIONS:

1. What is the distribution of reaction product at the light microscope level in liver, kidney, and small intestine?

2. What types of cells contain reaction product?

3. Is reaction product restricted to a particular area of the tissue or cells?

EXERCISE 24C

LOCALIZATION OF SUCCINIC DEHYDROGENASE AT THE LIGHT MICROSCOPE LEVEL BY A TETRAZOLIUM SALT OXIDATION-REDUCTION METHOD

John M. Robinson, Ph.D.

A. OBJECTIVES:

1. To localize the enzyme succinic dehydrogenase at the light microscope level with a tetrazolium salt
2. To compare the cytochemical localization of succinic dehydrogenase in several different tissues
3. To determine the specificity of the cytochemical reaction by use of control experiments

B. MATERIALS:

1. A laboratory animal (mouse or rat)
2. Dulbecco's phosphate buffered saline
3. Cryostat
4. Water-soluble tissue mounting medium for cryostat
5. Petri dishes
6. Clean razor blades
7. Clean glass cover slips (22 mm square)
8. Clean glass slides
9. Sodium succinate
10. Tetrazolium salt (nitro blue tetrazalium chloride, NBT or tetranitroblue tetrazolium chloride, TNBT)
11. Filter paper and funnel

12. Neutral buffered formaldehyde (See Exercise Number 24B)

13. Glycerin gelatin (See Exercise Number 24B)

C. RATIONALE AND DISCUSSION:

The rationale for various steps in enzyme cytochemical experiments given in Exercise number 24A also apply to localization of succinic dehydrogenase with tetrazolium salts. In this exercise fresh tissue (unfixed) will be used for the cytochemical reaction since succinic dehydrogenase is usually very sensitive to fixation. This, of course, means that morphological integrity is sacrificed for enzyme activity.

Tetrazolium salt methods have been shown to give reliable localization for a number of dehydrogenases. Cryostat sections will be used so that components of the cytochemical reaction will have access to cell compartments containing the enzyme. Control experiments will be used to assess the specificity of the cytochemical localization.

D. PROCEDURE:

1. Sacrifice the animal and quickly remove pieces of tissue.

   Remove a kidney, pieces of liver, pieces of small intestine, and pieces of cardiac muscle.

2. Transfer the tissue to a petri dish containing ice cold Dulbecco's phosphate buffered saline and rinse.

   Take special care to rinse the luminal contents from the small intestine.

3. Cut the tissue in small pieces (2-3 mm cubes) so that they will freeze rapidly.

4. Quickly blot excess moisture from tissue with filter paper.

5. Mount tissue in cryostat mounting medium and quickly freeze.

6. Prepare the <u>complete cytochemical medium</u> and the <u>control medium</u>. Directions for preparation of these media are given at the end of this section.

7. Cut cryostat sections at 10 μm.

8. Pick up sections on glass cover slips.

9. Bring sections to room temperature (do not allow them to dry).

10. Incubate the specimens in the complete or control medium at room temperature or in a 37°C incubator.

    The incubation may be done in a petri dish by placing the cover slips in the bottom of the dish and then covering with a few drops of medium.

    This will conserve on the amount of medium used.

NOTE: INCUBATE UNTIL COLOR DEVELOPS

11. Rinse in cold Dulbecco's phosphate buffered saline.

12. Transfer to cold neutral buffered formaldehyde

    Fix for 30-60 minutes

13. Rinse in distilled water.

14. Mount cover slips in melted glycerin gelatin on glass microscope slides.

PREPARATION OF SOLUTIONS

1. <u>Dulbecco's phosphate buffered saline</u>

   To make Dulbecco's phosphate buffered saline add:

   0.2 grams KCl

   0.2 grams KH_2PO_4

   8.0 grams NaCl

   1.15 grams Na_2HPO_4

   1,000 ml distilled water

   Dissolve all of the components then test pH and adjust to pH 7.4 if necessary.

2. Cytochemical Medium

To make complete medium add:

4.4 ml of Dulbecco's phosphate buffered saline

0.1 ml of 0.5 M sodium succinate

0.5 ml of a 1% solution of NBT or TNBT

To make control medium add:

4.4 ml of Dulbecco's phosphate buffered saline

0.1 ml of distilled water

0.5 ml of a 1% solution of NBT or TNBT

E. RESULTS:

The sites of succinic dehydrogenase activity are indicated by the blue deposits of the formazan reaction product.

Control sections should be devoid of reaction product.

F. EVALUATION:

Examine the slides under the light microscope and record whether reaction product is formed and if formed where it is located.

PRESENT SLIDES TO INSTRUCTION FOR FURTHER EVALUATION.

G. QUESTIONS:

1. What is the distribution of reaction product at the light microscope level in liver, kidney, small intestine, and cardiac muscle?

2. What types of cells contain reaction product?

3. Is reaction product concentrated in a particular area of the tissue or cells?

4. Which organelle contains succinic dehydrogenase activity?

5. What predictions about the metabolic activity of a tissue can be made from cytochemical localization of succinic dehydrogenase? What predictions can you make concerning the tissues you have examined?

NOTES

EXERCISE NUMBER 24D

LOCALIZATION OF ENDOGENOUS PEROXIDASE ACTIVITY AT THE ELECTRON MICROSCOPE LEVEL BY THE 3,3'-DIAMINOBENZIDINE (DAB) METHOD

John M. Robinson, Ph.D.

A. OBJECTIVES:

1. To obtain precise ultrastructural localization of endogenous peroxidase activity

2. To compare the ultrastructural localization of peroxidase in two different tissues

3. To determine the specificity of the cytochemical reaction by the use of control and inhibitor experiments

B. MATERIALS:

1. A laboratory animal (mouse or rat)

2. 2% glutaraldehyde in 0.1 M phosphate buffer (pH 7.2)

3. 2% osmium tetroxide in 0.1 M phosphate buffer (pH 7.2)

4. 0.2 M phosphate buffer, pH 7.2 (and proper dilutions of the buffer as the experiment requires)

5. 3,3'-diaminobenzidine

6. 1% H_2O_2 as a stock solution

7. Potassium cyanide (KCN)

8. Filter paper and funnel

9. Petri dishes

10. Tissue chopper (Smith-Farquhar type)

11. 10% agar solution

12. Clean razor blades

13. Glass tissue vials

14. A graded series of ethanol solutions for dehydration (See Exercise Number 12)

15. Embedding resin for specimen embeddment for electron microscopy (See Exercise Number 12)

C. RATIONALE AND DISCUSSION:

The various steps in enzyme cytochemical experiments given in Exercise Number 24A also apply to localization of endogenous peroxidase activity with DAB. In this exercise fixed tissue will be used to maintain cellular integrity. Furthermore, peroxidase is relatively resistant to inactivation by glutaraldehyde fixative.

The DAB cytochemical method for peroxidatic activity is perhaps the best cytochemical method currently available. Tissue chopper sections will be used so that components of the cytochemical reaction mixture can penetrate throughout the tissue section. Control experiments on an inhibitor of peroxidase activity will be used to assess the specificity of the cytochemical localization.

D. PROCEDURE:

1. Prepare the primary fixative (2% glutaraldehyde in 0.1 M phosphate buffer, pH 7.2) and chill on ice.

   Prepare the fixative as previously described (See Exercise Number 11).

2. Sacrifice the animal and quickly remove the thyroid gland and the submaxillary gland.

3. Transfer the tissue into separate petri dishes containing cold fixative.

4. Trim the tissue quickly into small pieces approximately 1 mm wide and 2-3 mm long.

5. Keep the tissue in fixative for 1-1/2 hour.

NOTE: SEE EXERCISE NUMBER 24A FOR THE COMPARISON OF IMMERSION FIXATION AND PERFUSION FIXATION.

6. Wash the tissue in ice cold 0.1 M phosphate buffer several times (4 changes of buffer over a 1 hour period).

   The wash may be extended to overnight for convenience.

7. Embed pieces of tissue in molten 10% agar (42-45 ^{o}C) on mounts designed for the tissue chopper being used.

   Transfer the mounts to a refrigerator to solidify the agar.

8. Cut tissue chopper sections at 70 µm

9. Transfer chopped sections to glass vials containing 0.1 M phosphate buffer.

10. Transfer some sections to separate vials containing the inhibitor solution (KCN).

    Incubate at room temperature for 1 hour.

    See the end of this section for directions on preparing the inhibitor.

11. Transfer sections to separate vials containing the complete medium at the control medium.

    There are three vials for each tissue type:

    a) Complete medium

    b) Control medium, and

    c) Tissue pretreated with inhibitor in complete medium with additional inhibitor

12. Incubate sections in the cytochemical media for 1-2 hours at room temperature or in a 37 ^{o}C incubator.

13. Wash tissue several times in 0.1 M phosphate buffer (4-5 times over a 1 hour period).

14. Postfix the sections in 2% OsO_4 in 0.1 M phosphate buffer for 1 hour at room temperature.

    The osmium solution is prepared as previously described (See Exercise Number 11)

15. Process the osmium treated sections for electron microscopy and examine thick and thin sections as described in previous exercises.

E. PREPARATION OF SOLUTIONS

1. Phosphate buffer

Stock solutions

Make Solution A:

0.2 M solution of sodium phosphate, monobasic by dissolving 27.8 grams in 1,000 ml of distilled water.

Make Solution B:

0.2 M solution of sodium phosphate, dibasic by dissolving 53.65 g in 1000 ml of distilled water.

Adjust to desired pH (7.2), by mixing a portion of Solution A (about 1/4) with a portion of Solution B (about 3/4).

If desired pH is not achieved titrate with acid or base as needed. Dilute with distilled water to give final desired concentration.

2. Cytochemical Medium

To make complete medium add:

9.9 ml of 0.1 M phosphate buffer (pH 7.2)
5 mg of DAB
0.1 ml of 1% H_2O_2

To make control medium add:

9.9 ml of 0.1 M phosphate buffer (pH 7.2)
5 mg of DAB
0.1 ml of 1% H_2O_2

To make inhibitor medium add:

9.9 ml of 0.1 M phosphate buffer (pH 7.2)
5 mg of DAB
6.5 mg of KCN
0.1 ml of 1% H_2O_2

To make inhibitor solution for preincubation add:

> 10.0 ml of 0.1 M phosphate buffer (pH 7.2)
> 6.5 mg of KCN

NOTE: KCN (CYANIDE) IS VERY POISONOUS, HANDLE WITH EXTREME CARE.

DAB MAY BE CARCINOGENIC, HANDLE WITH EXTREME CARE.

F. RESULTS:

Oxidized DAB reaction product is formed at sites of enzyme activity (peroxidase). The oxidized DAB is brownish red in color; the reaction product becomes very electron dense after reacting with osmium.

G. EVALUATION:

Examine thin sections under the electron microscope and record whether reaction product is formed and where it is located.

Present electron micrographs to instructor for further evaluations.

H. QUESTIONS:

1. What types of cells contain reaction product?

2. Which organelles contain reaction product?

NOTES

EXERCISE NUMBER 25

AUTORADIOGRAPHY

LIGHT MICROSCOPY AND ELECTRON MICROSCOPY

G. Colin Budd, Ph.D.

A. OBJECTIVES:

1. To prepare autoradiograms for light microscopy with 1 μm sections of epoxy-embedded radiolabeled tissue blocks or cell suspensions

2. To prepare electron microscopic autoradiograms with 100-120 mμ sections of epoxy-embedded, radiolabeled specimens

3. To evaluate the distribution of radioactive sources in autoradiograms by measuring the distribution of autoradiographic silver grains in relation to landmarks in the specimen

B. MATERIALS:

1. 3% glutaraldehyde in cacodylate buffer

2. Freshly prepared 2% buffered osmium tetroxide

3. A laboratory mouse

4. A 1 c.c. tuberculin syringe and 21 G needle

5. Tritiated thymidine or tritiated leucine (highest available specific activity; 5-10 μCi needed per g. body wt. of the laboratory mouse)

6. Surgical gloves

7. Plastic or stainless steel tray lined with absorbent paper

8. Dissecting kit

9. Glass vials with caps

10. Wooden applicator sticks

11. Clean pasteur pipettes with rubber bulbs

12. Disposable plastic petri dishes

13. Sucrose-cacodylate buffer for washing specimens

14. New single edge razor blades cleaned with absolute ethanol

15. Bottles containing the following solutions for dehydrating the specimens

    30% ethanol

    50% ethanol

    70% ethanol

    90% ethanol

    95% ethanol

    Absolute ethanol

    Acetone (dried over drierite)

16. Premixed epoxy resin containing all components required for infiltration and embedding

17. Capsules or embedding molds for final embedding of the specimens

18. In the darkroom: for preparing autoradiograms:

    a. Safelight with red safelight filter (Wratten #1 or #1A)

    b. Water bath at $45^{0}C$

    c. Absorbant paper on a clean plastic tray

    d. 30 ml beaker containing 20 ml distilled water (in water bath at $45^{0}C$)

    e. Empty 30 ml beaker

    f. Spatula for picking up emulsion gel

    g. Balance with a scale that is visible in red light

h. Plastic slide box in which to store the autoradiograms

i. Box of paper tissues

j. Light-proof box to contain the plastic slide box of autoradiograms during storage

k. Drying agent (e.g. drierite)

l. Nuclear emulsion for autoradiography (L4-Ilford)

m. Box of microscope slides

n. Light-proof drying box for autoradiograms

o. Light-proof drawer or cupboard

19. In the darkroom: for processing autoradiograms after storage:

a. Freshly prepared and filtered developer (D-170 for LM, D-19 or Microdol for EM)

b. 3% acetic acid stop bath

c. Photographic fixer (commercial rapid fixer or 20% sodium thiosulfate)

d. Distilled water for washing after each processing solution

e. Staining dishes or 30 ml beakers to hold autoradiograms during processing

f. Safelight with red safelight filter (Wratten #1 or 1A)

20. 5% hydrofluoric acid (5% HF)

21. Suction device attached to vacuum (see diagram)

22. EM grids (150 or 200 mesh)

23. 0.4% collodion in amyl acetate

24. Glass microscope slides with frosted ends

25. Vacuum evaporater set up for evaporating 40 A thick carbon layer onto collodion coated slides

C. RATIONALE AND DISCUSSION:

Autoradiography is a procedure for localizing and quantitating radioactive sources within sections of biological tissue or suspended cells. It is generally required that the radioactive material to be detected must be retained in the tissue during exposure to the fixatives, solutions, and dehydrating solvents used to prepare specimens for electron microscopic examination. This includes the epoxy resin infiltrating media in which the tissue is finally embedded.

Radioactive sources that are usually detectable with this technique include: ^{125}I-labeled iodide incorporated into thyro-globulin, ^{125}I-labeled antibodies attached to cell-surface antigens, ^{3}H- or ^{35}S- or ^{14}C-amino acids after incorporation into secretory or structural proteins, ^{3}H-nucleotides after incorporation into nucleic acids, or ^{3}H-monosaccharides after incorporation into polysaccharides or muco-proteins.

The radioactive precursors and intermediates from which the above radiolabeled macromolecules are synthesized are generally removed from the specimens during fixation and dehydration and are rarely detected in LM and EM autoradiograms.

By observing the localization of developed autoradiographic silver grains with respect to visible landmarks (e.g. organelles, membrane boundaries, extra-cellular fibers), it is possible to learn about the functions of specific cells and organelles in protein synthesis, nucleic acid metabolism, receptor turnover and other vital functions.

Light microscopic autoradiography should always be undertaken prior to or together with EM autoradiography because it provides a more extensive view of the distribution of a radiolabel in a specimen to complement the higher magnification images obtained with electron microscopy.

It is also possible to obtain information about the general distribution of radioactivity in a specimen some time before any information can be obtained from the corresponding EM autoradiograms. This is because semi-thin LM sections contain up to 10 times more radioactive sources per unit area of tissue surface than the thin sections used for preparing EM autoradiograms.

The EM autoradiograms do have the advantage that they provide more precise morphological information with which the identification of labeled structures can be made.

D. PROCEDURE: LABELED TISSUE PREPARATION

1. Weigh the mouse.

2. While wearing rubber gloves and working under a fume hood, load the tuberculin syringe with tritiated thymidine (DNA precursor) or tritiated leucine (protein precursor). Use 5 - 10 μCi/g body weight.

3. Inject the mouse intraperitoneally (ip) with the labeled compound. Place it in a clean cage under the hood, with adequate food and water.

4. After one hour kill the mouse by cervical dislocation or with an overdose of sodium pentobarbital injected i.p.

5. Open the abdomen and excise 1 cm. of small intestine or large intestine (if tritiated thymidine was the label) or 1 cm^3 of pancreas (if the label was ^{3}H-leucine).

6. Place the excised tissue in a few drops of 3% buffered glutaraldehyde in a petri dish and mince the tissue into 1-2 mm^3 fragments with razor blades.

7. Transfer the tissue fragments with applicator sticks into a clean glass vial containing 3% glutaraldehyde. Fix the tissue fragments for 1-2 hours.

8. Dispose of the animal, petri dish, razor blades and applicator sticks according to established procedures for disposal of radioactive materials.

9. After fixation in glutaraldehyde, remove the excess fixative from the tissue fragments and dispose of this solution as liquid radioactive waste.

10. Wash the tissue fragments three times with sucrose-cacodylate buffer.

11. Replace the last sucrose-cacodylate wash with 2% buffered osmium tetroxide solution. Post-fix in this solution for 1-2 hours.

12. Remove the osmium under the hood, and wash the tissue fragments three times with stock buffer during 30 minutes.

13. Dehydrate the tissue fragments by segmental immersion in a graded series of increasing concentrations of ethanol for 30 minutes each.

14. Dehydration should be completed by imersing the tissue fragments in absolute ethanol (2 changes of 30 mins. each) followed by dry acetone (2 changes of 30 minutes each).

15. The tissue fragments are then immersed overnight in a 50:50 mixture of dry acetone and the complete epoxy embedding medium. Preferably this should be done with continuous rotation of the tissue fragments on a rotary embedding apparatus.

16. Infiltration is completed by immersion of the tissue fragments in complete embedding medium at room temperature, with agitation, for 8 hours.

17. Selected small tissue fragments are finally embedded in plastic capsules in complete embedding medium and polymerized at 70^{0}C overnight.

PREPARATION OF LM AND EM AUTORADIOGRAMS

1. For each block of epoxy-embedded labeled tissue to be sectioned for autoradiography, clean six or seven microscope slides in hot water containing liquid detergent.

2. Rinse the slides in several changes of distilled water and wipe them dry with paper tissues.

3. Three of the slides should be coated with collodion by dipping each slide into 0.4% collodion in amyl acetate. The slides are placed on end with the narrow end opposite to the frosted end on clean absorbant paper so that the collodion solution drains as it dries.

4. When the collodion is dry, a 1 cm diameter circle is scratched with a glass-marking diamond pencil on the underside of each coated and uncoated slide about 1.5-2 cm from the unfrosted end of the slide.

5. The epoxy embedded tissue is then placed in the specimen holder of an ultramicrotome and sectioned with a glass knife to provide 1 μm sections of good quality (See Exercise Number 9).

6. Three to five 1 μm sections are transferred to a drop of water placed over the circumscribed diamond scratch on one of the slides which was not coated with collodion.

7. The slide is then placed on a hot-plate at 80^{0}C until the sections have dried and attached to the slide within the circumscribed area.

8. Similar sections are transferred in the same fashion to the remaining uncoated glass slides.

9. The ultramicrotome is then adjusted so that sections with a pale gold interference color (120 mμ thickness) are produced.

10. A small drop of water, which will tend to bead up, is placed on a collodion-coated slide within the circumscribed zone. A section or ribbon of sections is then transferred to this water drop with the aid of a wire loop.

NOTE: CARE MUST BE TAKEN NOT TO ALLOW THE WIRE LOOP TO TOUCH THE COLLODION

11. Additional sections may be transferred to the same slide until 4 or 5 groups of sections have been transferred.

12. Sections are transferred in a similar way to the other collodion-coated slides.

13. After the thin sections have dried onto the collodion, a thin layer (40 A) of carbon is evaporated onto each slide to cover as much of the collodion as possible.

14. If desired, the thin sections may be covered with 2% uranyl acetate or uranyl nitrate for 2 to 4 hours, and then gently washed and dried, prior to addition of the carbon layer.

15. The uranium salt helps to add electron dense material to some membranes and nucleic acids in the sections while the carbon layer helps to ensure a smooth distribution of nuclear emulsion over the thin sections.

COATING WITH NUCLEAR EMULSION

1. The slides with semi-thin (1 μm) and thin (120 mμ) sections are transferred to the darkroom.

2. Heat the water bath to 45^{0}C. Warm the beaker of water.

3. Under the red safelight, 4 g. of L4 (Ilford, Ltd.) emulsion are weighed out into a dry, 30 μl beaker.

   Warm water (20 μl) is added to the beaker containing weighed-out emulsion and this beaker is placed in the 45^{0}C water bath. After 5-10 minutes the emulsion will have melted.

4. Gently stir the emulsion in the beaker until it is thoroughly mixed.

5. Remove the beaker from the water bath and cool it (in a water bath if necessary) to 20^{0}C.

6. Using a pipette, spread a thick layer of emulsion on one side of a horizontally held microscope slide--<u>which does not have any sections attached</u>.

   Allow the emulsion to drain back into the beaker and then dry the slide by standing it on end while leaning it against the wall at the back of the bench.

7. Put the emulsion (both undiluted and diluted) into a light-proof drawer.

8. When the thin emulsion layer on the slide is dry, take it out of the darkroom and examine it in daylight. It should be possible to see a spectrum of colors on the slide.

These colors are related to variations in thickness of the dry emulsion layer. A purple-bordering-blue color corresponds to a closely-packed monolayer of silver halide crystals in the emulsion layer.

A blue color corresponds to a double layer of crystals in the layer. Mark both of these zones on the slide with a felt-tip marking pen.

9. Return to the darkroom with the test slide and coat the experimental slides with emulsion under the red safelight according to the procedure described above.

   Making certain that the 1 μm semi-thin sections are covered with a layer that would appear blue if exposed to white light. The 120 mμ thin sections should be coated with a purple-bordering-blue emulsion layer.

10. Allow an hour or more for the emulsion layers to dry in a light-proof drying cabinet before putting the slides away in a slide box.

    A small paper tissue packet containing drying agent should be included in the slide box. The slide box should be kept inside another, light proof box in a refrigerator until the slides are to be developed.

PROCESSING LM AND EM AUTORADIOGRAMS

1. Remove one or more slides of emulsion-coated 1 μm sections from the storage box under a red safelight. Immerse each slide in freshly prepared and filtered D-170 developer for 6 minutes at 20-22^{0}C.

2. Rinse the developed slides briefly in distilled water and transfer each slide to 3% acetic acid stop bath for 30-60 seconds.

3. Rinse each slide in distilled water and transfer to filtered fixer (rapid fixer or 20% sodium thiosulfate) for 2-3 minutes.

4. Wash each slide in several changes of distilled water for 5-10 minutes and then dry the slide. It is now ready for staining and light microscopic examination.

5. Slides containing 120 mu thin sections should be developed in freshly prepared and filtered D-19 developer or Microdol for 2 mins. at 20-22^{0}C.

   These slides should be treated with stop bath and fixer solutions as described above for LM autoradiograms. After thorough washing, these EM autoradiograms are dried before further processing.

EXAMINATION OF LM AUTORADIOGRAMS

1. The 1 µm sections are stained with toluidine blue prior to microscopic examination. Place a drop of freshly filtered stain (0.1% toluidine blue in 1% sodium borate) onto the sections and place the slide on a hotplate at 80 - 90^{0}C for 30 - 45 seconds.

   Carefully flood the slide with distilled water and drain the stain onto filter paper. Allow the slide to dry before examining it with the microscope.

2. Autoradiographic grains should be visible as small black spots overlying labeled areas of tissue.

   It will be necessary to use a high power dry objective (x70 to x80 magnification) or an oil immersion objective (x100) to see the grains clearly.

   It is not necessary to place a coverslip over the sections.

ATTACHMENT OF EM AUTORADIOGRAMS TO GRIDS

1. The 120 mµ thin sections are sandwiched between a collodion support and the developed emulsion layer. This entire sandwich must be stripped off the glass slide and attached to an EM grid.

2. Carefully scratch the emulsion and collodion in a wide circle around the area of sections with a needle.

While the slide is resting horizontally, place a few drops of 5% hydrofluoric acid on the edge of the scratched circle.

The solution will spread out under the collodion-section-emulsion sandwich and lift it from the slide.

3. By slowly immersing the slide at a shallow angle, carefully float the sandwich onto the surface of distilled water in a staining dish.

4. With the aid of a dissecting microscope, place EM grids onto the sections while the autoradiogram is floating on the dish of water.

5. Attach a piece of wet filter paper to the suction device and carefully lower the device onto the floating autoradiogram.

   Remove the suction device rapidly from the dish of water with the grids and attached autoradiogram.

   Peel away the filter paper from the suction device and allow the grids and attached autoradiogram to thoroughly dry.

   Drying may be facilitated by placing the filter paper into a 37^{0}C incubator.

6. When the filter paper is dry, each grid with its attached emulsion-section-collodion sandwich may be picked up with watchmakers forceps and examined in a transmission electron microscope, either with or without staining with lead or uranium salts.

ANALYSIS OF AUTORADIOGRAMS

1. The distribution of developed silver grains is a reflection of the distribution of radioactivity within the sections. To determine this distribution, count the number of grains over measured areas of the section and express the results as grain density (grains/unit area).

   It is always important to subtract background grain density before interpreting the data.

Background grain density can be estimated by counting the grains/unit area over areas of the section which do not contain any tissue.

2. Areas are most easily determined by placing a rectangular grid over a photomicrograph of the autoradiogram, and counting the number of grid intersections occuring over the tissue components.

   The number of grid intersections on a component divided by the total number of intersections on the tissue is equivalent to the relative area of the component in the tissue section. This can be converted to absolute area if the photomicrograph magnification is known.

3. For more detailed analyses of autoradiograms the reader should consult the references provided below.

E. EVALUATION:

Present one micrograph resulting from LM or EM Autoradiography to the instructor for evaluation.

F. REFERENCES:

1. Rogers, A.W. (1979). Techniques of Autoradiography. Third Edition. Elsevier-North Holland Biomedical Press.

2. Williams, M.A. (1977). Autoradiography and Immunocytochemistry. In: <u>Practical Methods in Electron Microscopy</u>. (Edited by Audrey M. Glavert). North Holland. Amsterdam, New York, London.

3. Salpeter, M.M. and McHenry, F.A. (1973). Electron Microscope Autoradiography. In: <u>Advanced Techniques in Biological Electron Microscopy</u>. (Edited by J.K. Koehler). Springer-Verlag. Berlin, New York.

4. Budd, G.C., E.A. Bernard, C. Porter and S.A. Mattimoe (1980). J. Histochem. Cytochem. <u>28</u>: 533-542. Fluorophosphate-sensitive esterases in mammalian liver. The radioautographic localization and measurement of fluoro- phosphate-reactive sites in adult rat liver.

G. QUESTIONS:

1. Describe a method that you could use to prepare autoradiograms for examination in a transmission electron microscope.

2. Describe a method of electron microscopic autoradiography in which the nuclear emulsion is uniformly deposited on thin sections to permit quantitative evaluation of the radioactivity in the sections.

3. You are provided with a sample of radioactively labeled biological tissue embedded in epoxy resin after suitable fixation and dehydration. Describe an electron microscopic autoradiography procedure that would permit you to demonstrate that the radioactivity is mainly located in the cell nuclei.

EXERCISE NUMBER 26

IMMUNOCYTOCHEMISTRY
LIGHT MICROSCOPY

LOCALIZATION OF BIOGENIC AMINES IN *HOMARUS AMERICANUS*

Betty R. Jones, Ph.D.

A. OBJECTIVES:

1. To expose the student to immunocytochemistry and immunofluorescent techniques at the light microscope level

2. To localize the biogenic amine, 5-Hydroxytrptamine (5-HT), in ganglia of *Homarus Americanus* (lobster)

B. MATERIALS:

1. 1 year old immature lobsters (7-10 cm in length, 25-50 gms in weight)

NOTE: THE LOCALIZATION OF 5-HT IN IMMATURE LOBSTERS IS USED HERE ONLY AS AN EXAMPLE. LOBSTERS WERE SELECTED BECAUSE THEY HAVE A SMALL GANGLIA AND ARE READILY PENTRATED BY ANTIBODY PREPARATIONS.

A GIVEN LABORATORY MAY WISH TO USE WHATEVER TISSUE IS AVAILABLE THAT HAS BEEN FOUND OR IS SUSPECTED TO CONTAIN 5-HT (I.E. MOUSE OR RAT BRAIN, PLANARIA, EARTHWORM, SCHISTOSOMES AND ETC.). FIXATION AND INCUBATION TIME MAY VARY WITH TISSUE TYPES.

2. Lobster saline (see procedure for composition)

3. 4% paraformaldehyde

4. 0.1 M phosphate buffer (pH 7.4)

5. 0.3% Triton X-100

6. 0.1% sodium azide

7. 1:200 dilution of antiserotonin antibodies (primary antibody)

8. Goat anti-rabbit IgG labeled with fluorescein isothiocyanate (FITC)

9. 80% glycerol in 20 mM carbonate buffer

10. Fluorescent microscope or a standard research light microscope set up for epifluorescence, transmitted light and bright field

11. cryostat

12. gelatin coated slides

13. dissecting kit

14. paraformaldehyde/lysine/periodate (PLP)

15. 20% sucrose

C. RATIONALE AND DISCUSSION:

This exercise was written with the intention of exposing the student to immunocytochemistry and immunofluorencent techniques at the light microscope level.

The procedure described is according to that of Dr. Barbara Beltz and colleagues, Department of Neuroanatomy, Harvard Medical School, Boston, Massachusetts (personal communication).

Students should experiment with carrying out the procedure using wholemounts or either sectioned material to demostrate the distribution of 5-HT in tissues or both methods.

Immunocytochemistry developed out of the field of immunology and still maintains and uses its methods of application and nomenclature. It can be defined as the study of the chemistry and interactions of antigens and antibodies and the basis of chemical immunity and chemical resistance to disease.

NOTE: SOME OF THE APPLICATIONS OF IMMUNOCYTOCHEMISTRY AS A VALUABLE AND SIGNIFICANT TOOL IN RESEARCH AND TEACHING ARE LISTED BELOW (KABAT, E.A. AND MAYER, M.M. (1961). EXPERIMENTAL IMMUNOCHEMISTRY, CHARLES C. THOMAS COMPANY, SPRINGFIELD, ILLINOIS.

a) used in the characterization of proteins and polysaccharides

b) in the detection and quantitative determination of minute amounts of certain impurities in biological preparations

c) in virology

d) in mammalian and bacterial protein biosynthesis

e) in tissue transplantation and graft rejection or tolerance

f) in bacterial genetics for the histochemical localization of antigens in tissues and microorganisms

g) in embryology and systematic biology

h) in the study of the mechanism of proteins associated with various diseases in man and in allergy

D. PROCEDURE:

NOTE: WHOLEMOUNTS

1. Dissect central ganglia with thoracic roots in cold lobster saline.

   The composition of lobster saline is as follows:

462 mM	Nacl
116 mM	KCL
26 mM	CaCL
8 mM	$MGCL_2$
11 mM	Glucose
10 mM	Tris
10 mM	Maleic Acid

NOTE: A SUFFICIENT AMOUNT OF NaOH IS ADDED TO ADJUST pH TO 7.4.

2. Fix tissues for 12-36 hrs. in 4% paraformaldehyde in 0.1 M phosphate buffer (pH 7.4).

3. Rinse tissues in several changes phosphate buffer and sodium azide for 6-8 hrs.

4. Incubate tissues at $4^{0}C$ in a 1:200 dilution of anti-serotin antibody (the primary antibody) for 15-20 hours.

NOTE: THE ANTI-SEROTIN ANTIBODY CAN BE PURCHASED COMMERCIALLY (IMMUNONUCLEAR CORPORATION, STILLWATER, MINNESOTA).

5. Following the primary antibody treatment, rinse tissues in phosphate/azide/triton for 6-8 hrs.

6. Then incubate with the secondary antibody (goat anti-rabbit IgG labeled with fluorescein isothiocyanate (FITC). (This antibody preparation can be obtained from Boehringer Mannheim Biochemicals, Indianapolis, Indiana).

7. Incubate tissues in IgG conjugated FITC at $4^{0}C$ for 15-20 hrs.

8. Remove excess secondary antibody by multiple rinses in 0.1 M phosphate buffer for 4-6 hrs.

9. Rinse tissues (X1) in 4 mM of sodium carbonate buffer (pH 9.5).

10. Mount in 80% glycerol in 20 mM g carbonate buffer.

11. View in a microscope with epifluorescent or in a fluorescent microscope.

NOTE: EXCITER-BARRIER FILTERS AND REFLECTOR COMBINATION CUBES CONTAINING A BLUE EXCITATION AT 440-490 nm AND A SELECTIVE BARRIER FILTER AT 520-560 nm CAN BE USED.

Sectioned Material

1. Fix tissues in 4% paraformaldehyde or PLP for 12-36 hrs.

2. Infiltrate tissue with 20% sucrose in 0.1 M phosphate buffer for 2-12 hrs.

3. Mount and freeze tissues in cryostat.

4. Cut (8-25 micron) sections using a cryostat.

5. Mount sections on gelatin coated slides.

6. Then process according to the previous described procedure for wholemounts.

NOTE: SECTIONS CAN BE INCUBATED AT ROOM TEMPERATURE AND INCUBATION TIMES CAN BE SHORTENED TO 1-1.5 HRS.

Specificity Controls

Absorption controls can be carried out by pre-incubating the anti-serotonin antibody at the working dilution, with the following antigens:

a) serotonin creatinine sulfate (1 mg/ml)

b) formaldehyde cross-linked serotonin (BSA in which the BSA concentration is 300 mg/ml, with a BSA: serotonin ratio of 10:1 (w/w)

c) BSA (500 mg/ml)

d) octopamine (1 mg/ml)

E. RESULTS:

Structures that show a yellow fluorescent color indicates the localization and distribution of 5-HT.

F. EVALUATION:

Students should experiment with carrying out the described procedure on a given available organism or tissue using wholemounts or sectioned material or both.

PRESENT THE RESULTS TO THE INSTRUCTOR.

G. QUESTIONS:

1. List the classes of immunoglobulins and their function(s).

2. Who is responsible for most of the pioneering work in immunocytochemistry and who first described the term?

3. Discuss the functional significance of the distribution of biogenic amines in invertebrate and vertebrate tissues.

4. What is the chemical structure for 5-HT?

5. What enzyme is responsible for the conversion of tryptophan to 5-hydroxytryptophan?

EXERCISE NUMBER 27

FREEZE-FRACTURE, FREEZE-ETCH TECHNOLOGY

Russell L. Steere, Ph.D.

A. OBJECTIVES:

1. To immobilize and preserve the ultrastructure of cells with a minimum alteration from the living state (preferably without chemical fixation)
2. To provide an alternate approach to thin sectioning for the preparation of specimens for electron microscopy
3. To evaluate the effects of various treatments, diseases, aging, etc. on the internal fine structure of membranes

B. MATERIALS:

1. A freeze-etch apparatus including appropriate specimen holders
2. Freon 22
3. Liquid nitrogen
4. A bench dewar
5. Wooden applicator sticks
6. Glycerol
7. Sucrose
8. Tweezers (Forceps)
9. TEM grids
   a) Freeze-etch grids, Pellco 5GC75FF or equivalent
   b) 400 to 500 mesh grids, Polaron type 400 HH or 500 HH or equivalent
10. Formvar support films on Pelco 5 GC75FF grids
11. Pt wire .005 inch diameter
12. 3 mm hard graphite rod

13. Chlorox (5.25% Sodium Hypochlorite)

14. Chromic acid sulfuric acid cleaning solution

15. Distilled water

16. Lexan (Polycarbonate)

17. Collodion or parlodian

18. Solvents including: acetone, chloroform, amyl acetate, methylene chloride, ethylene chloride

19. Beaker (1,000 ml)

20. Small beaker assortment

C. RATIONALE AND DISCUSSION:

Chemical fixation, by its very nature, must produce artifacts. It converts materials which are in a fluid flux state into a solid or fixed state after which the fixed components are presumed to be unable to move from their fixed position to another place. It is known, however, that some materials leak out of fixed cells during the subsequent processing.

Freeze-fracturing and freeze-etching permit the fixation of materials by a very different mechanism (freezing). Samples which are frozen by the recently developed "fast freezing" techniques reveal crystal free freezing to a depth of 15-20 μm. It is presumed that materials within this zone are frozen so rapidly that they are preserved in their "native" state and essentially in their original positions. Since water expands upon freezing, to occupy approximately 10% additional volume, one must expect a certain minimum displacement of materials in frozen samples. Perhaps there is less expansion of fast frozen parts in which the ice is presumed to be amorphous, than in regions where freezing is slow enough that true ice crystals develop.

An alternate approach to prevent the formation of ice crystals during freezing is to use cryoprotectants such as glycerol and glucose. A solution of 25% by weight of sucrose and 25% by volume of glycerol in buffer will freeze as an amorphous glass even with comparatively slow freezing rates. In this instance, also, adequate tests have not yet been made to know whether or not expansion occurs upon freezing.

Freeze-fractured samples in which the specimens are "shadowed" and replicated immediately after being fractured provides excellent specimens for examination of the internal structure of split membranes.

Freeze-etched specimens are customarily fractured at any temperature colder than -98^0C then "etched" by vacuum sublimation of ice from the fractured surface with specimen temperature somewhere between -90 and -110^0C. The etched surface is then "shadowed" and replicated. Freeze-etched specimens are useful for the examination of inner and outer surfaces of membranes and for the elucidation of fine structures within cells or fine details of suspended materials.

D. PROCEDURE:

1. Prepare the freeze-etch apparatus.

   a) Check the carbon source to be certain the point is properly sharpened and that contacts are good.

   b) Prepare the Pt source. Use a resistance gun mounted with source 7 cm from the specimen. Use 3 mm hard graphite rod (if the gun requires 1/4 inch rod use sharpening tool to make a 5 mm length of 3 mm diameter rod at the contact ends of both hard graphite rods).

      Sharpen one of the 3 mm diameter rods or ends to provide the same bevel as that used for an ordinary lead pencil.

      Flatten the point to a diameter of 1/2 mm. Flatten the contact end of the other graphite rod. Lock the two graphite rods into the gun with the point against the flat and with medium spring tension holding them together.

      Wrap 3 mm of Pt wire around the point of the pointed rod. Evacuate the chamber than melt the Pt.

      Open the chamber then rotate the pointed graphite rod (if necessary) so that the Pt bead faces the specimen. Wrap 3 additional turns of Pt wire around the point.

   c) Pump out unit again and start cooling the stage.

2. Cut specimens into proper size pieces to fit onto or into the specimen holders.

3. Use sharpened wooden applicator stick to transfer small pieces of tissue or samples of pellets or suspensions into or onto specimen holders.

If complementary specimen holders are used, make certain that the specimen slot is filled to both ends with the suspending liquid but that none of the liquid penetrates into the hinge.

4. As soon as each specimen is properly oriented in its specimen holder it should be frozen as rapidly as possible.

5. When all specimens for a single freeze-etch preparation are frozen, mount them in the appropriate specimen cap under liquid Freon or liquid nitrogen.

6. Cool the specimen stage of the freeze-etch unit to at least -100^{0}C while the chamber is under vacuum.

7. Break vacuum with dry nitrogen gas and, as rapidly as possible, transfer specimens in the specimen cap to the cold stage of the freeze-etch unit.

8. Reevacuate the chamber while maintaining a specimen temperature of -100^{0}C or colder.

9. When the vacuum reaches $1X10^{-6}$ Torr make certain: a) that specimen temperature is adjusted to that desired for fracturing, b) that electrode controls are ready for evaporation of the Pt shadow film, and c) that the monitor for control of Pt thickness is turned on and set at the desired position.

10a. For freeze-fracture specimens

a) Fracture specimens

b) Align the specimens with the evaporators

c) Evaporate the Pt

d) Return the powerstat to 0

e) Turn monitor off and set controls for carbon evaporation

f) Evaporate carbon

10b. For freeze-etch preparations

a) Fracture the specimens

b) Adjust the specimen temperature to the desired temperature for etching (usually -98^0C but may be anywhere between -90^0C and -110^0C)

c) Recool the specimen to -150^0C or colder (if possible without contaminating the specimen)

d) Evaporate the Pt

e) Return the powerstat to 0

f) Turn the monitor off and set all controls for carbon deposition

g) Evaporate the desired thickness of carbon

11. Close main valve to the diffusion pump.

12. Break vacuum with dry nitrogen gas.

13. Remove the specimens (cap and all) and transfer specimen holders with specimens to water or buffer so that the frozen samples are beneath the surface and melt under the liquid. -or-

14. For extremely rough fracture faces and specimens which fragment badly, place a drop or two of Lexan or collodion dissolved in the appropriate solvent on top of the specimens as soon as the chamber is opened.

    Transfer the cap to a freezer and transfer specimen holders to cold filter paper where solvent will evaporate. Remove the plastic coated samples and cut off plastic that is beyond the fractured surface.

15. Transfer the individual specimens from the holders to a well of a porcelain spot plate containing either chrome-sulfuric acid cleaning solution or chlorox (either one diluted to between 1/4 and 1/10 with water).

16. After 10 minutes, or longer (for some tissues more time may be required for the dilute solvent to penetrate all the tissue), transfer the specimens to full strength solution of the same kind as that used initially. Transfer again to clean full strength solution if necessary.

17. When the sample is thoroughly dissolved, transfer the replicas or replica pieces to a clean well of a spot plate filled with distilled water.

If tissue is dissolved in the chromic acid the replica will drop to the bottom of this first wash.

Use a Pt loop to stir the water and lift the replica to near the surface of the water but not floating.

Pick-up a drop of water containing the replica with the Pt loop and transfer it to a second well of distilled water.

Repeat this process two more times then use the edge of the pt loop to lift the replica to make it float on the surface of the water.

18. Hold uncoated grid (400-500 mesh) with tweezers, lower it horizontally onto the surface so that the specimen makes flat contact with the center of the grid.

    Tilt grid and lift away to pick-up a minimum of the water. Touch the edge of the grid to lint-free filter paper to remove most of the excess water. -or-

19. Insert coated grid vertically into the water beside the floating replica. Slowly move the grid toward the replica until contact is made between the two at the proper point on an opening.

    Slowly lift the grid allowing the replica to adhere to the support film over an opening.

    If the replica fails to stick to the grid, push the grid back down and keep trying. When the replica is finally attached as desired, let it dry in place. Once dry, a clean replica will adhere permanently to the support film allowing it to be submersed to pick-up additional pieces of replica.

20. If the Lexan or collodion coat was applied to the replica surface immediately after freeze-etching, it must now be removed by immersing the grid in the proper solvent, one which will dissolve the initial plastic film but not the support film on the grid.

21. Now that the clean dry replicas are ready for electron microscopy, one should attempt to obtain as much information as possible from them. This requires a bit more effort than most electron microscopists are accustomed to.

One should take stereo electron micrographs of the selected areas of interest. To do this, one takes one micrograph with the specimen in any selected position then tilts the specimen 5 to 10 in the desired direction, then takes a second micrograph of the same area.

In some instances best results of a given part of a specimen can be obtained only by tilting the specimen as much as 30 or more from the position normal to the electron beam.

Other areas of the same specimen may provide the most information when the beam is normal to the specimen for one of the two micrographs of a stereo pair.

Sometimes several stereo pairs of a single locus of a specimen are needed if one is to obtain enough information for accurate interpretation of certain structures.

E. RESULTS:

Unless proper cold traps are provided, or unless the Pt shadow film and C replica film are deposited immediately after fracturing, samples which are freeze-fractured at temperatures much colder than $-100^{0}C$ will be unetched but also will contaminate rapidly with H_2O and hydrocarbons which leave other surfaces within the vacuum chamber. Samples held at $-110^{0}C$ or warmer under conditions which encourage the trapping of H_2O onto colder surfaces will reveal etching (the removal of H_2O from the fractured sample by vacuum sublimation). This results in some enhancement of the relief of certain features, particularly cross fractured membranes; compare for example, the cross fractured membranes of a chloroplast (Fig. 1). A few split membrane faces are clearly observed and reveal the large number of membrane associated particles (MAP) on the P-face (PF) and depressions left by the MAPs on the E-faces (EF) regardless of whether the faces are on the freeze-fractured sample (Fig. 1a) or on the complementary freeze-etched sample (Fig. 1b). Deep etching may also result in the development of holes in the half membranes, particularly in the EF (small arrows Fig. 1b). The unetched cross-fractured membranes are barely discernable (large arrows Fig. 1a), whereas the complementary parts are clearly visualized (large arrows Fig. 1b). there was considerable etching of the contents of some vesicles (Fig. 1b) but one, presumably a lipid body (LB Fig. 1b), failed to etch.

Complementary freeze-fracture freeze-etch replicas provide the best evidence that membranes do split when the ice block is fractured. The half membranes complementary to the membrane faces of the freeze-fractured sample remain in place after etching (Figs. 1b and 2a). If the faces seen in freeze-fracture specimens were either inner or outer surfaces of the membranes, their complements would be only impressions and should etch out as does the surrounding cytoplasm or vesicular contents. The membrane faces with MAPs and corresponding depressions are present in the deeply etched specimens as well as in the unetched specimens (PF and EF Figs. 1 and 2). Fig. 2 is a particularly good example for observing depressions on one face complementary to the MAPs of the other face (narrow arrows).

One of the major reasons for etching is to expose true inner and outer faces of membranes. Often, because of the cellular or intracellular components which freeze-dry down onto these faces, little new information is revealed unless the cells are isolated and thoroughly washed. An exception is observed in deeply etched samples of plant tissue infected with potato yellow dwarf virus (PYDV). When the virus buds through the inner nuclear membrane into the space between the inner and outer nuclear membranes, it envelopes itself in a sheath of the membrane and spikes or particles form on these membrane envelopes. Freeze fractured specimens usually reveal only split membrane faces (Fig. 3a). With light etching, sharp spikes appear at the outside of the outer half membrane. With deep etching, the rows of spikes look like corn on the cob (Figs. 3b and c). In these figures we see the smooth faces of split membranes of the virions (small arrows Figs. 3b and c) and the outside membrane surfaces where etching has removed H_2O and exposed the spikes (large arrows Fig. 3b and c). One virion was completely beneath the fracture plane and has only the outer surface exposed (S Fig. 3b and c). Because of the deep etching, the spikes on the virions are partially coated with other components and are less sharp than we see on lightly etched specimens.

F. EVALUATION:

It must be remembered that the final product of the freeze-fracture and freeze-etch methods is a very thin replica film supporting a much thinner metallic shadow film. Therefore, one can visualize only those surface features exposed by fracturing and etching. Except for those situations in which cellular components are cross fractured, interal structures of such components will not be revealed. When examining objects such as the larger virus particles, surface structures can be revealed more accurately than when the same particles are deposited by air drying or freeze-drying onto a support film then shadowed. This is so because, in the freeze-etched preparations, the particles are not present to scatter electrons and reduce contrast. One must examine these thin replicas carefully and usually take stereo electron micrograph pairs to be certain that the area of interest has not been distorted by the cleaning procedure or by drying onto the support grid.

The freeze-etch procedure is particularly useful in evaluating the effect of various treatments on membrane structure. It has been discovered that most membrane faces exposed by freeze-fracturing are faces of split membranes rather than true inner or outer surfaces which are revealed only if the specimens are deeply etched. Thus, these techniques are the only means of visualizing the internal structure of membranes.

PRESENT ONE MICROGRAPH TO THE INSTRUCTOR FOR EVALUATION.

Fig. 1. Complementary freeze-fracture, freeze-etch preparation of a portion of a tobacco chloroplast in 25% glycerol, 6% sucrose. Fig. 1a, shadowed and replicated immediately after fracturing. Fig. 1b, fracture face, complementary to that of Fig. 1a, etched for 3 min. at -98^{0}C then cooled to -150^{0}C before being shadowed and replicated. Micrographs match along connecting line. Magnification X50,000.

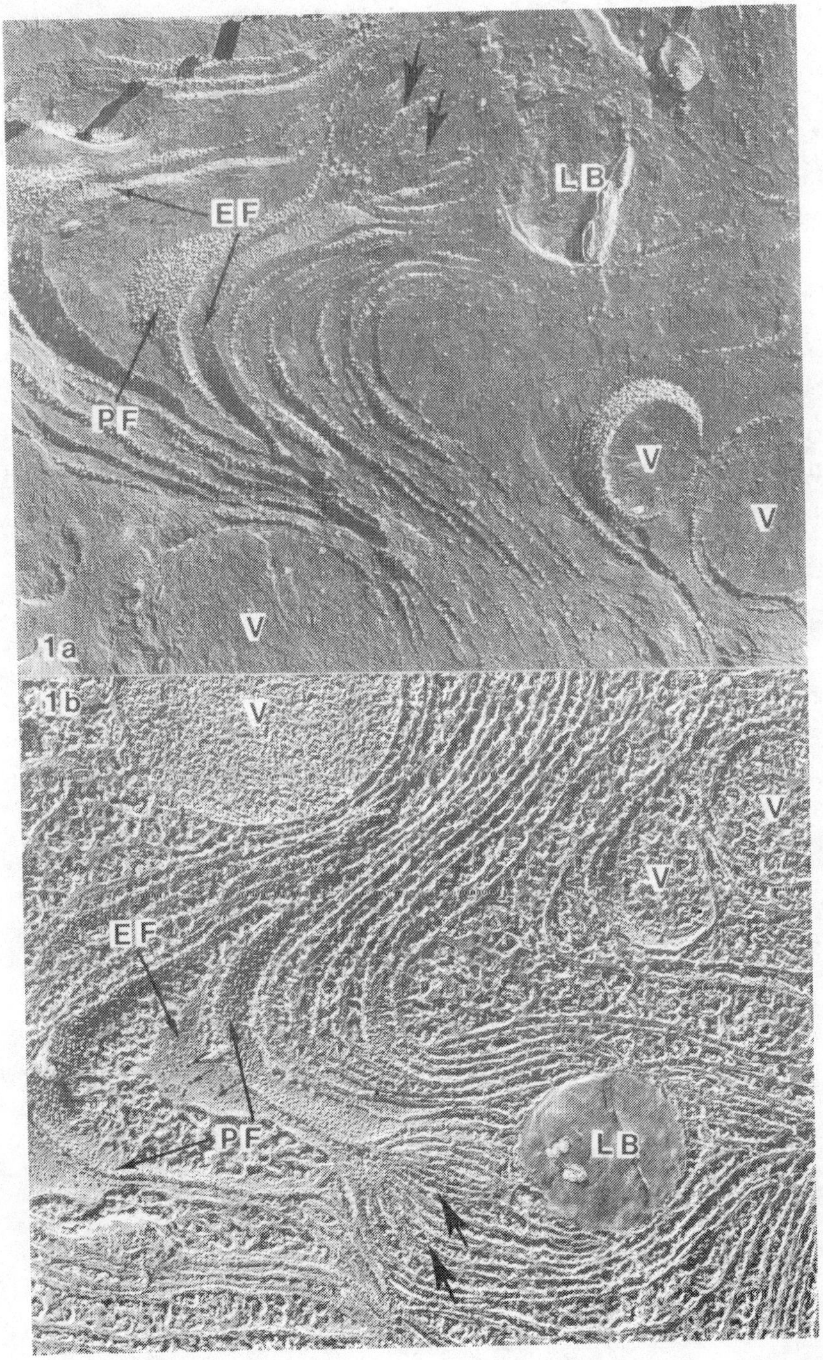

LB
EF
PF
V
V
V
1a
1b
V
V
V
EF
PF
LB

Fig. 2. Complementary freeze-fracture freeze-etch preparation of vesicles in tobacco tissue. Preparation same as for Fig. 1 but freeze-etch preparation is Fig. 2a. Magnification X100,000.

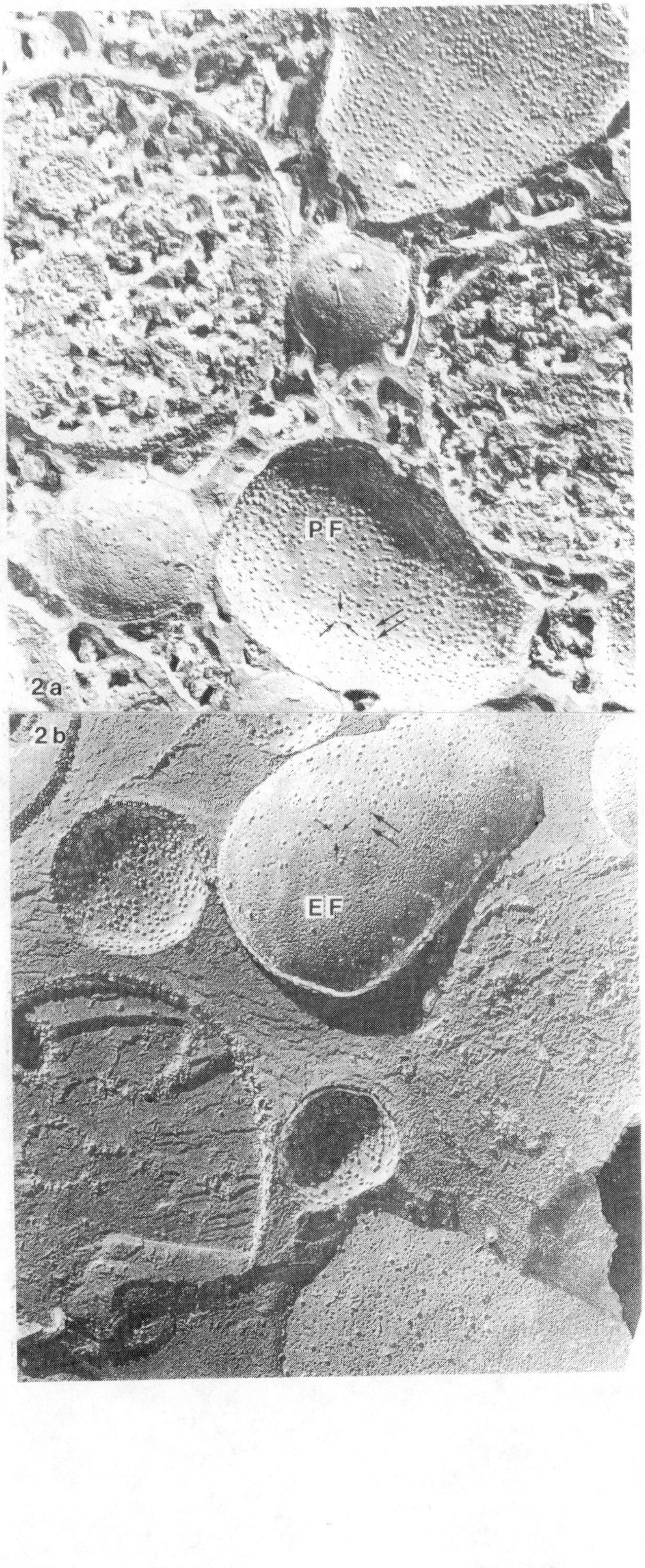
PF
2a
2b
EF

Fig. 3. Freeze-fracture and freeze-etch preparations of potato yellow dwarf virus in *Nicotiana rustica*. Fig. 3a, stereo electron micrographsof freeze-fracture preparation. Magnification X50,000.

Fig. 3b. Stereo electron micrograph pair of freeze-etched preparation. Sample deeply etched 5 min. to reveal subunit spikes on outer surface of membrane envelope. Magnification X50,000. Fig. 3c, portion of Fig. 3b enlarged to reveal greater detail. Magnification X100,000.

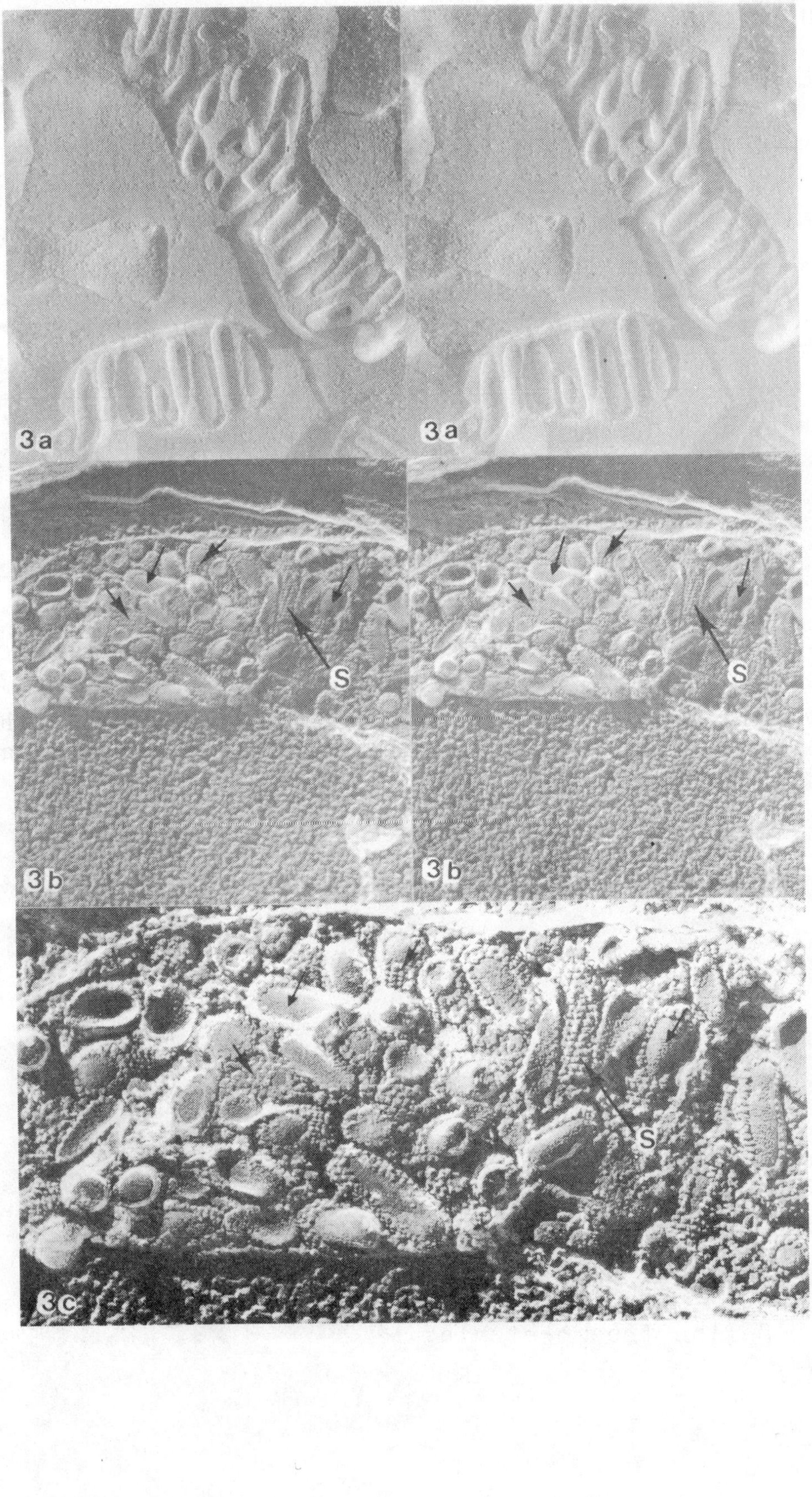
3a
3a
S
S
3b
3b
S
3c

G. *QUESTIONS:

1. Briefly discuss the experiment and the investigator responsible for the pioneering work on freeze-fracture-etch replication.

2. What other investigators are given credit for the first successful freeze-fracture replicas?

3. Discuss some of the problems that may occur in preparing specimens before and after producing a replica.

4. How does one evaluate a good replica?

5. Define freeze-etching.

6. What are the advantages of etching?

7. What do you visualize as being some future perspective ways the freeze-fracture-etch technique may be used other than studying membrane structure?

*Questions written by Betty Ruth Jones.

NOTES

EXERCISE NUMBER 28A

PREPARATION OF SPECIMENS FOR HIGH VOLTAGE ELECTRON MICROSCOPY

Preparation of Cell Whole Mounts With The Critical Point Drying Apparatus

Hans Ris, Ph.D.

A. INTRODUCTION:

The conventional 100 KV electron microscope requires specimens that are exceedingly thin, i.e. less than about 100 nm thick. This is due in part to limited penetration of the electron beam, but more importantly because of chromatic aberration of the electromagnetic lenses.

Electrons passing through a specimen lose energy which changes their wave length. The thicker the specimen the greater the spread of wave lengths in the electron beam. Since the lenses focus electrons of different energies in different image planes, the images become blurred and resolution is decreased.

As a rule of thumb, the resolution at 100 KV is about 1/10 of specimen thickness. At higher accelerating voltages, this blurring due to chromatic aberration of the lens is greatly diminished. For instance, at 1 million volts (1 MeV), the resolution becomes approximately 1/200 of specimen thickness.

High voltage electron microscopy (HVEM) therefore can provide good images of much thicker specimens. This is of course a great advantage in the analysis of complex three-dimensional structures or where reconstruction from serial sections is necessary. An ordinary tissue cell requires from 110 to 200 consecutive sections for conventional EM. With 0.5 μm sections and HVEM this is reduced to about 30 sections.

Many cells grown in cultures will attach and spread on formvar-coated grids. They are usually thin enough (0.25 to 2 μm) to be imaged intact with HVEM. Such preparations are particularly useful to study cytoplasmic structures such as microtubules, intermediate filaments and actin filaments and their role in cytoplasmic organization and cell motility.

The following exercises describe the preparation of stereoscopic micrographs of intact unsectioned mammalian cells grown on an electron microscope support grid.

High voltage electron microscopes are expensive installations. At present, three instruments are supported by the Biotechnology Resources Program of NIH as National Facilities available to biomedical researchers (Boulder, Colorado; Madison, Wisconsin and Albany, New York).

The cells described in this exercise are usually thin enough at the periphery to provide useful pictures at 100 KV. Most electron microscopes have stereo stages so that stereoscopic micrographs of intact cell whole mounts can also be obtained with conventional electron microscopes, though with reduced resolution.

B. OBJECTIVES:

1. To preserve cells with minimum alterations from the living state

2. To make cell structures and their chemical components insoluble and resistant to all of the procedures used in preparing the cells for electron microscopy

3. To dehydrate cells in ways that avoid distortions due to surface tension phenomena

4. To stain cell components with heavy metal salts to provide proper contrast in electron micrographs of all cell structures

C. MATERIALS:

1. 2.5% glutaraldehyde in 0.1 M Hepes buffer containing 0.05% saponin and 0.2% tannic acid

2. 0.1 M Hepes buffer

3. 0.1% OsO_4

4. 1% uranyl acetate in distilled water

5. Graded series of ethanol: 20%, 30%, 50%, 70%, 80%, 95%

6. Absolute ethanol containing molecular sieve to remove traces of water.

NOTE: THE MOLECULAR SIEVE IS WASHED IN DISTILLED WATER UNTIL THE SUPERNATANT IS CLEAR AND THEN DRIED IN AN OVEN AT 100 ^{o}C.

Add dry molecular sieve to absolute ethanol and keep well sealed for at least 24 h before use.

7. Liquid carbon dioxide in a siphon tank

Drying filter to remove residual moisture in the CO_2

8. Critical point drying apparatus

9. Cultures of any mammalian cell strain with cells grown on formvar-covered gold grids

10. Small petri dishes to hold fixatives

11. Appropriate dishes to receive the grid holder for the critical point drier during dehydration in the graded ethanol concentrations

12. Vacuum dessicator containing a flat dish with phosphorus pentoxide

D. RATIONALE AND DISCUSSION:

Cells are grown on formvar-coated gold grids in appropriate media and checked periodically with an inverted phase microscope. When there are a few free cells in most grid squares and before confluence the cells are fixed.

The fixative contains, in addition to glutaraldehyde, tannic acid and saponin. (Maupin and Pollard, J. Cell Biol. 95(2), 309a). The tannic acid was shown to protect cytoplasmic filaments against breakage and distortion during post-fixation in OsO_4 and critical point drying and to give them more contrast in the electron beam.

The low concentration of saponin makes the plasma membrane permeable to tannic acid without visibly affecting cell structure. Brief exposure to OsO_4 vapor before immersion into the glutaraldehyde-tannic acid fixative has also been used to make the membrane permeable to tannic acid.

The drying of the cells is the most critical step in the preparation of unembedded whole mounts. Because of surface tension effects, drying in air from an aqueous phase or from ethanol distorts and flattens the specimens.

The critical point drying procedure was designed to avoid such surface tension distortions. The specimens are thoroughly dehydrated in ethanol with molecular sieve in the final absolute ethanol to remove all traces of water. They are then immersed in ethanol in a pressure vessel attached to a liquid carbon dioxide siphon tank.

The ethanol in the pressure chamber is thoroughly flushed out with liquid CO_2 until the specimen is suspended in pure CO_2. Now the temperature of the pressure vessel is raised above the critical temperature, i.e. the temperature above which CO_2 is in the gas phase at any pressure.

During this heating, the CO_2 changes from liquid to gas gradually without any phase boundaries. The gaseous CO_2 is then released and the dry specimen stored in a dessicator over phosphorus pentoxide until it is inserted into the microscope.

E. PROCEDURE:

1. Gold grids are washed in distilled water for several hours and covered with a 0.5% formvar film.

2. After sterilization with a UV lamp, the grids are immersed in a plastic dish containing tissue culture medium and seeded with the cells to be studied.

3. After 2-3 days in a CO_2 incubator, the grids should be covered with several cells per grid square and are ready for fixation.

4. The following fixation steps should be performed in a fume hood.

With forceps, pick up grids at edge and rapidly transfer them to a dish with approximately 10 ml of fixative.

Swish them around a few times before depositing the grids with cells facing up.

Cover the dish.

5. After 15 minutes, the grids are rinsed in 0.1 M Hepes buffer and left for 30 minutes in a covered dish in fresh buffer.

6. Place the grids into a 0.1% solution of osmium tetroxide for 10 minutes.

7. Rinse grids and leave them in a dish with 20 ml distilled water for 5 minutes.

8. Filter 1% uranyl acetate with a millipore filter and immerse the grids, always with cells facing up, in the staining solution for 10 minutes.

9. Rinse grids briefly in distilled water and transfer them to a dish containing 20% ethanol.

10. Load the grids into the grid holder for the critical point dryer.

11. After 2 minutes, place grid holder for 2 minutes each into 30%, 50%, 70%, 80%, 95% ethanol.

12. Place grid holder in absolute ethanol that was dehydrated with molecular sieve. Leave for 5 minutes in a covered dish.

    Replace with fresh ethanol and leave for another 5 minutes.

13. Prepare the critical point drying apparatus:

    a. Close inlet and exit valves to pressure chamber, open valve at CO_2 tank.

    b. Open inlet and exit valves of chamber.

    c. When liquid enters the chamber, close the inlet valve.

d. Release the gas in the chamber and close exit valve.

e. Warm chamber to room temperature to prevent condensation of water in the chamber when opened.

f. Now cover to pressure chamber, fill the chamber about two thirds with absolute ethanol and place the grid holder into the pressure chamber.

   Make sure that the grids are well covered by ethanol.

g. Replace the cover to the chamber. Cool the chamber to about $10^{\circ}C$.

   If there is no cooling device for the chamber, place a piece of dry ice on the chamber cover, until the chamber content is at $10^{\circ}C$.

h. Now open the inlet valve and fill the chamber. Wash out the ethanol with CO_2 by filling and emptying the chamber 6-10 times.

NOTE: THE LIQUID LEVEL MUST NEVER BE ALLOWED TO GO BELOW THE GRIDS.

   To replace the ethanol from the cavities of the grid holder, it is necessary to stir the liquid in the chamber, for instance by tilting and shaking the apparatus.

   Total removal of all ethanol is essential for preservation of structure.

i. The chamber is again filled with CO_2 and let stand for 10 minutes.

j. After a few more fillings and emptyings, the chamber is filled to about two-thirds, the inlet valve and tank valves are closed and the heater is turned on.

k. After the chamber has reached a temperature of $43^{\circ}C$, the outlet valve is slightly opened and the gas very slowly released.

l. Evaporate a thin layer of carbon on the backside of the grids.

m. Critical point dried specimens should be stored in a vacuum dessicator over phosphorus pentoxide.

F. QUESTIONS:

1. The use of ultrathin sections (less than 100 nm thick) in conventional EM is obviously a disadvantage for understanding the 3-dimensional architecture of cells. Why are such thin specimens required with the 100 KV instruments?

2. What are the advantages of using a million volt electron microscope in biological research? Describe a research problem where application of the million volt EM could provide new information or facilitate the analysis.

NOTES

EXERCISE NUMBER 28B

PREPARATION OF STEREOSCOPIC MICROGRAPHS

Hans Ris, Ph.D.

A. OBJECTIVES:

1. To prepare stereoscopic electron micrographs which have good depth effect and are comfortable for viewing

2. To mount prints of electron micrographs to be viewed with a stereo pocket viewer (with two lenses enlarging x2 or x4)

B. MATERIALS:

1. A conventional 100 KV electron microscope with a tilt stage that can be used to prepare stereoscopic micrographs of the thin edge of critical point dried cells.

NOTE: FOR BEST RESULTS, A EUCENTRIC GONIOMETER STAGE IS RECOMMENDED.

The tilt axis is in the specimen plane and tilting will not change magnification appreciably.

2. Darkroom with photographic enlarging paper, developer, stop bath, fixer, print washer, print drier, paper trimmer, dry mounting press, tacking iron, mm ruler, mounting cardboard.

C. RATIONALE AND DISCUSSION:

In an ordinary electron micrograph, structural information from all levels of the specimen is projected onto one plane. With ultrathin sections, this provides little difficulty but with thicker specimens, the images become confusing and information on the arrangement of structures in three-dimensional space is lost. The electron microscope, because of the small illuminating angle, has great depth of field, i.e. structures up to several um thick are imaged in focus with a resolution better than 5 nm.

Because we have two eyes which look at the world from slightly different angles, the brain receives two views of an object which are fused into a single image providing depth perception. We can obtain such three-dimensional views of microscopic objects by taking two micrographs, the second one after tilting the specimen by a predetermined angle using a tilt stage. When these images are viewed so that each one can be seen only by the appropriate eye, a three-dimensional image is obtained.

For comfortable fusion, the differences between the images should not be greater than with an object viewed with two eyes at a distance of 250 mm. In the electron microscope, the disparity between the two images depends not only on the tilt angle, but also on the specimen thickness and magnification. At higher magnifications, the tilt angle must be smaller to give the same depth perception. Similarly, as the specimen thickness increases, the tilt angle must be diminished.

Since our eyes are positioned on a horizontal line, corresponding object points in a stereo image pair must lie on a horizontal line for proper fusion. This condition is met if the orientation of the object to the tilt axis is the same in the image as in the stage. An elongated object oriented in the tilt axis should be vertical in the image.

In older microscopes, the image is rotated relative to the specimen. In this case, this angle must be determined and the prints rotated through this angle before trimming and mounting. In modern microscopes, this image rotation is compensated for, so that the long axis of the negative is always parallel to the tilt axis and prints can be mounted vertically side by side.

For viewing with a pocket stereo viewer, corresponding points in the two prints should be separated by the interocular distance, i.e. 60-65 mm.

This applies also to the vertical edges of the prints. Each print is therefore 60-65 mm wide. Image points which are this distance apart appear to be in the plane of the table. Points above the table lie closer together, those below the table farther apart.

D. FOR TAKING STEREO MICROGRAPHS:

1. Select an appropriate thin area near the edge of the cell.

2. Determine the stereo angle. Near the edge the cell may be from 0.25 µm to 0.5 µm thick. At a magnification of x10,000 on the screen, a tilt angle of 20° is appropriate. Tilt the stage 10° on one side of the horizontal position.

3. Prepare a sketch marking the position of a prominent object on the fluorescent screen. This will help placing the object into exactly the same position after tilting for the second picture.

4. Focus and expose.

5. Tilt the stage 10° on the other side of the horizontal position giving a total stereo angle of 20°.

6. Using the sketch, place the object into the exact identical position as on the first picture. Focus and expose.

If the thickness of the specimen is unknown, one can take a thorough tilt series, for instance 20° and 10o to the left and 10° and 20° to the right. After printing, the pictures with the best stereo effect can be selected for mounting.

E. PROCEDURE FOR MOUNTING STEREO PRINTS:

1. Print the negatives so that the area of interest lies within a strip 63 mm wide. Since the stereo viewers magnify 2-4 times, an enlargement between 1-2 times is sufficient.

2. Trim one side of the print parallel to the long axis of the negative (which is parallel to the tilt axis) and along a line so that the area to be viewed is within 63 mm of it.

3. The position of the corresponding edge in the second print will determine what level of the specimen will appear at the height of the table when viewed in stereo.

   Whole mounts of cells on a formvar film appear most pleasing if the formvar film appears in the plane of the table with the cell rising from it.

   To give this effect, corresponding structures on the formvar film must be the same distance from the lateral edges in both prints.

4. Select some marker on the film near the upper and lower edges of the prints and trim the edge of the second print so that it is the same distance from the marker as in the first print.

   With thick sections, the stereo effect is more pleasing if the top layer of the section appears in the plane of the table. One sees the section as if looking through a window into the specimen.

5. To create this effect, place the prints side by side and observe with the stereo viewer.

6. Select a few prominent points that lie in the top surface of the section.

7. Trim the side edges so that these points are equidistant from either the left or right edge in both prints.

8. Mark left and right print.

9. Trim both prints to a width of 63 mm.

10. Trim the top and bottom of the prints so that exactly the same area is included in both.

11. Place prints side by side and view with stereo viewer. The formvar film should appear in the plane of the table with the cell either above or below it.

12. If it lies below, reverse left and right. The cell should now be sitting on the film.

13. Mark the prints left and right and mount them side by side in the correct position on a piece of cardboard using dry mounting tissue and a dry mounting press.

F. EXAMPLES OF CRITICAL POINT DRIED CELLS:

The stereomicrographs should be viewed with a stereo viewer such as the stereo pocket viewer PS-2 from Air Photo Supply (Yonkers, NY) or cat. #758 from Pelco Co. (P.O. Box 510, Tustin, CA).

Inexpensive cardboard viewers are available (Model 707) from Taylor-Merchant Corp. (25 West 45th Street, New York 10036).

Figure 1. PtK$_1$ cell grown on formvar coated gold grid, fixed in 2.5% glutaraldehyde in 0.1 M HEPES buffer (pH 7.0) containing 0.05% saponin and 0.2% tannic acid. Postfixation in 0.1% OsO_4, staining with 1% uranyl acetate. Critical point dried. X 7500, tilt angle 30°.

This low magnification micrograph shows the transition between cytoplasm containing endoplasmic reticulum, mitochondria, polysomes and the thin lamella at the edge of the cell which contains mainly cytoskeletal elements such as microtubules, intermediate filaments and actin filaments.

This picture illustrates the complex relationship between cytoskeletal filaments at the cell border.

Figure 2. PtK$_1$ cell prepared as in Figure 1. Higher magnification of the lamella at the cell border. Actin filaments form a network with attached particles of unknown composition. The mesh of the network is smallest at the cell edge and increases towards the interior. X 60,000, tilt angle 10°.

Figure 1

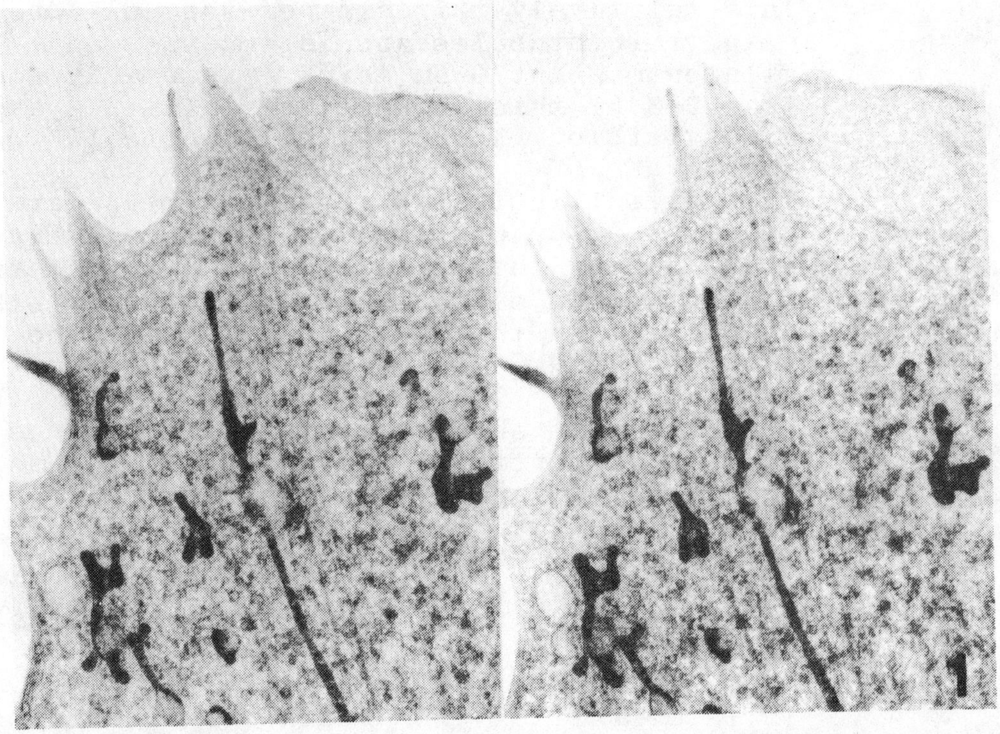

Figure 2

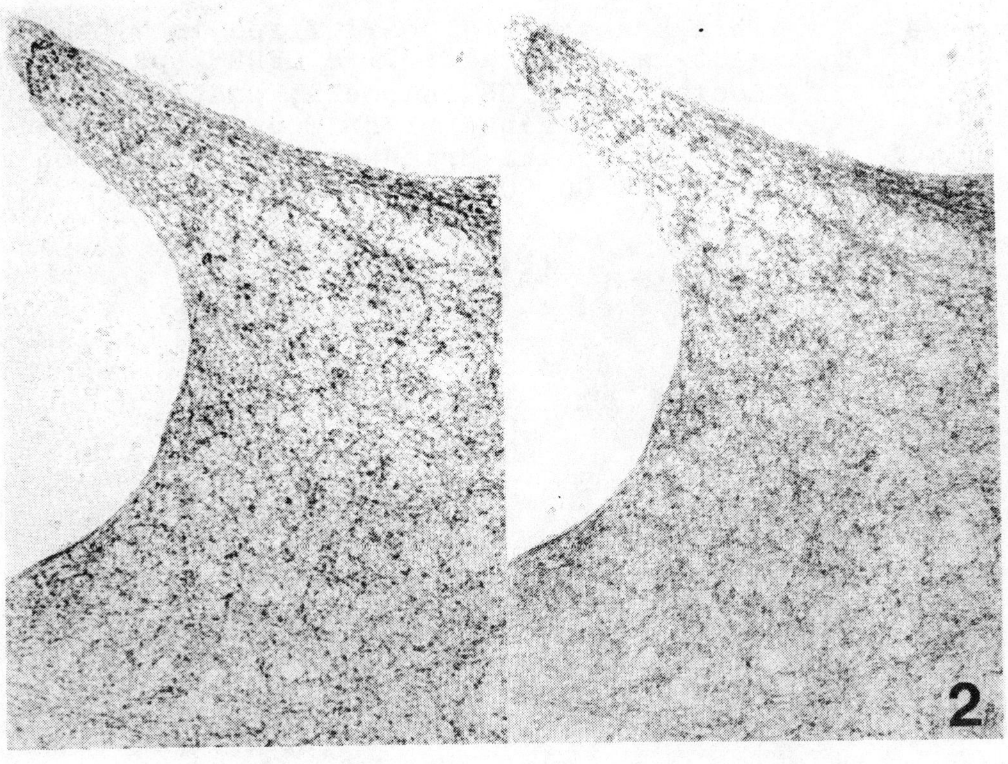

Figure 3. PtK_1 cell prepared as in Figure 1. Higher magnification of the cytoplasmic interior. In this relatively thin area of the cell one sees mainly microtubules and interconnected filaments about 6 nm thick (presumably actin). Some 2-3 nm thick connecting filaments are also visible. X 60,000, tilt angle 10°.

The cells illustrated above had been carefully dehydrated and critical point dried. Figures 4 and 5 show the appearance of cytoskeletal elements when either residual water or ethanol is present in the CO_2. In this case the filter on the CO_2 tank was omitted.

The filaments show characteristic surface tension distortions, they vary in thickness and fuse into an alveolar network.

This structure has been described as a new cell organelle ("microtrabecular lattice") by Wolosewick and Porter (J. Cell Biology 82, 114).

It is most likely an artifact induced during preparation. As seen in Figures 1-3 this structure is not present in well dried preparations.

Figure 4. Thin edge of PtK_1 cell fixed in 2.5% glutaraldehyde in 0.1 M HEPES (pH 7.0) containing 0.05% saponin, postfixed in 0.1% OsO_4 and stained with 1% uranyl acetate. Critical point dried without filter on the CO_2 tank. X 60,000, tilt angle 10°.

Figure 3

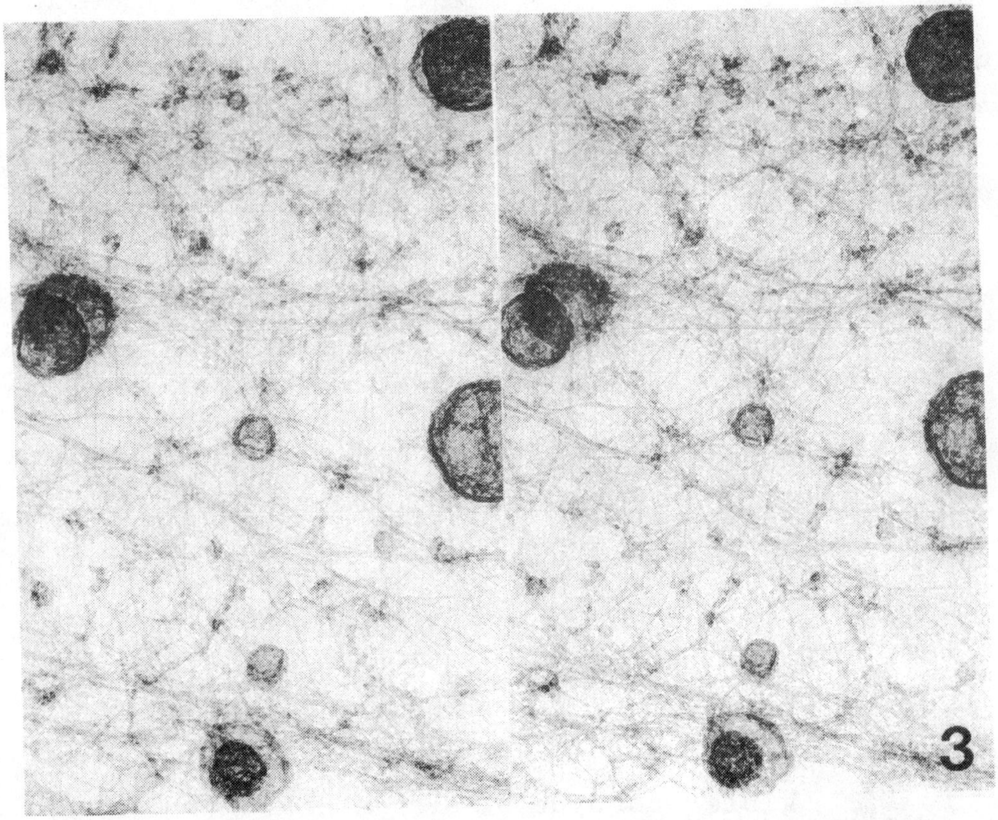

Figure 4

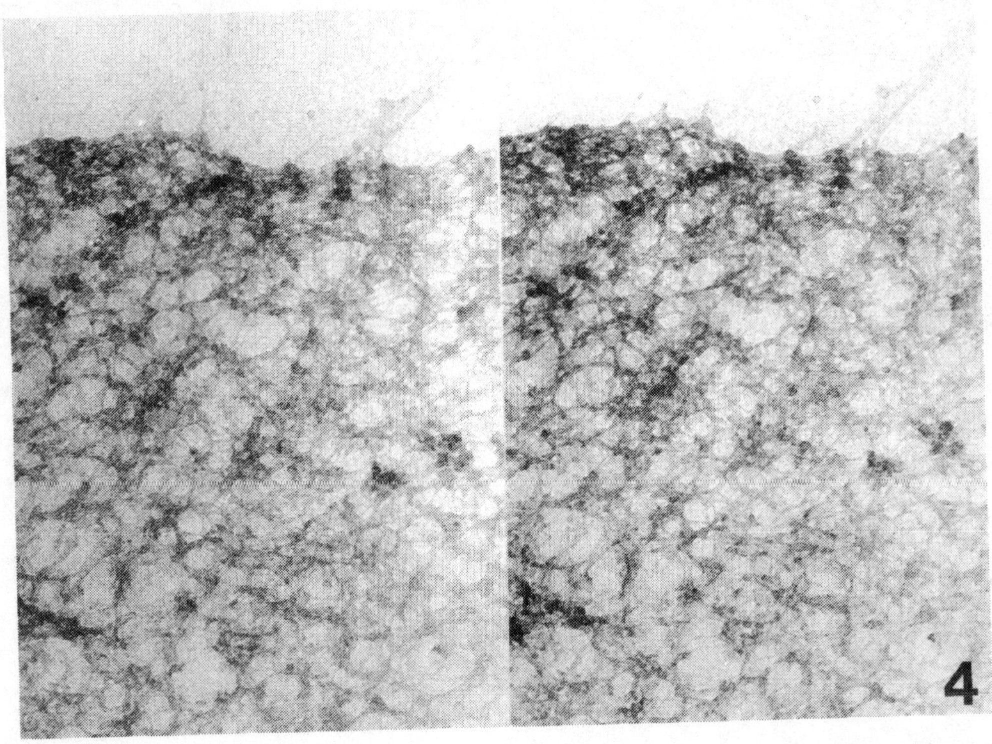

Figure 5. Interior of PtK_1 cell fixed in 2.5% glutaraldehyde in 0.1 M HEPES (pH 7.0), postfixed in 0.1% OsO_4 and stained in 1% uranyl acetate. Critical point dried without filter on the CO_2 tank. X 60,000, tilt angle 10°.

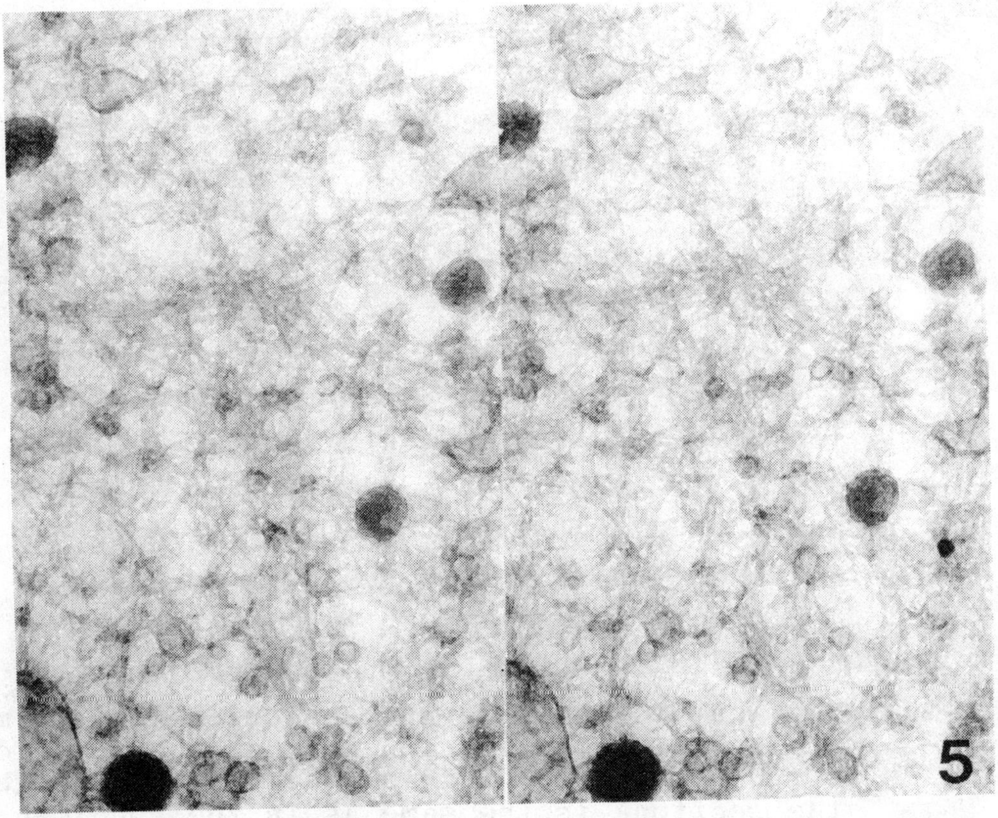

G. QUESTIONS:

1. You have a single lens reflex camera with a close-up attachment and want to take a stereoscopic photograph of a flower. Describe in detail how you would proceed to obtain such a photograph and how you would view it to get a 3-dimensional effect.

2. You want to obtain a stereoscopic electron micrograph of a thick section so that the position of structures in the Z-axis (the relative height in the section) is the same as in the specimen. In other words, what is in the top plane of the section should appear on top in the stereo micrograph. What factors must be considered during preparation of the stereo micrograph to obtain this effect?

EXERCISE NUMBER 29A

PREPARATION OF VIRUSES FOR ELECTRON MICROSCOPY

PSEUDOREPLICA TECHNIQUE FOR PREPARATION AND NEGATIVE STAINING OF VIRAL SPECIMENS FOR ELECTRON MICROSCOPY

G. William Gary, Jr., M.P.H., Dr. P.H.

NOTE: THIS EXERCISE IS INTENDED TO BE PERFORMED ONLY IN A LABORATORY WHICH IS EQUIPPED FOR THE PROPER SAFE HANDLING OF VIRUSES AND VIRUS-INFECTED CELL CULTURES.

FURTHER, THIS EXERCISE SHOULD BE PERFORMED USING ONE OR MORE OF THE LEAST HAZARDOUS ANIMAL VIRUSES; OR, PREFERABLY, A PLANT VIRUS IF ONE CAN BE OBTAINED WHICH CONTAINS A SUFFICIENT NUMBER OF PARTICLES PER UNIT VOLUME, I.E., 10^6 PARTICLES/ml.

Acknowledgements: The pseudoreplica technique is from work done by Sharp and Overman (1958) and Sharp (1960); and described in more detail by Smith and Melnick (1962).

The technique as specifically described here and illustrated in Figure 29-a is that introduced into the Virology Division of the Centers for Disease Control by Fred Murphy, D.V.M., Ph.D. in 1964.

The technique has been used extensively in the Virology Division since that time and has been taught to a large number of national and international visitors to the Division.

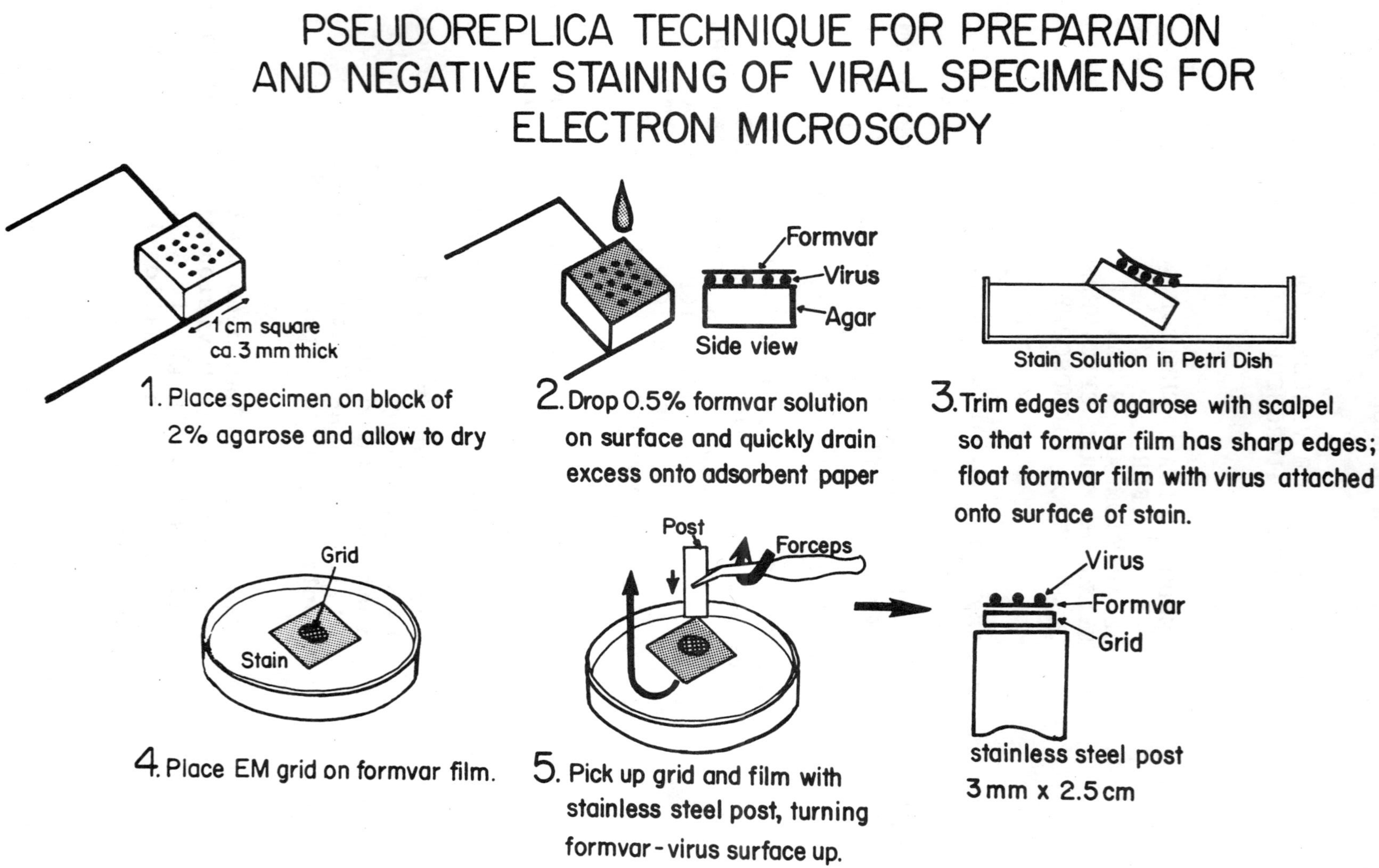
PSEUDOREPLICA TECHNIQUE FOR PREPARATION AND NEGATIVE STAINING OF VIRAL SPECIMENS FOR ELECTRON MICROSCOPY
1 cm square
ca. 3 mm thick
1. Place specimen on block of 2% agarose and allow to dry
Formvar
Virus
Agar
Side view
2. Drop 0.5% formvar solution on surface and quickly drain excess onto adsorbent paper
Stain Solution in Petri Dish
3. Trim edges of agarose with scalpel so that formvar film has sharp edges; float formvar film with virus attached onto surface of stain.
Grid
Stain
4. Place EM grid on formvar film.
Post
Forceps
5. Pick up grid and film with stainless steel post, turning formvar-virus surface up.
Virus
Formvar
Grid
stainless steel post
3 mm x 2.5 cm

FIGURE 29-a.

A. OBJECTIVES:

1. To demonstrate one method for transferring viruses from their natural or artificial habitat to a grid for examination in the transmission electron microscope

2. To utilize crude material, such as fecal suspension, or very purified material, such as density gradient purified polio viruses

B. MATERIALS:

1. Agarose, 2%, in petri dishes. The desired thickness is approximately 3 mm and can be achieved by pouring 25 ml of melted agarose into a 100 mm petri dish

2. Formvar solution, 0.5% in ethylene dichloride

3. Grids, copper, 400 mesh, of appropriate diameter for microscope to be used

4. Filter paper

5. Forceps

6. Negative stain, 0.5% uranyl acetate or 1.5% Phosphotungstic acid (or stain of choice)

7. Petri dish, 60 x 15 mm (for staining)

8. Stainless steel rods (posts), 3 mm x 2.5 cm (used to pick up grid and film with virus) (The diameter of the posts [3 mm] is appropriate for the diameter of grids used by the author; you may need to use a different diameter for the rods)

9. Support for the rods (posts), plexiglas block measuring 15 cm x 7 cm x 2 cm and drilled to hold 12 rods (See Figure 29-b)

10. Glass microscope slides, 1 x 3"

11. Suspension of cell culture medium, or other suspension, containing virus in a concentration of at least 10^6 particles/ml (Use an innocuous plant virus for practice and this exercise if at all possible)

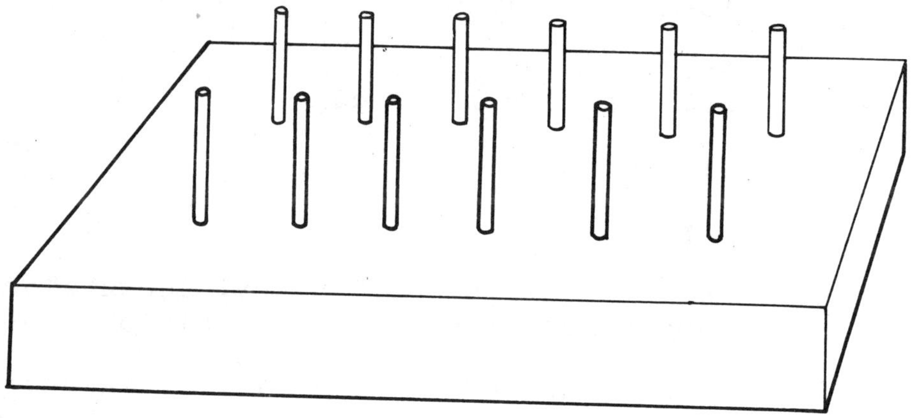

FIGURE 29-b. Plexiglas block with stainless steel posts.

12. Scalpel (disposable, sharp pointed if possible)

C. RATIONALE:

There are a number of techniques by which one may transfer virus from fluid or suspension to a grid for electron microscopic examination. The method described here is rather simple, works very well with readily-obtainable supplies, allows some separation of salts from the virus, and affords some degree of concentration of the virus. A number of investigators have found it to be their method of choice.

The method works well for rapid diagnosis; it has been used to differentiate varicella from smallpox, detect cytomegalovirus in urine, and, in the author's hands, to detect rotavirus in fecal specimens.

This method has also been used for a number of studies of virus morphology, including the preparation for many of the photomicrographs in the atlas by Palmer and Martin (1982).

The virus suspension is placed on top of an agarose square and allowed to dry. This allows salts and fluid from the suspension to diffuse into the agarose while leaving the virus on the surface. Formvar solution is flooded over the surface, allowed to dry, trimmed with a scalpel, and floated onto the surface of the appropriate stain.

The grid is then placed over the formvar film and a post held with forceps is then pushed down through the grid and film, flipped over, and excess stain removed from the grid with the edge of a piece of filter paper. The post and grid (with film and virus) are placed in the plexiglass holder to further dry and await examination in the electron microscope.

D. PROCEDURE:

1. Make sure that all materials are immediately accessible.

2. Place a 1 cm^2 square of agarose on one corner of a glass microscope slide (See Figure 29-a).

3. Place a drop of virus suspension on the agarose square and spread fluid across the surface with the side of a pipette (pasteur preferrable).

4. After the virus suspension has dried, the upper surface of the agarose square is flooded with 0.5% formvar solution and immediately the slide is tilted sideways so that the formvar will run off onto an absorbent paper towel. This step will leave a very thin layer of formvar to which the virus is attached.

5. Trim all four vertical edges of the agarose square with a scalpel so that the film of formvar has sharply defined edges.

6. Carefully immerse the glass slide (on which rests the agarose square) into a dish of negative stain at approximately a 30^0 angle. The formvar film will separate from the agarose and float on top of the stain.

   Occasionally a specimen will contain substances that cause the film to adhere tightly to the agarose.

NOTE: ONCE THE FILM HAS BEEN SUBMERSED BENEATH THE STAIN IT IS GENERALLY IMPOSSIBLE TO SAVE THAT PREPARATION AND YOU MUST BEGIN AGAIN.

7. Gently lay a grid onto the surface of the film. (Staining processes take place almost immediately; it is not necessary to delay for staining to take place).

8. Take a stainless steel post with forceps and push one end of the post down on top of the grid, into the stain, and flip over the forceps bringing the grid and film out on top of the post (See Figure 29-a).

   The grid will sit on top of the post and the excess film around the grid will adhere to the post. Remove excess stain with the edge of a piece of filter paper.

9. Set the post with grid into the plexiglas holder to further dry. The grid may then be picked off the top of the post with forceps and without damage to the film.

10. Examine the grid in the transmission electron microscope.

E. RESULTS:

A very nice preparation will result almost every time. On occasion a film will contain too much extraneous material from the specimen such that the interaction of the electron beam and this material will cause severe distortion of the film due to heat buildup in the film. Dilution of the specimen or further purification may be necessary in these instances.

F. EVALUATION:

The appearance of the film on the surface of the stain will be the first evidence of success or failure to prepare a good film. A shredded film may result from inadequate drying of either the specimen or the formvar, or may result from something in the specimen that reacts adversely with the formvar solution.

The next evaluation occurs when the grid is examined in the electron microscope. At this step the most obvious faults observed are (1) use of too heavy of a virus suspension and (2) poor film due to inadequate drying at one or both drying steps. The method to correct either fault is obvious.

NOTE: THE PSEUDOREPLICA TECHNIQUE CAN BE PERFORMED EASILY AND EFFICIENTLY AFTER A LITTLE PRACTICE.

G. REFERENCES:

1. Sharp, D.G. and J.R. Overman 1958. Enumeration of vaccinia virus particles in crude extracts of infected tissues by electron microscopy. Proc. Soc. Biol. Med. 99: 409-413.

2. Sharp, D.G. 1960. Sedimentation counting of particles via electron microscopy. Proc. Intern. Congr. Electron Microscopy 4th Congr. 1958, Berlin, pp. 542-548.

3. Smith, K.O. and J.L. Melnick. 1962. A method for staining virus particles and identifying their nucleic acid type in the electron microscope. Virology 17: 480-490.

4. Palmer, E.L. and M.L. Martin. 1982. An atlas of mammalian viruses. CRC Press, Inc. Boca Raton, Florida, 154 pp.

H. QUESTIONS:

1. How does this technique differ from replication techniques used in metallurgy electron microscopy?

2. What is the advantage of using an agarose block?

3. What specimens other than viruses can be examined by this technique?

EXERCISE NUMBER <u>29B</u>

FIXATION OF VIRUS-INFECTED CELL CULTURES FOR EXAMINATION BY ELECTRON MICROSCOPY

G. William Gary, Jr., M.P.H., Dr.P.H.

NOTE: THIS EXERCISE IS INTENDED TO BE PERFORMED ONLY IN A LABORATORY WHICH IS EQUIPPED FOR THE PROPER SAFE HANDLING OF VIRUSES AND VIRUS-INFECTED CELL CULTURES.

FURTHER, THIS EXERCISE SHOULD BE PERFORMED USING ONE OR MORE OF THE VIRUSES AVAILABLE WHICH HAVE THE LEAST HAZARD TO MAN.

A. OBJECTIVES:

1. To demonstrate one procedure for taking virus-infected cells from their *in vitro* habitant in a cell culture flask through a series of steps to fixation, and ultimately through embedding steps, sectioning and examination in the transmission electron microscope

2. To allow the student to observe ultrastructural changes caused by virus infection of a cell and the replication of the virus within the cell

B. MATERIALS:

1. A 75 cm^2 cell culture flask (See Figure 29-c) containing a monolayer of cells that have been infected with a strain of virus

2. A 75 cm^2 cell culture flask containing a monolayer of cells that *have not* been infected with virus

NOTE: THIS IS THE "CONTROL" PREPARATION THAT WILL SERVE TO COMPARE THE ULTRASTRUCTURE OF CELLS INFECTED

WITH VIRUS TO THAT OF CELLS NOT INFECTED.

3. An appropriate container containing a 10% solution of sodium hypochlorite (1:10 dilution of common liquid household bleach) in which to discard media from the cell culture flask

   Other materials that may have been contaminated with virus-containing fluids

NOTE: THIS PAN MUST BE PLACED IN A STEAM AUTOCLAVE FOR AN APPROPRIATE TIME AT THE END OF THE EXERCISE, OR DAY, TO STERILIZE OR DECONTAMINATE THE PAN AND ITS CONTENTS.

4. Rounds-bottom, polypropylene 50 ml centrifuge tubes (not conical)

5. Rubber policeman, 2 each with glass-rod handles (i.e., American Scientific Products #R5110-2)

6. Centrifuge, low speed

7. 2.5% glutaraldehyde, freshly made from 25%, diluted with "EM PO_4 Buffer"

8. EM PO_4 Buffer, 0.15 M

   $NaH_2PO_4 \cdot H_2O$ 5.85 g

   Na_2HPO_4 15.25 g

   qs to 1 liter with distilled water

9. Pipettes, 10 ml and pasteur

10. Pipetting bulb

11. Applicator sticks (without cotton)

12. Osmium solution, 1%

    1 part 2% osmic acid in distilled water
    1 part "EM PO_4 Buffer"

C. RATIONALE AND DISCUSSION:

Most viruses can be grow in one or more cell lines <u>in vitro</u> and provide an excellent means of studying the replication of viruses in cells. Investigators can harvest and fix cells at selected intervals after inoculation of the cell cultures using a number of cell cultures flasks, all inoculated at the same time. By this method one can observe attachment, penetration, uncoating, and replication of viruses.

This technique is one that was developed in the Division of Viral Diseases in the middle '60s and has been used with a number of viruses since that time. There are undoubtedly a number of similar techniques in existence, but this one has worked well for us.

The primary purpose is to get cultured cells from their growing state through fixation, embedding, sectioning, and staining to examination in the electron microscope with as little damage as possible to their ultrastructural integrity.

This method utilizes 50 ml round-bottom centrifuge tubes and low-speed centrifugation to gently concentrate cells from culture into a thin, flat pellet that can be rapidly fixed with glutaraldehyde.

After fixation with glutaraldehyde and washing with EM PO_4 Buffer, appropriate sized pieces of the pellet can be post-fixed with osmium tetroxide and embedded through standard procedures. This exercise does not include embedding procedures.

NOTE: FOR THOSE UNFAMILIAR WITH CELL CULTURES AND THEIR USE IN VIROLOGY, SOME EXPLANATION IS NECESSARY. FOR A NUMBER OF YEARS INVESTIGATORS HAVE BEEN ABLE TO GROW LIVING CELLS IN A VARIETY OF GLASS OR PLASTIC CONTAINERS.

The most common technique is to "seed" a flat flask, either plastic or glass with a suspension of cells in a chemically defined medium which supports growth of the cells. The initial concentration of cells thus seeded is calculated to allow the multiplication of cells as they grow to eventually cover one side of the flask.

This monolayer of cells is then inoculated with an optimum concentration of virus in a small volume of fluid, the virus is allowed to attach to the cell surface, and a volume of maintenance medium is added to the flask to cover the cells and provide sufficient nutrients to keep the cells alive but not promote their growth.

D. PROCEDURE:

This exercise assumes that someone with expertise in virus culture techniques is involved with teaching this course or is available to provide a virus-infected cell culture and a non-infected control culture.

1. The medium on the control and infected flask is poured off or decanted into a container containing 10% sodium hypochlorite.

2. Each flask is then "washed" with approximately 50 ml of EM PO_4 Buffer to remove residual culture medium.

3. Approximately 40 ml of EM PO_4 Buffer is then added to each flask.

4. A rubber policeman is then used to gently scrape the cells into the buffer solution.

NOTE: THE CELLS ARE NOT FIRMLY ATTACHED TO THE FLASK WALL AND ARE THEREFORE NOT USUALLY DAMAGED BY THIS PROCEDURE.

5. The suspension of cells resulting from step 4 is then poured into respective 50 ml centrifuge tubes and centrifuged for 20 minutes at 700-800 x g.

   A thin, intact layer of cells will be pelleted to the bottom of the tube.

6. The supernatant EM PO_4 Buffer is decanted carefully, leaving the cell pellet intact.

   This supernatant is discarded carefully into the solution of sodium hypochlorite as the supernatant from the infected bottle may contain residual virus.

7. Fifteen ml of freshly prepared 2.5% glutaraldehyde is very slowly and carefully layered on top of the cell pellet.

   Both tubes are then placed in a refrigerator or ice bath for 20 minutes to allow fixation of the cells.

8. At the end of 20 minutes the glutaraldehyde is carefully decanted, leaving the fixed pellet in the tube.

9. Approximately 45 ml of EM PO_4 Buffer is gently added to the tube and then decanted to wash away the excess glutaraldehyde.

10. The resulting pellet is fairly firm and can be teased into pieces approximately 1 mm^2.

11. These pieces are then fixed in an osmium solution (See Materials, #12) for 20 minutes.

CAUTION: OSMIUM FUMES HAVE THE POTENTIAL TO FIX THE CORNEAL CELLS OF THE PERSON USING THE OSMIUM AND THEREFORE THIS SOLUTION MUST BE USED IN A PROPERLY OPERATING FUME HOOD.

UNEXPERIENCED PERSONS SHOULD NOT USE OSMIUM WITHOUT PROPER SUPERVISION.

12. After post-fixation with osmium the pellets are then ready for embedding by any acceptable technique.

E. RESULTS:

These procedures are relatively easy to perform and give excellent results if the proper equipment is available. In the author's environment the steps through glutaraldehyde fixation were handled in the virology laboratory and the osmium fixation steps and further were handled in the electron microscopy laboratory.

F. EVALUATION:

True evaluation can only occur after the embedded cells are sectioned, stained, and examined in the electron microscope. At that time, the degree of preservation of cell structure can be determined.

The instructor should observe the various stages of the exercise to be sure that the directions are being followed.

The critical point is the appearance of the pellet before and after glutaraldehyde fixation; there must be intact pieces of "chunks" of clumped cells, not totally dispersed individual cells.

Figure 29-c

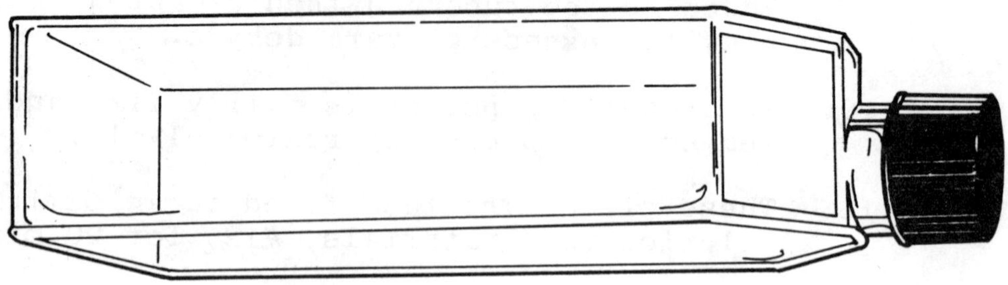

G. QUESTIONS:

1. Speculate as to, or relate from your own knowledge, why this use of cell cultures is frequently preferable to using tissues from infected animals.

2. Why is rapid fixation necessary?

3. Why do we use osmium under a fume hood?

4. Why do we decant or pour fluids from virus inoculated cell cultures into sodium hypochlorite solutions?

EXERCISE NUMBER 30

PREPARATION OF BACTERIA FOR ELECTRON MICROSCOPY

Frederic A. Giere, Ph.D.

A. OBJECTIVE:

1. To prepare bacterial specimens for examination by transmission electron microscopy

B. MATERIALS:

1. Bacterial culture: petri dish, slant, or liquid culture
2. Distilled water
3. Vials or small culture tubes
4. Inoculating loop
5. Filter paper
6. Pipet, dropping
7. Jeweler's forceps
8. Storage box for prepared grids
9. 2% phosphotungstic acid

Optional materials:

10. Clinical centrifuge and tubes
11. Paraffin block or parafilm
12. Nebulizer
13. 5% Bovine serum albumin

C. RATIONALE AND DISCUSSION:

It is obvious to you that the ultrastructure of bacteria can be revealed only by the sectioning of the bacteria. Procedures for that would be similar to the preparation of unicellular plants and animals and would involve their concentration, fixation, dehydration, embedding, sectioning, etc. However, considerable information can be gleaned from the observation of their shape, size, presence of flagellae and pili etc of the whole organism.

This involves the application of the bacteria to an electron transparent support film and a subsequent step which will enhance the contrast of the specimen with respect to the background (i.e. the application of an electron dispersing material - a negative stain or metal shadowed from an oblique angle). Metal evaporation in a vacuum will be considered in another exercise.

It will be necessary to mount the particulate matter in such a way that the particles appear as individuals rather than as aggregates where overlap will cloud the appearance of individuals. Methods which facilitate that application are the float, drop and spray methods.

D. PROCEDURE:

1. Transfer a loop of culture grown on a solid substrate (plate or slant) to a few milliliters of distilled water.

   This suspension will likely contain more than 0.1 mg/ml and will be appropriate for application to the support film by the float method.

NOTE: IF YOU START WITH A LIQUID CULTURE, THAT SOLUTION MUST BE CENTRIFUGED AND WASHED WITH DISTILLED WATER SEVERAL TIMES TO REMOVE CELL DEBRIS AND BUFFERS WHICH MAY PRECIPITATE DURING DRYING AND THEREFORE INTERFERE WITH YOUR OBSERVATIONS.

   After several washes a loopful of the bacteria may be transferred as above. Grids can be prepared by the drop, float or spray method.

2. To apply the sample to the support film you may follow one of the following sets of instructions:

a) Drop Method:

A drop of the particulate solution is transferred to a support film with a pipette or a platinum loop. The drop remains about 30 seconds or more and the excess fluid is removed with a shred of torn filter paper. The grid should then be negative stained or air-dried for metal shadow casting.

b) Floating Method:

A drop of prepared culture solution is placed on a hydrophobic surface such as dental wax, parafilm, or a paraffin block and a support film is floated on the drop with the support film attached to the grid in opposition to the drop.

After contact in excess of 30 seconds, the grid and support film are removed to ashless filter paper to drain the excess fluid by capillarity from the edge of the grid. If the specimen is to be negative stained it must be done before drying. If the sample is to be metal shadowed, allow it to dry in a dust free environment.

Carbon and Formvar support films are more hydrophobic than is Parlodian and some difficulty may be experienced in application of the particulate matter to those films.

A number of investigators have found that the addition of a drop(s) of a dilute (5%) Bovine Serum Albumin to the culture suspension will have sufficient wetting action to apply the sample and not permit the particles to aggregate.

c) Spray or Nebulizer Method:

This method employs an atomizer which deposits microdroplets of the culture solution to the support films. The atomizer consists of a large glass bulb which contains a diffuser head.

About 0.1-0.2 ml of sample is placed in the glass chamber. The opening is

i) Fill the bag immediately with water checking for leaks, and store it immersed in water.

IV. Preparation of Stock Buffer - Uranyl Acetate Stain

Ammonium Acetate (1 M) pH 7.5

77 g ammonium acetate (M.W. 77.08)

dissolved in 1000 ml H_2O

pH adjusted with acetic acid or ammonium hydroxide

Ammonium Acetate (1 M) + EDTA (0.002 M) pH 7.0

77 g ammonium acetate (M.W. 77.08)

0.74 g EDTA (M.W. 372.24) disodium ethylenediamine tetraacetate

dissolved in 1000 ml H_2O

pH adjusted with acetic acid or ammonium hydroxide

Tris (1 M) + EDTA (0.1 M) pH 8.0

121.1 g Tris (hydroxymethyl)-aminomethane (M.W. 121.1)

37.2 g EDTA (M.W. 372.24)

dissolve in 700 ml H_2O and adjust pH to 8.0 with HCl, bring volume to 100 ml with H_2O

Tris-HCl (1.8 M) + Tris-base (0.2 M) pH 6.8

28.35 g Tris (hydroxymethyl)-aminomethane hydrochloride (M.W. 157.5)

2.42 g Tris (hydroxymethyl)-aminomethane (M.W. 121.1)

dissolve in 100 ml H_2O

closed with a stopper, and the rubber bulb compressed to nebulize (spray) microdroplets onto the support films which are placed about 5-10 cm from the exit of the instrument.

The negative stain may be mixed with the specimen and sprayed simultaneously. This method will reduce artifacts attendent with specimen drying when there are significant changes in surface/volume ratios.

3. Negative Staining:

This step must be accomplished before there is significant air drying of the sample.

Negative staining makes use of the electron-scattering property of heavy metals. In its simplest form, the stain forms a thick, electron opaque continuum between the individual particles (bacteria or viruses). The particles largely unstained will then stand out against the dark background.

The negative stain is a solution of a heavy metal salts which upon drying form an almost structureless solid. The commonly used materials include: uranyl acetate, ammonium molybdate, and sodium silicotungstate. We shall use the original method.

Place a drop of a 2% Phosphotungstic Acid (PTA) adjusted to the pH of the medium in which the specimen was suspended. Float the particulate-coated support film in contact with the stain for 30 seconds of more. Longer staining (frequently up to 10 minutes) may be required.

It is imperative that the specimen grid remain wet between the time of mounting the sample and the negative staining. Failure in that progression will result in nonuniform staining.

4. Dry the grids in a dust free environment and examine in the transmission electron microscope.

E. QUESTIONS:

1. For what reasons is the procedure described above called "negative" staining?

2. What role does the wetting agent play in the:

   a) application of the particulate matter to the support?

   b) deposition of the "negative stain" around the particles?

3. What precautions should be taken when dealing with pathogens?

4. Would you expect that killed, fixed bacterial specimens have the same appearance as those prepared by the methods above?

5. What effects might the absence of adjusting the pH have on the action of the negative stain?

6. What morphology can be determined by the procedures described above?

EXERCISE NUMBER 31

PREPARATION OF SUPPORT FILMS FOR ELECTRON MICROSCOPY

Frederic A. Giere, Ph.D.

A. OBJECTIVE:

1. To construct electron transparent supports films for particulate samples to be observed by transmission electron microscopy

B. MATERIALS:

1. Clean 1x3 inch glass microscope slides
2. Beaker 50 ml or 100 ml
3. Parlodian (Collodion) 0.5% in ethyl or amyl acetate
4. Formvar 0.25% in ethylene dichloride
5. 75x150 mm culture dish or finger bowl
6. Demineralized water
7. Glass stirring rod or orangewood applicator stick
8. Desk lamp
9. Razor blades
10. Glass petri dish
11. Acetone

*Original illustrations were prepared by Betty R. Jones and Frederic A. Giere.

12. 300 - 400 mesh athene type grids

13. Filter paper

14. Jeweler's forceps

15. Large test tube (about 30 mm in diameter)

16. Covered dish for storage of grids with the support films

C. RATIONALE AND DISCUSSION:

Electron microscope grids are analogous to the glass slides used for specimen support in light microscopy. These grids come in a variety of sizes and shapes of their perforations. Many have a regular coordinate grid pattern, e.g. 200 mesh which have 200 'holes' per inch and about 60-70% open space.

Grids with fewer holes have more open space and those with more holes (400 mesh) have a lesser open space for electron transmission. Other grid patterns encountered appear as long slots or ovals.

You can readily see that whenever a section or ribbon of sections is smaller than the slot, or when particulate samples are smaller than the hole, some electron transparent material (film) will be required to maintain the specimen in position on the grid.

Suitable electron transparent support films can be constructed of carbon evaporated on a freshly cleaved sheet of mica or a more easily prepared plastic film. There are two plastics which are commonly used for support films: Collodion and Formvar (polyvinyl formal). Both can be produced by the same technique and with reasonable ease.

Collodion is less hydrophobic and less stable in the electron beam than is Formvar. Therefore, the specimens will aggregate more on a Formvar film, but that film will sublime less in the electron beam.

While there are a number of techniques for the preparation of these plastic support films, you are asked to prepare suitable examples by the 'cast on glass' method. These support films will be used in a subsequent exercise to examine bacterial specimens. Please read these directions in their entirety before commencing the exercise.

D. PROCEDURE:

1. Assemble all material items at your work station.

2. Clean several glass microscope slides, but allow a soap film to dry on the slide. Some prefer to wipe it clean with the fingers which allows the deposition of a slight film of oil on the slide.

   This soap or oil facilitates the stripping of the plastic film from the glass.

3. Fill a small (50 or 100 ml) beaker with the Formvar or Parlodian solution. Dip the glass slide into the solution and air dry the slide in a dust free environment under cover (See Figure 31-a-b).

   You may wish to return the solution of plastic to its container at this time. Check with the instructor. Evaporation of the solvent will lead to an increase in the concentration of the plastic.

   The drying time will increase with an increase in the relative humidity. The lower edge of the slide will have the thickest film and will be the last part to dry. A dessicator might be of considerable help on a humid day.

4. Fill the 75x150 mm culture dish or finger bowl to overflowing with demineralized water. Sweep the surface with a glass stirring rod or applicator stick to remove any dust particles.

   Oblique illumination from a desk lamp will assist your evaluation of the next steps.

5. Select a clean sharp razor blade to cut the film about 1 mm from edge of the slide as well as the bottom and the top of the film (See Figure 31-c).

6. Breathe on the film you have just scored and introduce the slide into the water at an angle of from 30° to 45° (See Figure 31-d). This manuever should be slow and steady.

   The surface tension of the water will float the film as the slide is introduced into the water. When the film is free, allow the slide to drop to the bottom of the dish.

Examine the film for variations in thickness. If bright interference colors, dust, or wrinkles are present, you should discard the film and try again (See Figure 31-e).

7. Athene type grids are cleaned by rinsing them in a glass petri dish with acetone. The solvent removes the oils associated with the etching process and also reduces the static charge on each grid.

   When the grids are removed from the acetone bath they are dried on filter paper. As they are transferred with jeweler's forceps you should make the matte side slightly convex.

8. The grids are then transferred individually to the plastic film 'raft' floating on the water. They are placed matte (convex) side down on the film. Slight pressure may 'dimple' the film. Do not crowd the grids on the film (See Figure 31-f).

9. Select a clean large (about 30 mm diameter) test or centrifuge tube and bring it down upon one end of the film, simultaneously depressing the film into the water and rolling the tube. This winds the 'raft' onto the test tube (See Figure 31-g-h).

10. Stand the tube in a dust free environment to dry.

11. Evaluate the grids and remove them to a petri dish with filter paper and double stick tape for storage.

12. These support films can be stablized by the evaporation of a carbon film on their surface (See Exercise Number 32); or they may receive particulate specimens immediately (See Exercise Number 23 or 30 for negative staining; for metal shadowing (See Exercise Number 32).

13. Special 'holey films' are often employed for the diagnosis of astigmatism in the microscope.

    The introduction of small amount of water into the liquid Formvar will cause many holes of various diameters in the film because water droplets interfere with the polymerization.

This can be accomplished in several ways, the easiest is to breathe on the wet film as indicated in step 6.

Microdroplets from the warm humid breath will condense on the cooler slide and form small holes.

Subsequent steps remain unchanged, although the film should be shadowed to make it more permanent.

<u>A Few Words of Caution:</u>

Plastic films should be viewed within a month of preparation. The films are unstable in the electron beam and a low beam current and intensity should be used until the photography is ready to be done because there may be film and specimen drift as the energy of the electron beam is absorbed by the film.

If a higher resolution is required for your work, you will then require a carbon support film which is described in many other more complete manuals. Your instructor may want you to cast films on water. Again those instructions will be found in other laboratory manuals.

E. EVALUATION:

PRESENT TEST TUBES CONTAINING GRIDS TO THE INSTRUCTOR FOR EVALUATION. THEN REMOVE GRIDS AND STORE IN PETRI DISH ON FILTER PAPER OR DOUBLE STICK TAPE.

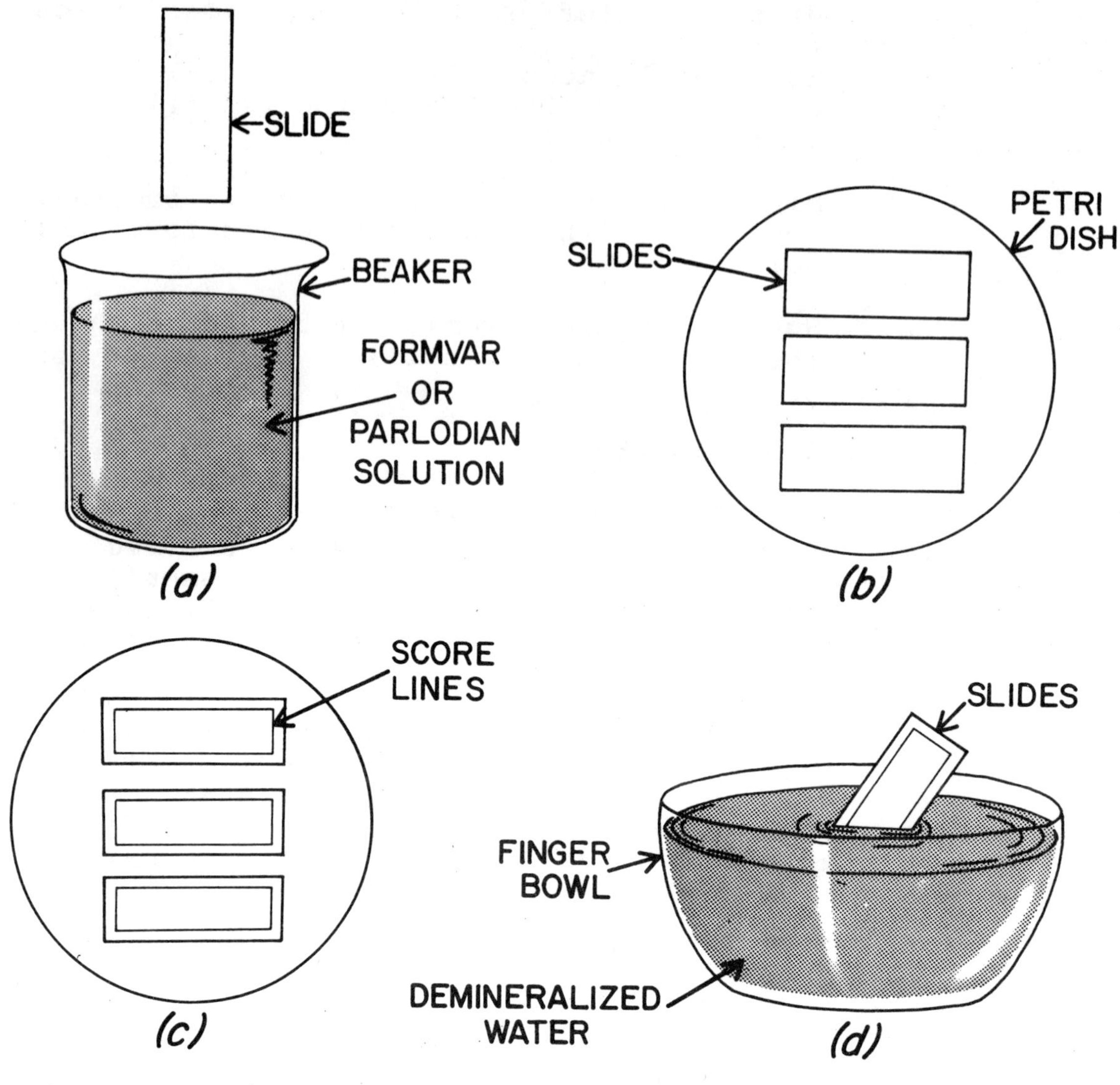

FIGURE 31-a-d. Preparation of support films for EM grids.

(a) Glass slide dipped into formvar or parlodian solution.

(b) Slides air drying in open petri dish in dust free environment.

(c) Scoring of plastic film with razor blade.

(d) Breathing on slide containing film and lowering it into demineralized water at a 30° to 45° angle.

TOP VIEW OF FINGER BOWL

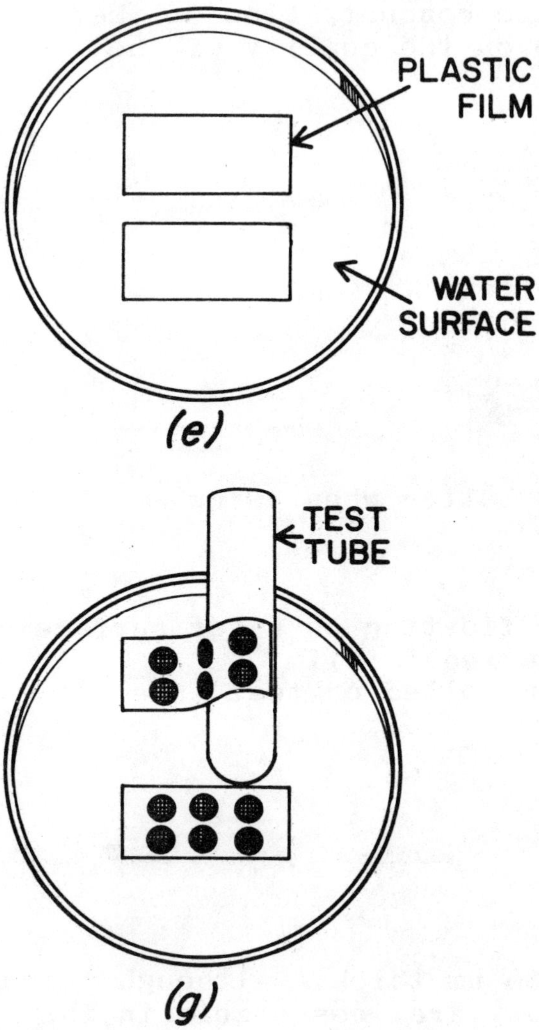

FIGURE 32-e-h. Transferring and collecting grids.

(e) Plastic film floating on water surface.

(f) Grids transferred to film.

(g) Grids on film rolled on test tube.

(h) Test tube with grids stored in a dust free environment.

F. QUESTIONS:

1. What effect will solute concentration of the plastic solution have on the quality of the support film?

2. Why is the resolution better when using a carbon support film?

(a) Plastic film floating on water surface
(b) Grids transferred to film
(c) Grids on film rolled on test tube

3. Plastic films are 20-30 nm thick. Although thinner films are possible, they are less stable in the electron beam. Why is this true?

4. What role does the hydrophobicity of the support film play in the selection of an appropriate film such as carbon, Formvar, or collodion?

EXERCISE NUMBER 32

PRINCIPLES OF VACUUM EVAPORATION

Frederic A. Giere, Ph.D.

A. OBJECTIVES:

1. To evaporate carbon and a metal in vacuum
2. To produce a carbon stabilized plastic support film
3. To metal shadow a particulate specimen

B. MATERIALS:

1. Grids with a plastic support film
2. Plastic support films with specimens of particulate matter (bacteria or viruses)
3. Vacuum evaporator equipped with:
   a) spectroscopic grade carbon rods
   b) tungsten baskets
   c) metal wire to evaporate: platinum, platinum-palladium, gold, etc.

C. RATIONALE AND DISCUSSION:

A carbon film is the best support that can be used in high resolution electron microscopy. The films are amorphous and very strong. In this exercise a carbon coat will be applied to a plastic support film. Many investigators will subsequently dissolve the plastic support from under the carbon. That takes a long time and will not be done here.

As mentioned earlier, a carbon film deposited on a

freshly cleaved sheet of mica is the best support for high resolution microscopy. Although the carbon film is amorphous, in our case the imperfections of the plastic film will be faithfully reproduced in the carbon film.

The apparatus to be used consists of a vacuum bell jar which is evacuated by rotary and oil diffusion pumps. High voltage cables enter the bell jar and connect the power supply to electrodes of spectroscopically pure carbon rods. One electrode is spring-loaded so that the carbon arc is maintained. It is often pointed and contacts the blunt immobile electrode.

The target films are placed 10-15 cm below the electrodes. Place a clean porcelain plate (cup, crucible) with a drop of diffusion pump oil or vacuum grease on it near the target. No carbon will adhere to the oil while the porcelain will accept the carbon. This provides a good visual index to the thickness of the carbon film being deposited.

The chamber should be evacuated to $1x10^{-5}$ Torr (follow the operating directions for your instrument since the sequence of valve closings and openings is critical) and 30 amp at 15 VAC is passed through the electrodes. The carbon rods will heat to incandescence. DO NOT LOOK AT THE ARC WITHOUT A PIECE OF COBALT GLASS AS A SCREEN!!

The thermal activity will volatilize some carbon atoms and they will travel in a line of sight to be deposited on a cooler surface (bell jar, plate and targets). A continuous film should be formed in less than 1/2 minute. Some technicians prefer pulses of current rather than a continuous flow claiming that the deposition rate can be controlled with greater precision.

When a satisfactory film has been deposited the current shall be turned off and the system returned to atmospheric pressure. Again, FOLLOW THE DIRECTIONS FOR VALVE CLOSURE AND OPENING provided by the manufacturer.

The results obtained in metal shadowing can be likened to the view one might have in flying over a village or landscape covered with fresh snow, a morning or afternoon sun and the observer looking straight down at the scence. Houses can be identified as ranch, two story, split-level, and the landscape plantings can be identified all on the bases of the shadows cast by those structures.

Similarly, electron transparent material, e.g. bacteria, can be coated at an oblique angle with a metal film which is relatively electron opaque with respect to the area in the shadow of the particle. The size and shape of the bacterium can then be determined by assessing the shadow of the specimen.

In this procedure the cable from the power supply enters the evacuated bell jar and terminates at two posts which in turn are connected by a length of tungsten wire fashioned into a cup or cone. The metal to be volatilized is placed in that cup and the vaporized metal will travel in a line of sight to be deposited on the cooler surface in the chamber.

Any elevation from the surface will intercept the line of sight and will create a 'shadow' where no metal is deposited. When the grid is placed in an electron beam there will be mosaic of electron opaque areas (where the metal film is present) and an electron transparent area (the shadow). Much information about size, shape, the presence of flagellae or pili can be garnered with the use of this technique.

D. PROCEDURE:

a) Carbon films

1. Start the mechanical and diffusion pumps to evacuate the vacuum system. Follow the directions on your instrument as the sequence of valve openings and closings is critical!

2. Place the plastic coated grids (the target) 10-15 cm below the carbon electrodes. They should be affixed to some support attached to filter paper on a Petri plate with double stick tape) so that they are not displaced when the bell jar is evacuated.

   The filter paper they are supported on can serve as a reference for the amount of carbon deposit or a 2nd piece of filter paper can be placed adjacent to the grids (see figure 32-a).

3. Check that the power supply is connected to the carbon electrodes.

4. Adjust the carbon electrodes so that the points touch or that the spring-loaded pointed carbon is against the immobilized blunt

FIGURE 32-a. Diagram of carbon rod arrangement for the deposition of carbon films.

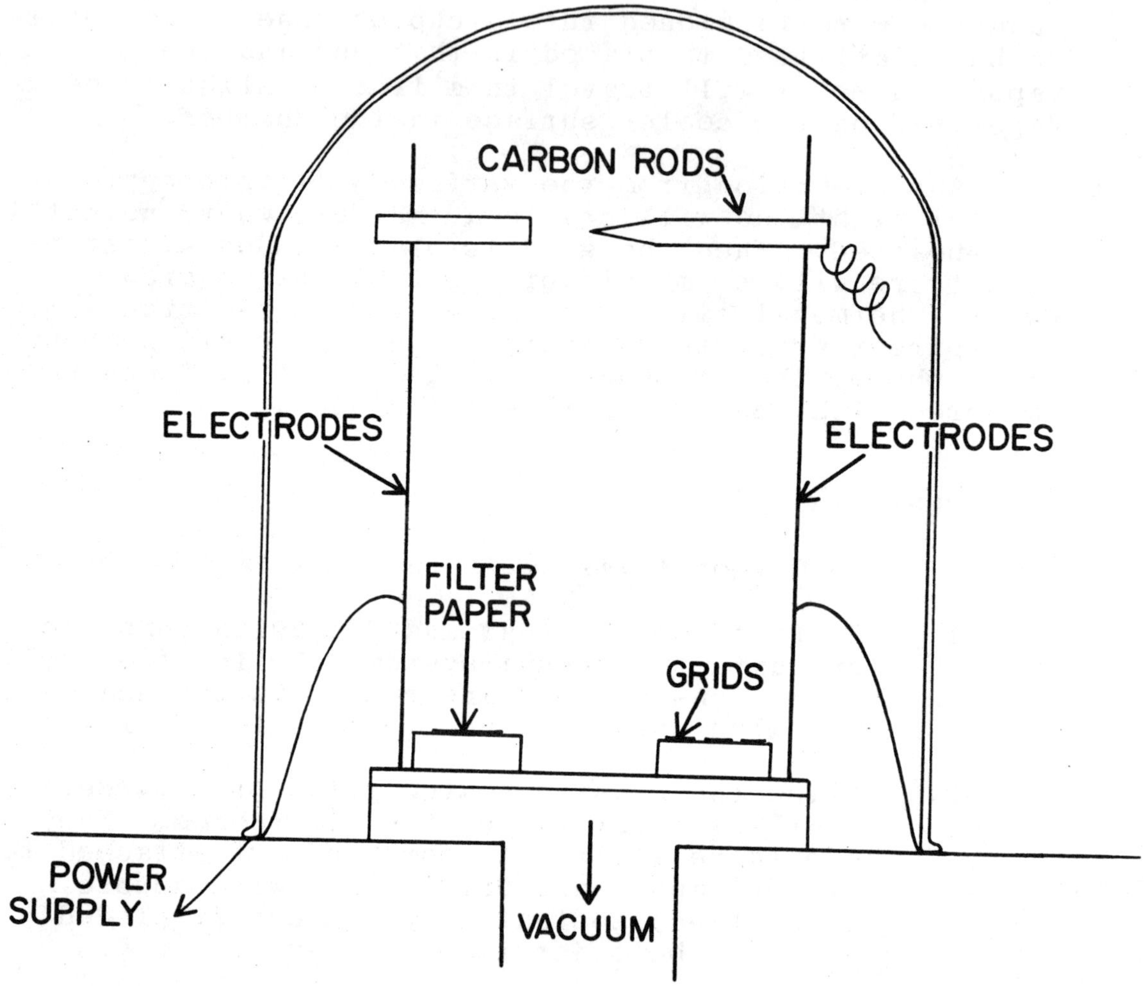

carbon. Check the porcelain reference for film thickness.

5. Evacuate the system to 1×10^{-5} Torr.

6. Energize the electrodes. Pass 30 amps at 15 VAC for about 30 seconds, although more or less may be required. The actual deposition should be monitored visually. DO NOT LOOK AT THE ARC EXCEPT THROUGH A COBALT GLASS PLATE.

7. When the film thickness is adequate, disconnect the power to the electrodes and bring the bell jar to atmospheric pressure.

8. Particulate samples may be placed directly on these carbon-stabilized grids immediately by the drop, float, or spray method described elsewhere in this manual.

b) Metal Shadowing

1. Start the mechanical and diffusion pumps to evacuate the vacuum system. Follow the directions on your instrument as the sequence of valve openings and closings is critical.

2. Place the grids with specimens (target) about 10-15 cm lateral to the metal source and inclined about 30° (25° - 35°) toward the source.

   This distance is required to avoid heat damage to the specimens during the evaporation of the metal. The grids must be held in place by double stick tape to a porcelain plate (to be used for the index of film thickness as with the carbon deposition).

   This will prevent them from being displaced by suction when the rotary pump valve is opened or by vibration from the pump.

3. A 10-15 mm length of wire (0.25 mm diameter) to be evaporated is coiled and placed in the tungsten wire basket.

4. Evacuate the system to at least 1×10^{-4} Torr. The greater the vacuum, the better the thin film.

5. When a current is passed through the tungsten

basket, the temperature of the basket will increase toward the melting point of the contained metal wire.

The metal atoms will vaporize and travel in a line of sight to a cooler surface where they will lose their excess thermal energy and eventually form a continuous thin film.

Ideally, the film thickness should be minimal but continuous (See Figures 32-b-c).

6. The white hot source is very bright.

   DO NOT LOOK AT THE SOURCE WITHOUT PROTECTING YOUR EYES WITH COBALT GLASS.

7. When the film thickness is sufficient as judged by the contrast between the diffusion pump oil drop on the porcelain and the oil-less area, disconnect the power supply and bring the system to atmospheric pressure.

   FOLLOW THE MANUFACTURERS INSTRUCTIONS FOR VALVE OPENINGS AND CLOSINGS.

8. The specimens may be observed in the transmission electron microscope immediately.

FIGURE 32-b. Metal shadowing technique.

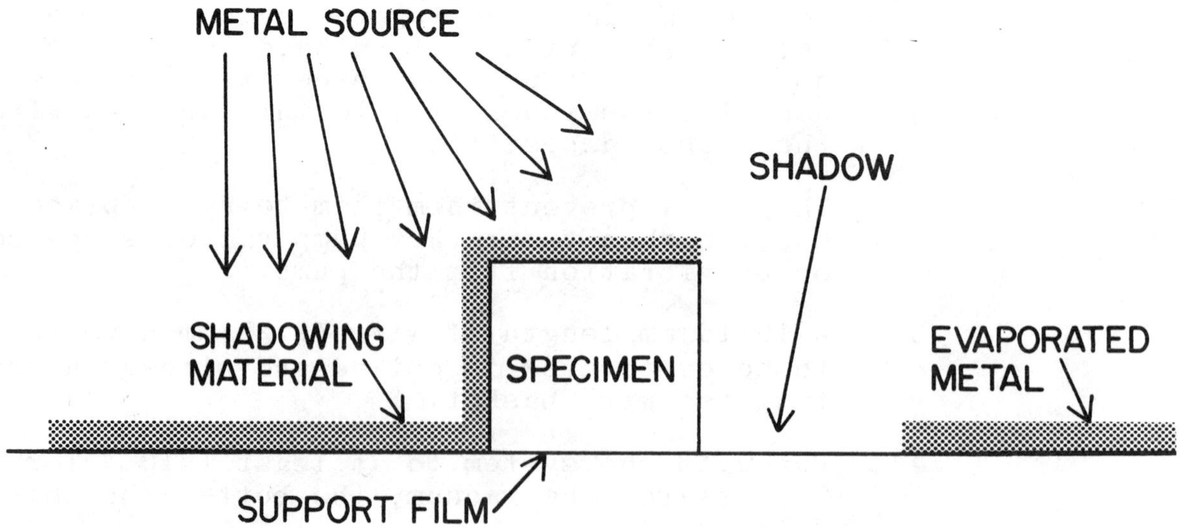

FIGURE 32-c. Metal shadowing technique with coiled tungsten wire basket attached to electrodes.

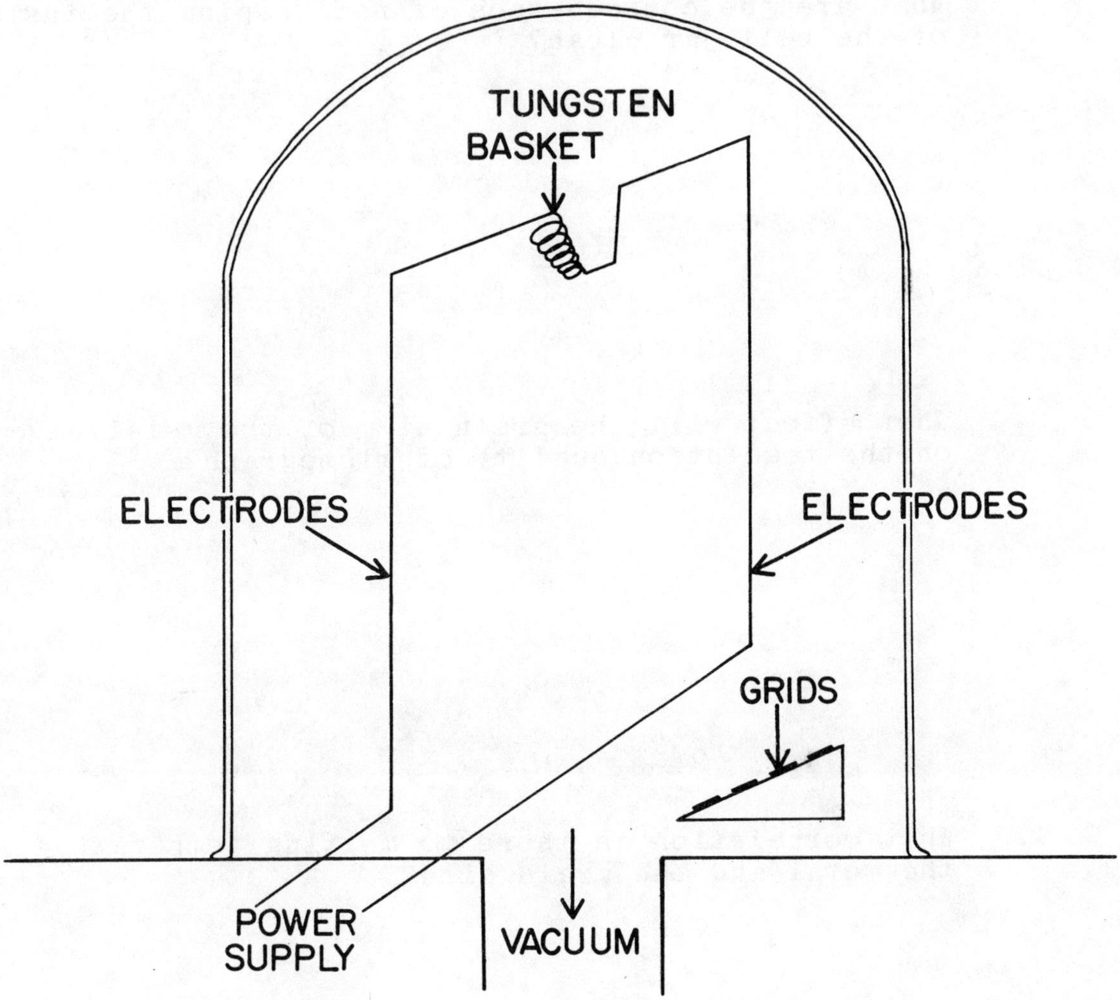

E. QUESTIONS:

1. What would be the appearance of a grid which was shadowed using a current and short exposure relative to one shadowed with an adequate current for volatilizing the metal but a longer shadowing time?

2. The bell jar will build up layers of metal-carbon-water vapor from the atmosphere etc. unless there is a frequent cleaning of the inside of the jar. What are the consequences of not keeping the inside of the bell jar clean?

3. What effect will the grain size of the metal have on the resolution quality of micrograph?

4. What correlation is there to melting temperature of the metal and the grain size?

5. Why does one use spectroscopic grade carbon as the electrode?

6. What are the relative merits of metal shadow casting _vis a vis_ negative staining?

EXERCISE NUMBER 33

PREPARATION OF DEOXYRIBONUCLEIC ACID (DNA) FOR ELECTRON MICROSCOPY

Dave P. Allison, Ph.D.

A. OBJECTIVES:

1. To visualize the molecular configuration of DNA, (i.e. circular or linear)
2. To measure the length of linear DNA or the circumference of circular DNA molecules
3. To identify adenine-thymine rich regions in DNA molecules
4. To construct heteroduplex DNA molecules and identify single stranded unpaired regions

B. MATERIALS:

I. Dialysis Bags

a) Parlodion (strips) Mallinckrodt

b) Ethyl ether

c) Ethyl alcohol (100% undenatured)

II. Nitrocellulose Film on Grids

a) Copper grids (400 and 200 mesh) (Fullam)

b) 10% nitrocellulose solution (Fullam)

c) Amyl acetate (Mallinckrodt)

d) Buckner Funnel (9 or 10 cm)

e) Whatman filter paper #1, 9.0 cm

f) Pasteur pipettes 5-3/4 inches long

III. Spreading DNA

a) Formaldehyde 37% (Mallinckrodt)

b) Formamide 99% (Mallinckrodt)

c) Cytochrome C, equine heart (Cal Biochem)

d) HCl

e) EDTA

f) Tris base (hydroxymethyl) aminomethane

g) Sodium carbonate

h) Glass microscope slides

i) Tris-HCl (hydroxymethyl) aminomethane hydrochloride

j) Glass scintillation vials

k) Glass test tubes disposable (10 x 75 mm)

l) Amylamine

m) Plastic petri dishes 100 x 15 mm (Falcon 1001)

n) Ammonium acetate

IV. Staining and Shadowing DNA

a) Uranyl acetate (Fisher)

b) Ethyl alcohol (100% undenatured)

c) 2-Methylbutane (isopentane) (Eastman Kodak)

d) Platinum wire 8 mil diameter (Fullam) (Ladd)

e) Tungsten wire 65 mil diameter preferred red, however 30 mil wire will suffice (Fullam)

V. Miscellaneous Supplies

a) Map measurer, Keuffel and Esser catalog #620300

b) Glass petri dishes (Kimble, Corning, Pyrex) three sizes that will hold 4.5, 9.0, and 12.5 filter papers Whatman #1 or 40

c) Glass disposable micro pipettes 10, 20, 50 λ made by Corning, order through Scientific Products

d) Tape - both sides sticky, 2 inch wide

e) Forceps - Pelco catalog #510 self-closing

VI. E.M. Supply Catalogs

(See List of Suppliers in Appendix A of this manual).

C. RATIONALE AND DISCUSSION:

The electron microscopic visualization of nucleic acids cannot be reconciled to a single experimental technique. Therefore, the method used to prepare a DNA sample will depend upon the information to be gained from the sample. In this laboratory exercise four basic techniques are described beginning with the simplest and therefore the most likely to succeed and proceeding to the most complex where failure will be more probable for the beginning student. It is left to the discretion and charitability of the instructor offering the course to select the technique or techniques appropriate for the level of student being reached.

1. Aqueous Technique

The aqueous variation of the basic protein film technique (1) is the simplest and most fail safe method for looking at double-stranded DNA. The DNA filament is coated with a basic protein (usually cytochrome C), spread onto a water surface (cytochrome C film), picked up on electron microscope grids, stained with uranyl acetate and shadowed with a heavy metal. The normally 2 nm (20 A^0) wide DNA molecule may attain a width of 30 nm (300 A^0) due to coating with cytochrome C, staining with uranyl acetate and shadowing with a heavy metal.

2. Formamide Technique

The formamide technique (2) is a modification of the basic protein film technique that has the advantage of rendering areas of single stranded DNA visible. In the aqueous technique single-stranded DNA condenses or collapses because of random base interactions. The addition of formamide has the effect of melting out random base interactions so that the single-stranded DNA appears much like double-stranded DNA except that it may be somewhat thinner and will usually exhibit more abrupt changes in direction such as right angle bends. The spreading solution or hyperphase contains 40% formamide in addition to buffered DNA and cytochrome C. This solution is spread onto a hypophase containing 10% buffered formamide.

3. Inman-Schnös Technique (3)

This DNA spreading method is another modification of the basic protein film technique. The addition of formaldehyde to the hyperphase prevents renaturation of partially denatured DNA. Therefore, this is the method of choice for doing partial denaturation mapping of DNA using either heat or pH as the denaturing agent. Another advantage of this technique is that the DNA does not become as heavily coated with cytochrome C as in both the aqueous and formamide techniques. This makes differentiating between double and single-stranded DNA much easier than with the formamide technique. Also, because of this lighter coating, denatured areas as small as 60 base pairs can be resolved.

4. Dubochet Technique (4)

This method does not involve a protein film which means that the width of the DNA molecule is not increased by the deposition of cytochrome C or other basic protein additives. Therefore, naturally occurring protein-DNA complexes such as histones, nucleosomes and polymerase molecules can be seen associated with the DNA. The difficult part of this technique involves rendering the carbon-coated electron microscope grids hydrophilic by some means. Otherwise, DNA placed directly onto the grids will form clumps and aggregates instead of spreading out onto the carbon film. The Dubochet technique employs the use of glow discharge under vacuum in the presence of amylamine for the deposition of basic radicals onto carbon coated electron microscope grids. These charged radicals then absorb the DNA to the grid in a spread condition.

D. METHODS:

I. Spreading DNA in a Plastic Petri Dish

NOTE: THIS METHOD IS USED WITH ALL OF THE BASIC PROTEIN FILM TECHNIQUES (AQUEOUS, FORMAMIDE, INMAN-SCHNÖS).

a) Use any 100 x 15 mm circular petri dish.

Use a glass microscope slide that has been cleaned in chromic acid, rinsed extensively in distilled water, and stored in distilled water.

The slide will serve as a ramp for the deposition of hyperphase containing DNA.

b) Pipette 30 ml of hypophase into the plastic petri dish.

NOTE: THE TERM HYPERPHASE APPLIES TO THE SOLUTION CONTAINING THE DNA, WHILE HYPOPHASE APPLIES TO THE LIQUID THE HYPERPHASE WILL SPREAD ONTO.

BOTH OF THESE ARE LITERATURE TERMS USED EXTENSIVELY BY DNA MICROSCOPISTS.

For the aqueous technique the hypophase is 0-25 M ammonium acetate pH 7.5, while for the formamide technique in addition to 0.25 M ammonium acetate buffer the hypophase contains 10% formamide; and finally using the Inman-Schnos technique the hypophase is distilled water.

c) Insert the clean glass slide into the hypophase, resting one end at the bottom center of the petri dish and the other end on the side of the dish, thereby forming a ramp at a roughly 30^0 angle with the surface of the hypophase.

d) Allow the slide to air dry for about 5 minutes prior to spreading the DNA.

If the progress of the DNA protein film is to be monitored, a light dusting of talc should be made to form a line from one side of the petri dish to the other at about half a centimeter from the slide/hypophase boundary.

Apply the talc by dipping a Q-tip into talcum powder and depositing the talc onto the hypophase much like tapping the ashes off a cigarette.

e) With a micropipette deliver about 40 ml (0.04 ml) of the hyperphase spreading solution to the slide in a back and forth motion roughly 5 millimeters above the surface of the hypophase.

NOTE: THE SPREADING SOLUTION WILL GLIDE DOWN THE SLIDE RAMP AND THE PROTEIN-DNA FILM WILL SPREAD ONTO THE HYPOPHASE SURFACE.

IF TALC IS USED, THIS SPREADING FILM CAN BE SEEN BY THE MOVEMENT OF THE TALC.

f) Pick up a nitrocellulose-filmed grid (Methods II) by its edge with self-closing tweezers.

g) With the nitrocellulose side down the grid is touched to the DNA-protein about 1 to 2 grid widths from the slide/hypophase boundary.

h) Rinse the grid by touching its meniscus to a beaker filled with 50% ethanol, then 95% ethanol.

i) Stain the grid by immersion for 30 seconds in uranyl acetate staining solution (Methods IV), then rinse in 90% ethanol by rapid up and down movement.

NOTE: FOR BEST RESULTS THE GRID SHOULD BE ALLOWED TO DRY UNTIL AFTER THE FINAL RINSE IN 90% ETHANOL.

j) Rotary shadow the grid with platinum in a vacuum evaporator (Methods V).

II. <u>Nitrocellulose Film on Grids (Buckner Funnel Method)</u>.

a) Place a 9-cm Buckner Funnel fitted with a 15 cm length of rubber tubing loosely into a ring stand set up in the sink.

b) Wet the funnel with distilled water and place a 9-cm Whatman #1 or #40 filter paper at the bottom.

c) Place 2 cm of water into the funnel, seat the filter paper by allowing some water to run through the funnel, and clamp off the rubber tubing.

d) With clean tweezers, place 200 mesh copper grids, shiny side up, on filter paper.

e) With a pipette, introduce one drop of 2.5% nitrocellulose in amyl acetate onto the water surface.

This forms a film which dries in 3-5 minutes.

f) Release the clamp allowing the water to drain out.

NOTE: THE FILM WILL DESCEND ONTO THE GRIDS AND THE FILTER PAPER CAN BE PICKED UP AND STORED IN A GLASS PETRI DISH WITH THE GRIDS UP.

IT IS IMPORTANT TO PREPARE NITROCELLULOSE-COVERED GRIDS IN THIS MANNER FOR ALL PROTEIN FILM TECHNIQUES.

III. Preparation of Dialysis Bags

a) To make the dialysis bags dissolve 7% parlodion in a 1:1 mixture of ethyl alcohol: ethyl ether.

b) Chop the parlodion into small pieces about 2 mm square and dissolve a little at a time on a magnetic stirrer over a 2 day period.

c) Any size test tube can be used as a mold for dialysis bags. For DNA work a 10 x 75 mm tube is adquate.

d) Dip the test tube bottom first into the nitrocellulose solution to within 2 cm of the tube top.

e) Withdraw the tube and invert it onto a peg, such as a pipette held in a test tube rack.

f) When dry, cut around the top of the tube with a razor blade.

g) Immerse the coated test tube in 95% ethanol.

h) Withdraw and peel off the dialysis bag like a sock.

EDTA (0.1 M)

3.72 g disodium ethylenediamine tetraacetate (M.W. 372.24)

dissolve in 100 ml H_2O

Uranyl Acetate (0.05 M) Stain (5)

1.94 g uranyl acetate (M.W. 388.12) dissolved in 100 ml 0.05 N HCl.

Must be kept in light proof bottle. Stain diluted 1:1000 into 90% ethanol (i.e. 0.01 ml uranyl acetate stain into 10 ml 90% ethanol in a glass scintillation vial).

V. Rotary Shadowing with Heavy Metal

NOTE: REFER TO EXERCISE NUMBER 32 OF THIS MANUAL FOR A SLIGHTLY DIFFERENT PROCEDURE AND ADDITIONAL INFORMATION.

a) After staining, enhance specimen contrast by shadowing with platinum, platinum:palladium 80:20, or palladium:gold 40:60.

b) Bend a 8 to 10 cm length of tungsten wire cleaned by rubbing with fine emery paper into a v shape at its center.

c) Wrap the point of the v with 2 to 3 cm of 8 mil platinum wire or other suggested alternates.

d) Place the wrapped tungsten wire in petroleum ether for 10 minutes to remove any grease or dirt.

e) Secure the tungsten-platinum filament in a holder in a vacuum evaporator.

f) Place the grids to be shadowed on the rotary turntable in the vacuum evaporator and shadow at an angle of 7^0 to 10^0 at a distance of 10 cm from the v source to the center of the rotary turntable.

g) Attain a vacuum of 10^{-5} to 10^{-6} torr to ensure goods shadowing.

VI. Denaturation and Reannealing of DNA (6)

In this procedure double stranded DNA is denatured or separated into two complementary single strands. Under controlled conditions the separated strands are allowed to reanneal or renature.

If the DNA is homogeneous i.e. from one virus, bacteriophage, episome, etc. then the reannealed DNA is a homoduplex. If two similar but somewhat different DNAs are denatured and renatured, then some of the molecules constructed will be heteroduplex molecules containing a single strand of each of the related DNAs.

Where these constructed molecules are the same, base-pairing will occur and the molecules will appear double stranded. However, in areas where the two single strands are not complementary, the bases will not pair and the DNA will appear single stranded.

a) Denaturation

100 μl of 0.1 M EDTA pH 8.0

250 μl of H_2O

50 μl of DNA [total concentration of heteroduplex DNA is 5 μg of each of two similar DNAs] total volume at this step 400 μl

50 μl 1.0 N NaOH

↓

DENATURE 10 min at room temperature

↓

b) Renaturation

50 μl of 1.8 M Tris-HCl + 0.2 M Tris-base

500 μl Formamide

↓

RENATURE 1 to 2 hours at room temperature

Further renaturation is prevented by dialysis against

Tris (0.01 M) + EDTA (0.001 M) pH 8.0

VII. Preparation of Carbon Films

NOTE: FOR ADDITIONAL INFORMATION AND A SOMEWHAT DIFFERENT PROCEDURE REFER TO EXERCISE NUMBER 32 OF THIS MANUAL.

a) Carbon films for EM can be prepared by evaporating carbon onto freshly cleaved mica.

Cleave a rectangular piece of mica roughly 3 cm wide and 6 cm in length and 0.2 to 0.5 mm thick at its edge by working a sharp object such as a razor blade, pen knife or even a fingernail between the mica layers.

b) Bend the thin layers in opposite directions, so the mica sheets will snap apart.

c) Place the mica in the vacuum evaporator cleaved side up, at a vacuum of 10^{-5} torr, and evaporate a thin film of carbon onto it.

d) Float the carbon off the mica and onto a water surface very slowly inserting the mica into a vessel of water at an angle of not greater than thirty degrees.

A pyrex container with adequate room to work is sufficient.

If a good film is formed it should be reasonably strong and gently blowing on it should not break it apart.

NOTE: WE FIND THAT RAINY OR OTHERWISE HIGH HUMIDITY DAYS SEEM TO BE BETTER FOR MAKING CARBON FILMS.

e) Place this film onto copper grids individually by picking the edge of a grid up with tweezers, submerging it to one side of the floating carbon film, and then lifting the grid up under just enough of the carbon film to completely cover it.

f) Place each grid with the carbon film side up on filter paper to dry.

Perhaps a hundred carbon grids can be collected from one good film.

g) Use 400 mesh or small grids.

NOTE: THESE FILMS ARE FRAGILE!

THE BEST CARBON FILMS ARE THIN TO THE POINT OF BEING DIFFICULT TO SEE ON A WATER SURFACE.

VIII. Glow Discharge in a Vacuum (4)

Glow discharge in a partial vacuum of amylamine vapors makes carbon films suitable substrates for mounting preparations of nucleic acid and proteins.

The vacuum evaporator must be equipped with a glow discharge unit. Basically an AC power supply capable of generating 500 volts is needed.

Usually the hardware within the evaporator consists of two circular electrodes 6 to 10 cm in diameter positioned one above the other 5 to 10 cm apart (most units will have an adjustment option).

a) Place the carbon films within the lower electrode.

b) By using the mechanical vacuum pump of the evaporator, evacuate the vacuum chamber to 10^{-2} torr, then seal off from the vacuum pump, and valve the amylamine vapors into it.

NOTE: THE VACUUM SHOULD DECREASE AND THE PRESSURE SHOULD INCREASE TO 0.5 TORR.

c) Use the mechanical pump to bring the vacuum down to 0.1 to 0.15 torr, and then isolate it from the system.

d) Apply 500 AC volts across the electrodes for 30 seconds.

NOTE: THE GRIDS SHOULD NOW BE RECEPTIVE TO DNA-PROTEIN COMPLEXES.

IX. Inman-Schnös Denaturation Buffer

This buffer is used for doing partial pH denaturation studies on DNA.

4 ml of stock Formaldehyde (37%)

0.4 ml of 0.126 M EDTA

0.32 ml of 1 M Na_2CO_3

The pH of this stock buffer is 9.4. Therefore, in order to be useful for denaturation experiments, the buffer should be titrated together with the usual vector buffer for the DNA.

This is done by taking 3 ml of stock Inman-Schnös buffer + 7 ml of the DNA vector buffer and titrating with 1 M HCl to go from pH 9.4 to pH 6.5. In titrating toward a pH that will partially denature DNA from pH 9.4 to pH 11.5, use 5 N of NaOH.

NOTE: DNA WILL USUALLY BEGIN TO DENATURE AT PH 10.5 TO 11.2. CONSTRUCT A COMPLETE TITRATION CURVE. AS AN EXAMPLE, USE A DNA VECTOR BUFFER OF 0.01 M TRIS + 0.001 M EDTA PH 8.0 (DNA DILUTE OR DIALYZE INTO BUFFER).

Inman buffer	vector buffer	5 N NaOH	pH
3 ml	7 ml	0.02 ml	10.0
"	"	0.06 ml	10.5
"	"	0.12 ml	10.9
"	"	0.18 ml	11.1

X. DNA Length Measurement

The molecular lengths of either circular or linear DNAs are calculated from length measurements made by tracing DNA molecules either directly from micrographs or from projected negatives with a map measurer (curuimeter). See Materials - Miscelleneous Supplies.

The final magnification of the DNA molecule must be known. As an example, if the negative of the DNA preparation is taken at a magnification of 6,000X in the electron microscope and the final micrograph is printed at an enlargement of 5X in the enlarger the final magnification of the micrograph is 30,000X.

In a like fashion, if the negative is projected and this projection magnification is 8X then the final magnification of the projected DNA molecules would be 6,000 x 8 = 48,000X.

a) Trace the length of DNA with the curvimeter.

NOTE: USUALLY THIS WILL BE RECORDED IN CENTIMETERS AND DNA LENGTHS WILL NORMALLY BE GIVEN IN MICRONS.

b) Since 1 centimeter is equal to 10,000 microns (micrometer), the traced DNA recorded in centimeters must be converted to microns by multiplying by 10,000.

c) To get the actual length of the DNA, the measured length of the DNA must be divided by the total magnification of the micrograph or the projected DNA molecules.

$$\text{DNA Length (In Microns)} = \frac{\text{DNA Measured In Microns}}{\text{Magnification}}$$

E. EXPERIMENTAL PROCEDURE:

I. Protein Spreading Techniques for DNA

The DNA should be free of detergents or anything else that will compete with the formation of a protein film. Dialysis (Methods III) against 0.01 M Tris + 0.001 M EDTA pH 8.0 is highly recommended.

a) Aqueous Method: to look at double-stranded DNA.

Spreading solution or Hyperphase

50 μl of 1 M Ammonium Acetate + 0.002 M EDTA pH 7.0

30 μl of H_2O

10 μl of DNA (concentration of DNA 3 to 5 μg/ml)

10 μl of 1.0 mg/ml aqueous cytochrome C

↓ 5 min at room temperature, then spread DNA (Methods I)

Hypophase

30 ml of 0.25 M ammonium acetate pH 7.5

b) Formamide Method: to look at double-stranded, single-stranded, or heteroduplex DNA

Spreading solution or Hyperphase

30 μl of H_2O

40 μl of Formamide 99%

10 μl of 1 M Tris + 0.1 M EDTA pH 8.0

10 μl of DNA (concentration of DNA 3 to 5 μg/ml)

10 μl of 0.5 mg/ml aqueous cytochrome C

↓

5 min at room temperature, then spread DNA (Methods I)

Hypophase

30 ml 10% Formamide in 0.01 M Tris + 0.001 M EDTA pH 8.0

c) Inman-Schnös Method: to look at both double and single-stranded DNAs --- the method of choice for doing partial denaturation mapping of DNA using either pH or heat as the denaturing agent

Spreading solution or Hyperphase

Ice bath

30 μl Inman buffer (adjusted to desired pH) (Methods IX)

70 μl of 0.01 M Tris + 0.001 EDTA containing 0.05 μg of DNA

↓

Incubate 10 min at room temperature - return to ice bath for 5 min

100 µl of cold Formamide 99%

20 µl of cold 1.0 mg/ml aqueous cytochrome C

↓ Room temperature 10 min, then spread DNA (Methods I)

Hypophase

distilled H_2O

d) Dubochet Technique: to look at DNA-protein complexes and for larger samples of DNA such as disrupted cell nuclei.

1. Use carbon filmed 400 mesh grids (Methods VII).

2. Make grids receptive to DNA-protein complex by glow discharge in amylamine (Method VIII).

3. Place a drop of the DNA-protein or disrupted nuclei either directly onto a grid, held by its edge with a self-closing tweezer, or float the grid carbon-filmed side down onto a drop of the DNA-protein complex.

4. Allow 1 to 5 minutes for absorption.

5. Wick off excessive DNA-protein by touching the edge of the grid to Whatman #1 filter paper.

6. Stain the grid with a 1% aqueous solution of uranyl acetate. This can be done either by placing a drop of stain onto the grid held in tweezers or by floating the grid onto a drop of stain. Remove excess stain by wicking with filter paper.

7. For added contrast grids may be shadowed (Methods V).

F. REFERENCES:

1. Kleinschmidt, A.K., D. Lang, D. Jackerts, and R.K. Zahn 1962. Biochem. Biophys. Acta 61: 857-864.

2. Westmoreland, B.C., W. Szybalski, and H. Ris. 1969. Science 163: 1343-1348.

3. Inman, R.B., and M. Schnös, M. 1970. J. Mol. Biol. 49: 93-98.

4. Dubochet, J., M. Ducommun, M. Zollinger and E. Kellenberger. 1971. J. Ultrastructure Research 35: 147-167.

5. Davis, R.W., and N. Davidson. 1968. Proc. Nat. Acad. Sci. 60: 243-250.

6. Davis, R.W., and J.W. Parkinson. 1971. Mol. Biol. 56: 403-423.

G. GENERAL REFERENCES:

1. Inman, R.B. 1967. J. Mol. Biol. 25: 209-216.

2. Studier, W.F. 1965. J. Mol. Biol. 11: 373-390.

3. Ferguson, J., and R.W. Davis. 1978. Quantitative Electron Microscopy of Nucleic Acids. In: Advanced Techniques in Biological Electron Microscopy edited by J.K. Koehler, Springer-Verlag, Berlin, Heidelberg, New York.

H. QUESTIONS:

1. You have a sample of presumed circular plasmid DNA that has been isolated from the bacterium *Escherichia coli* and you wish to determine both the relative degree of homogeneity and the molecular length. What steps would you take?

2. You find that the aforementioned plasmid DNA is indeed a 15-micron circular DNA and that your purification technique is excellent as judged by the relative homogeneity of the samples you have looked at. A friend of yours in Molecular Genetics Department tells you that she has successfully inserted the gene that codes for tallness in turkeys into this plasmid DNA using genetic engineering. She has also figured out that this gene codes for at least 5 different proteins and must have a DNA sequence 4 microns long. Very simply, how would you determine if she has successfully inserted this DNA into the plasmid DNA?

3. You now have the *E. coli* plasmid DNA and the *E. coli* plasmid DNA with the tall turkey gene inserted. You also have an endonuclease CT-1 that you know cuts the DNA at a single site. How would you go about physically locating the CT-1 site in relation to the inserted tall turkey gene?

4. You now have a fair amount of data about your circular plasmid and your plasmid + the tall turkey gene. Your major professor wants to publish an E.M. study on the origin and direction of DNA replication in the plasmid DNA. Really two models are likely. Either the rolling circle (starting from a fixed origin a linear piece of replicated DNA evolves and upon completion of replication it circularizes) or the Cairns circle (starting from a fixed origin the circular DNA is replicated in either one or both directions resulting in two complete circles). First, how would you determine which type of replication you have? Secondly, what E.M. experiments might you design to determine the origin and direction of replication in relation to the CT-1 endonuclease site?

EXERCISE NUMBER 34

ELECTRON MICROSCOPY OF TISSUE CULTURE

Sara E. Miller, Ph.D.

A. OBJECTIVES:

NOTE: THE USES FOR CELLS IN CULTURE ARE MANY AND VARIED. SOME GENERAL OBJECTIVES ARE AS FOLLOWS:

1. To study metabolic pathways and genetics
2. To study the interaction of pathogens such as viruses with host cells
3. To study the interaction of different treatments such as drugs or toxins on cells
4. To study transformed cells as models for cancer
5. To study the amplification of viruses from suspected specimens in diagnostic virology
6. To study attachments, communications and interaction between cells or cell types such as between neurons and muscle cells or immunological responses

B. MATERIALS:

NOTE: SEE RECIPES IN APPENDIX A FOR PREPARATION OF SOLUTIONS.

1. Cells in suspension

   a) 2% Glutaraldehyde in PS buffer

   b) Centrifuge and tubes

   c) Small spatula

   d) Filter paper

   e) 2% Agar

   f) Phosphate-sucrose (PS) buffer

g) Pasteur pipettes and bulbs

h) 1% Osmium tetroxide in PS buffer

i) Veronal acetate (VA) buffer

j) 1% Uranyl acetate

k) Graded ethanols

l) Epoxy resin

m) Oven

2. Cells in monolayer

a) 1% Glutaraldehyde in PS buffer

b) Items f-m in (1) above

3. Whole cells

a) Cells on formvar-coated gold grids

b) 1% Glutaraldehyde in PS buffer

c) Items f-k in (1) above

d) Critical point drier

e) Carbon coater

4. Cultured cells for SEM

a) Cells on 9 mm x 9 mm coverslips

b) Items b-d in (3) above

c) Sputter coater or vacuum evaporator

C. RATIONALE AND DISCUSSION:

Techniques for growing cells in culture are numerous and vary for different cell lines, thus the subject is far too broad to be covered here. Many books are available on the maintenance of tissue culture (1-8). Briefly, adherent cells (cells which attach themselves to a substrate) are grown on flat-sided flasks or Petri dishes (Corning, Fisher, Linbro, See Appendix B) at 37^0C in an incubator with an atmosphere of 5% CO_2. They are bathed in a buffered medium which supplies all the necessary nutrients and growth factors.

A typical medium is minimal essential media (MEM) (9; GIBCO, See Appendix B) including 10% fetal bovine serum. Once every week or 10 days (depending on the density and growth cycle), cells should be fed (old media replaced with fresh) or passaged (detached from the substrate, usually with trypsin and/or ethylenediaminetetraacetic acid (EDTA), divided into 4-6 new containers and refed). Non-adherent cells are cultured in similar conditions except that feeding and passaging may require sterile centrifugation to remove the cells from the old media.

Since most cells are finicky and require precise, and aseptic techniques, one should consult a good manual or an experienced tissue culture technician before attempting to grow them. However, assuming one is able to obtain cultured cells, preparing them for electron microscopy (EM) is similar to working with tissue.

D. PROCEDURE:

I. Transmission Electron Microscopy of Cells in suspension

1. Melt 2% agar

2. Add a volume of 2X fixative at 37^{0}C equal to the volume of the media. Mix gently.

3. Fix for 15 min. (Cells can be refrigerated and stored for several days if necessary).

4. Centrifuge at 2000 g 15 min.

5. Remove supernatant with a Pasteur pipette.

6. Trim a piece of Whatman #1 filter paper or paper towel so that the width will fit into the test tube.

   The length should reach the cells, and the tip of the paper should be pointed. Fold a crease down the length to give it strength.

7. Lay the tube on its side and just barely touch the top edge of the cell pellet with the paper (See Figure 1).

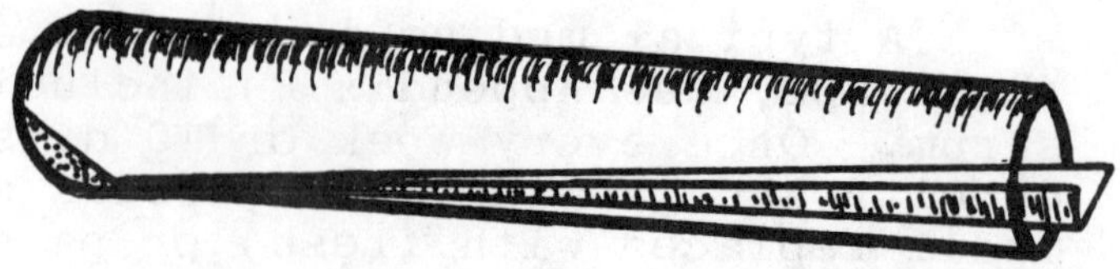

Figure 1.

8. Allow the tube to rest undisturbed until the liquid front stops moving up the strip (about 15 min).

   The cells at this stage should be a thick paste.

NOTE: DO NOT PERMIT THEM TO DRY OUT COMPLETELY.

   We have been able to work with as few as 10^6 cells, resulting in a clump of approximately 0.5 mm^3.

   However, more cells (10^7 - 10^8) give a pellet which is not easily lost.

9. Scrape cells with a small spatula or beveled applicator stick into 1-3 clumps about 0.5-1.0 mm^3.

10. With a pasteur pipette, remove some of the molton agar. Touch the pipette to assure that it does not feel hot.

    If it is still steaming or is too hot to touch comfortably, allow it to cool until it feels barely warm to the fingertips.

    At this point do not waste time because the agar will solidify at about 37^0C.

11. Form a small drop about 1 mm^3 at the pipette tip and gently touch a cell clump. Repeat with all clumps.

Do not allow the drop to fall from the pipette, or the agar surrounding the clump will be large and will slow penetration of additional solutions.

NOTE: IF EXCESS AGAR DOES FORM AROUND THE CELLS, SHAVE IT AWAY WITH A CLEAN RAZOR BLADE AFTER REMOVAL FROM THE TUBE.

12. After the agar solidifies (1-2 min), the pellets can be gently dislodged with a spatula and further processed in the tube or moved to a disposable vial.

13. Proceed washing and fixing the pellets as though they were pieces of tissue.

    Take care to remove solutions gently with a pipette to avoid crushing the agar.

14. Wash pellets 3 times for 15 min each in PS buffer (0.1 M phosphate buffer pH 7.4 containing 1.8% sucrose).

    SEE NOTE ON OSMOLARITY IN "RECIPES".

15. Fix 1 hr in a fume hood in 1% osmium tetroxide in the same buffer.

16. Wash 2 times, 15 min each in PS buffer.

17. Wash once for 15 min in 0.11 M veronal acetate buffer.

18. Stain 1-2 hr (not critical) in 1% uranyl acetate in VA buffer.

19. Rinse briefly in distilled water.

20. Dehydrate in a graded series of ethanols:

30%	15 min
50%	15 min
70%	15 min
95%	15 min
100%	15 min ea (3 changes)

    Make certain the 100% ethanol is anhydrous (See Note in Appendix A).

21. Infiltrate with resin at room temperature:

Spurr

50% resin - 50% ethanol	30 min
100% resin	30 min ea (2 changes)

Place in capsule with freshly made 100% resin.

Epon or Epon Equivalent

30% resin - 70% ethanol	30 min
70% resin - 30% ethanol	30 min
100% resin	30 min ea (2 changes)

Place in capsule with freshly made 100% resin.

22. Bake Spurr at 70^0C for 8 hr.

NOTE: DO NOT OVERBAKE OR SPECIMEN WILL BECOME BRITTLE.

IF USING AN OVEN ON A TIMER, SET IT FOR 7.5 HR.

BAKE EPON AT 60^0C FOR 24 HR; MORE TIME WILL NOT CAUSE DAMAGE.

II. Transmission Electron Microscopy of Attached Cells

A. <u>As Pellets</u>

Monolayer cells can be processed through uranyl acetate simply by bathing them in the dish with the appropriate fixative.

NOTE: FIXATION TIMES ARE THE SAME AS THOSE DESCRIBED BELOW IN "<u>IN</u> <u>SITU</u> EMBEDMENT".

If orientation is not important they can then be scraped with a cell scraper (Bellco, Costar, See Appendix B), a rubber policeman, or a piece of plastic tubing, pelleted, surrounded with agar, and embedded, as described in "Cells in Suspension".

B. In Situ Embedment

Many papers have been published on in situ preparation, involving growing cells on "cutable" substrates (10, 11) and film-lined dishes (See Appendix B), or separation of resin from plastics (12-17).

Most of these require a large amount of cells and reagents. We have developed an easy and economical technique which makes use of chamber slides (See Figure 2) which can be obtained with 1, 2, 4 or 8 wells (Lab Tek, See Appendix B).

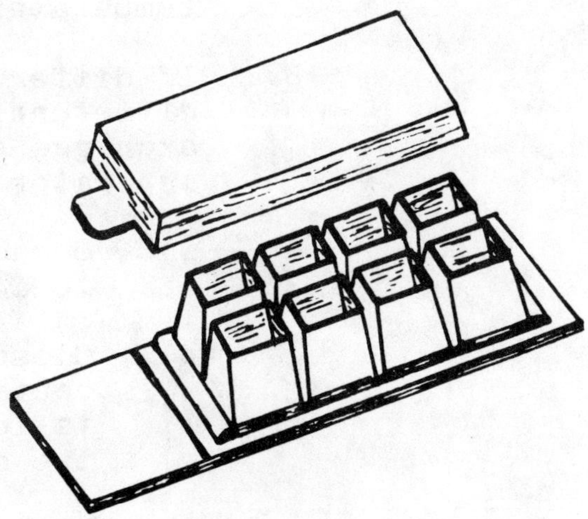

Figure 2.

These permit treatment of essentially the same cell population under several different conditions (fixations, virus concentrations, antibody reagents, etc.).

The cells can be embedded in situ, the sheet of resin peeled up while warm, and the squares cut with a razor blade and glued onto old blocks.

1. Monitor cell growth using an inverted phase light microscope.

2. When cells are confluent (touch one another in a monolayer with few areas devoid of cells, perform the desired experimental treatments (e.g. add different reagents or virus concentrations, etc.).

3. Fix at 37^0C by adding a volume of (warm) 2% glutaraldehyde in PS buffer (See recipe and note on osmolarity in Appendix A) equal to that of the media in the well(s).

4. Allow fixation for 10 min.

5. Decant and add 1% glutaraldehyde in PS buffer. Continue fixation on the bench at room temperature for 10 more min.

   Finish processing at room temperature.

6. If different wells are receiving different fixatives, a good way to exchange solutions is to use an aspirator in the following manner:

   a) Connect a side-armed Erlenmeyer to a vacuum with vacuum tubing.

   b) Close the top with a stopper containing a glass tube which is long enough to reach below the evacuation arm.

   c) With another piece of vacuum hose, connect a longtipped pasteur pipette to the glass tubing in the stopper.

NOTE: WHEN EVACUATING LIQUIDS FROM THE WELLS, DO NOT TOUCH THE CELLS WITH THE PIPETTE, RATHER, TIP THE SLIDE SO THAT SOLUTIONS RUN TO CORNERS AND HOLD THE PIPETTE IN THE WELL CORNER.

d) Immediately, without allowing the cells to dry (VERY IMPORTANT!), gently add the next solution by allowing it to run down the side of the well.

7. If all wells are receiving the same fixative after they have been treated differently, pop off the plastic well form.

   Leave the rubber gasket attached to the side.

8. Quickly immerse the slide into the appropriate solution in a Petri plate or Coplin jar.

9. Wash wells 3 times for 5-10 min each wash in PS buffer.

NOTE: SOME LABS HAVE ON HAND STOCKS OF PHOSPHATE BUFFERED SALINE (PBS).

THIS IS SUITABLE, BUT TEST THE OSMOLARITY TO INSURE THAT IT IS INDEED ABOUT 300 mOsm.

10. In a fume hood, fix slide in 1% osmium tetroxide in PS buffer.

NOTE: DO NOT OVERFIX !!

11. Rinse briefly, then wash once for 5 min in PS buffer.

NOTE: IF NECESSARY, CELLS CAN BE LEFT HERE OR IN VERONAL ACETATE BUFFER OVERNIGHT IN THE COLD.

12. Wash 2 times in 0.11 M veronal acetate (VA) buffer.

NOTE: OSMOLARITY ISN'T AS CRITICAL AFTER OSMIUM. SOME RESEARCHERS WASH WITH WATER FOLLOWED BY AQUEOUS URANYL ACETATE. SOME EVEN SKIP URANIUM; HOWEVER, WE OBTAIN BETTER PRESERVATION IF IT IS USED.

13. Fix 20-30 min with 1% uranyl acetate in VA buffer.

14. Dehydrate in graded ethanols:

30%	5 min
50%	5 min
80%	5 min
95%	5 min
100%	5 min ea (3 changes)

NOTE: BE EXTREMELY CAREFUL AT THIS STAGE NOT TO LET THE CELLS AIR-DRY.

ALSO, MAKE SURE THE ETHANOL IS ANHYDROUS. SEE NOTE IN APPENDIX A.

SAVE THE WASTE FOR CLEAN-UP.

15. While cells are being dehydrated, mix Epon 812, Embed 812, or Poly/Bed 812 components (See Appendix A).

16. Turn on oven to 55^0C. Monitor with a thermometer!

If the resin is baked at a higher temperature, it will harden completely and be difficult to separate from the slide.

17. Infiltrate with resin in ethanol and then resin:

30% resin - 70% ethanol	5 min ea (2 changes)
70% resin - 30% ethanol	5 min ea (2 changes)
100% resin	5 min ea (4-6 changes)

18. Decant the resin-ethanol mix into a waste container (<u>not down the sink</u>!).

That remaining at the corners of the gasket can be aspirated into a long-tipped pasteur pipette.

19. When the pipette becomes 1/3 - 1/2 full of foamy resin, discard and start a fresh one.

In this way the vacuum hose doesn't get clogged.

NOTE: IF RESIN DOES GET INTO THE HOSE, ASPIRATE SOME OF THE WASTE ETHANOL SAVED FROM DEHYDRATING.

20. The number of changes of 100% resin can be determined by observing the diffraction patterns of the interface between the fresh addition and that remaining on the slide.

When new resin looks homogeneous after swirling with the residue, change once more and make up fresh resin.

The increased numbers of changes is due to the lack of agitation as could be given a piece of tissue.

21. Make 1/2 the recipe if there are only 1 or 2 slides.

Allow one more 5 min change with the new resin, remove and add enough to fill the wells even with the gasket.

NOTE: DO NOT FILL SO THAT THE WELLS ARE ROUNDED OVER.

22. Aspirate all the excess outside the gasket and wipe the bottom thoroughly.

23. If you wish to embed a label in each well, record it in pencil, not pen, (ink will run) on white paper.

Cut it to fit into the well and push it underneath the surface.

It will not hurt the cells. However, we omit labels since the well eventually gets cut into many small pieces.

24. Lay slides onto an inverted plastic petri dish and place into the oven at 55^0.

25. The resin must be warm to peel up.

Before removing the slide from the oven, thoroughly clean a single-edged razor blade with acetone or ethanol.

Clear a space on a desk that is lower than waist-high (for leverage).

26. Next morning (16-20 hr later) test the resin for hardness.

NOTE: YOU SHOULD NOT BE ABLE TO LEAVE A FINGER-PRINT, BUT YOU SHOULD BE ABLE TO MAKE A DENT WITH A FINGERNAIL, EASILY.

a) If the resin is still sticky, bake it longer. Monitor its doneness every hour by touching it.

If you are in a hurry or cannot watch it all day, turn up the heat to 90^0C.

Further baking will take 1/2-1 hr depending on the heating speed of the oven.

NOTE: DO NOT OVERBAKE.

b) If the resin is too hard already (the oven got over 55^0C, or you forgot and left it too long), turn up the heat to 95^0C or place slide on a hot plate.

When it is very warm, but not too hot to pick up, quickly peel up the slab (See Number 27 below).

27. Place the slide with the ground glass tab toward you and hold it tightly against the desk with one hand on the tab.

Hold the razor blade with the other hand at a 30-45^0 angle and move it away from you under the resin slab inside the gasket.

NOTE: THE RESIN WILL BEND SLIGHTLY AS IT IS PRIED, NOT CUT, AWAY FROM THE GLASS (SEE FIGURE 3). FLATTEN IT OUT AGAIN BEFORE IT COOLS.

THIS PROCEDURE SHOULD TAKE NO MORE THAN 30 SECONDS, OR THE RESIN WILL COOL AND CRACK AS IT BENDS. (SEE FIGURE 3).

Figure 3.

28. Once the slab is free from the slide, if no labels have been embedded in the wells, scratch the well number on the top side (cells are on the bottom next to slide).

29. Then turn the slab upside down (cells up) on the slide and return it to the oven for another 6 hr at 60^{0}C or 1/2 hr at 95^{0}C.

30. When completely cured and cooled, break slab and remove squares from gasket.

31. Place pieces immediately and separately into labeled containers.

32. Then one at a time, scarify the top (away from the cells) sides of the small slabs so that you will be able to tell where the cells are not if they flip over.

33. When ready for use, cut the small squares into 2-3 mm^2 with a new clean razor blade.

NOTE: THIS CAN BE ACCOMPLISHED WITHOUT THE PIECES FLYING OVER THE LAB BY PLACING A FOREFINGER AND THUMB ON OPPOSITE SIDES OF THE SQUARE AND POSITIONING THE RAZOR BLADE BETWEEN THEM.

Keeping the blade vertical with the edge flat against the surface of the square, apply firm pressure, and it will snap in two, easily.

Carefully remove the fingers so that the pieces don't flip over.

If they do, look for the scratched side (opposite the cells).

34. With rapid curing epoxy glue, mount 1-2 mm squares onto old blocks which have been trimmed flat. (See Figure 4). Press them down so that they are level. Even though the glue may be "rapid-cure", blocks cut better if allowed to harden overnight.

Figure 4.

a) For sectioning parallel to the plane of growth, mount scratched surface down, cells up.

b) When dry, to determine the location of cells, place block into trimming chuck and focus the binoculars at the highest magnification on the surface.

c) Place a forefinger just above the block between it and the light source so that its shadow covers 1/3 of the block face.

At the junction of the shadow and the shiny area of the face will be a gray area where the light is diffracted (penumbra).

In this area, fibroblasts will appear rounded in the centers and stretched out at the edges.

Other cell lines may appear plump and more cuboidal.

In this way compact cell area can be selected (See Figure 5).

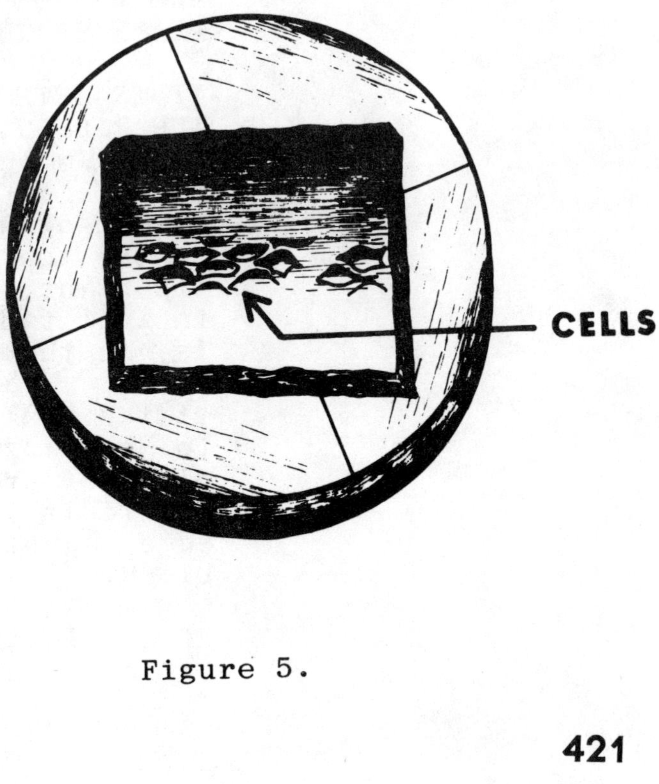

Figure 5.

35. Locate an area of dense cells and hand trim with a razor blade the excess surface away.

Either by hand or with a glass knife on the microtome, complete trimming of the trapezoid.

NOTE: THE BOTTOM EDGE SHOULD BE 0.5-0.75 mm.

a) Align and cut ultrathin sections being careful not to "lop off the end of the block".

NOTE: THE FIRST 10-20 SECTIONS CUT WILL BE OF THE BOTTOM SIDE OR AREA OF ATTACHMENT TO THE GLASS.

MANY MICROTUBULES AND CAVEOLAE WILL BE VISIBLE.

THE NEXT 20 OR SO SECTIONS THROUGH CELL CENTERS WILL SHOW CLOSELY PACKED CELLS, PERHAPS WITH CELL JUNCTIONS, LARGE NUCLEI, GOLGI, RIBOSOMES, AND MITOCHONDRIA.

THE LAST FEW SECTIONS THROUGH THE CELL TOPS WILL HAVE WELL SEPARATED CELLS, MANY MICROVILLI OR RUFFLES, AND FEW OR SMALLER NUCLEI DUE TO THE PLANE OF THE SECTION.

IT IS OFTEN HELPFUL TO KEEP TRACK OF THE RIBBONS WITH RESPECT TO THE ORDER IN WHICH THEY ARE CUT.

36. If orientation of cells with respect to the top or attachment site is important, re-embedding the square in a flat mould or a BEEM capsule is, in the long run, easier.

If the square is mounted on its side for cutting perpendicular to the plane of growth, it is more difficult to section since a clean edge cannot be cut when trimming the block.

If sectioning in the above manner, try placing the cell edge on a side of the trapezoid rather than the bottom or top (See Figure 6).

Figure 6.

NOTE: THE AREA OF INTEREST (CELLS) IS ON THE VERY EDGE AND IS RATHER UNSTABLE IN THE BEAM.

IT FOLDS OVER IF SECTIONS ARE NOT PICKED UP ON CARBON COATED FORMVAR GRIDS.

(SEE EXERCISE NUMBERS 31 AND 32 FOR COATING GRIDS).

C. Whole Cell Viewing

Determining the spatial organization of cell components is difficult in ultra-thin sections without careful reconstruction of serial sections.

The most advantageous method for examining thick material is stereo high voltage electron microscopy (See Exercise Number 29B).

However, one can obtain some helpful information from the periphery of whole cells using a conventional, 100 KV electron microscope.

To accomplish this, one must culture cells on a substrate thin enough to allow passage of the electron beam.

a) Clean gold grids (other metals are toxic to cells) in 1 M nitric acid and wash thoroughly.

b) Coat with 0.75% formvar (collodian will work but is less stable).

c) Collect grids on a cover slip.

d) Sterilize both sides under a UV lamp at a height of 7 cm for 20 min each.

e) Aseptically place coverslip, grids up, into a Petri dish and add cells and media.

f) When cells have attached and grown to about 80% confluency on the grid (do not allow them to become too thick), fix and dehydrate as described in "*In Situ* Embedment".

g) Critical point dry as described under "SEM of Tissue Culture".

h) Do not coat with metal, rather coat with carbon at least on the cell side. If stability is a problem, coat the underside lightly with carbon.

i) At 100 KV, cell centers including nuclei will be too dense, but stereo micrographs can be made of the thinner edges.

III. Scanning Electron Microscopy of Cultured Cells

Scanning electron microscopy is described in Exercise Numbers 20A-C, 21 and 22 of this manual. Below are a few hints to aid in viewing tissue cultures:

1. SEM of Attached Cells

Sawing plastic vessels or cutting with a hot spatula damages cells near the edges.

Cells can be more easily processed if grown on 9 x 9 mm cover slips (Bellco, See Appendix B).

a) First, notch the coverslip on one edge -- away from the middle. By always placing the chip in the same corner, e.g. bottom right (See Figure 7) you will be able to tell on which side the cells are, if it should get turned over during processing.

Figure 7.

NOTE: THIS MAY TAKE SOME PRACTICE BEFORE YOU CAN CHIP WITHOUT BREAKING THE GLASS IN TWO.

(We have about 70-80% success rate, but it is well worth it later on).

b) Hold the glass gently between the thumb and forefinger in one hand.

c) With the tips of an old pair of EM forceps, grasp at an angle about 0.5 mm of the edge of the glass and pull the tips downward over the edge (See Figure 8).

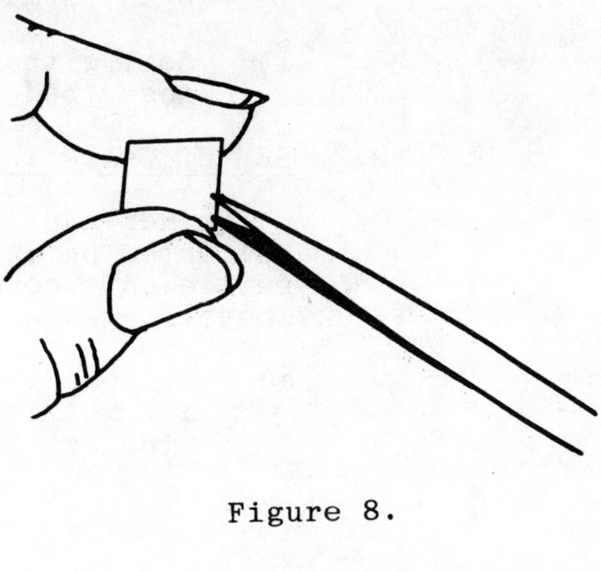

Figure 8.

d) Sterilize the coverslips by autoclaving, UV irradiating <u>both</u> sides (20 min each at a distance of 7 cm) or by covering with ether in a glass Petri dish with a lid and allowing the ether to evaporate in a fume hood.

e) With a sterile pair of EM forceps, place several coverslips into a culture dish, noting the orientation of the chip.

f) Add cells and media and allow growth to about 90% confluency.

NOTE: BE CAREFUL THAT COVERSLIPS DO NOT SLIDE OVER ONE ANOTHER.

g) Fix, including uranyl acetate, and dehydrate with ethanol as described in "TEM of Attached Cells". Make sure the buffer is isotonic regardless of the fixative concentration.

h) Critical point dry as described in Exercise Number 20B of this manual.

NOTE: DO NOT ALLOW THE COVERSLIPS TO LIE UPSIDE DOWN IN THE BOMB OR BE IN DIRECT FLOW OF THE INCOMING CO_2.

If a 4 chambered basket is available (our instrument shop made us one from stainless steel screen), 8 coverslips can be processed back-to-back, simultaneously.

i) Coat with gold-palladium as described in Exercise Number 20C.

2. <u>SEM of Cells in Suspension</u>

Individual non-attached cells can be collected on a filter or allowed to settle on a coverslip and processed as above.

a) Be sure to snip the filter in an asymmetric pattern so that the cell side can be recognized (See Figure 9).

NOTE: FILTERS HAVE A DISADVANTAGE OF OFTEN CURLING WHILE DRYING IF THEY ARE NOT MADE OF METAL.

b) For cell deposition on a notched coverslip, first coat the glass overnight in a moist chamber with 1% collagen dissolved in 0.1 M HCl or with 0.5% poly-L-lysine.

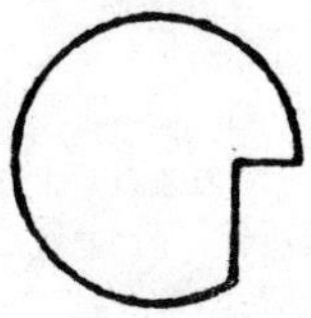

Figure 9.

c) Rinse the coverslips until the wash water is a neutral pH, and rinse once with 0.05 M phosphate buffer. Drain and dry.

d) Pipette cells onto coverslips taking care not to allow suspension to run off the sides.

e) Allow settling for 30 min in a covered moist chamber.

Add 1 drop or about 30 μl of 4% glutaraldehyde in PS buffer.

f) Allow the cells to fix to the glass undisturbed for another 30 min.

g) Gently remove the excess suspension with a Pasteur pipette.

h) Add PS washing buffer to the dish and process as described under TEM of attached cells.

Take care to remove and add fresh solutions gently and not directly over coverslips.

i) Dehydrate, critical point dry, and metal coat as previously described.

E. REFERENCES:

1. Barz, W., E. Reinhard, M.H. Zenk. 1977. Plant Tissue Culture and Its Bio-technological Application. Springer-Verlag, New York, 419 p.

2. Fogh, J. 1973. Contamination in Tissue Culture Academic Press, New York. 288 p.

3. Harris, C.C., B.F. Trump, and G.D. Stoner. 1980. Methods in Cell Biology. Vol. 21. Normal Human Tissue and Cell Culture. A. Respiratory, Cardiovascular, and Integumentary Systems. Academic Press, New York. 323 p.

4. Harris, C.C., B.F. Trump, and G.D. Stoner. 1980. Methods in Cell Biology. Vol. 21. Normal Human Tissue and Cell Culture. B. Endocrine, Urogenital and Gastrointestinal Systems. Academic Press, New York. 512 p.

5. Kuchler, R.J. 1977. Biochemical Methods in Cell Culture and Virology. Dowden, Hutchinson, and Ross, Inc. Stroudsburg, PA. 331 p.

6. Masamorosch, K., and H. Hirumi. 1979. Practical Tissue Culture Applications. Academic Press, New York. 426 p.

7. Paul, J. 1975. Cell and Tissue Culture. Churchill Livingstone, New York. 484 p.

8. Wasley, G.D. 1972. Animal Tissue Culture. Advances in Technique. Buttersworth, London. 192 p.

9. Eagle, H. 1959. Amino Acid metabolism in mammalian cell cultures. Science 130:432-437.

10. Dalen, H. 1973. Vertical sectioning. A. Millipore filters. In: Tissue Culture; Methods and Applications. P.F. Kruse and M.K. Patterson, eds. Academic Press, New York. pp 443-462.

11. Nopanitaya, W., R.K. Charlton, R.L. Turchin, and J.W. Grisham. 1977. Ultrastructure of cells cultured on polycarbonate membranes. Stain Technol. 52:143-149.

12. Duchateau, A., P. Zeitoun, F. Escaig, A. Leclerc, and J.-F. Guillerm. 1980. Separation of epoxy-embedded cell cultures from plastic flasks for electron microscopy. Stain Technol. 55:225-228.

13. Douglas, W.H.J., and J.E. Elser. 1972. A method for in situ embedding of cultured cells grown on plastic surfaces. In Vitro 8:26-29.

14. Johnson, J.E. 1977. A view on techniques of preparation of cell and tissue culture specimens for electron microscopy. Lawrence Berkeley Laboratory Reports LBL, 5638 pl.

15. Johnson, J.E. 1978. Transmission and scanning electron microscope preparations of the same tissue culture. Stain Technol. 53:273-278.

16. Kushida, H., and K. Suzuki. 1970. Evaporated tungsten oxide as a stripping agent for monolayer-cultured cells embedded in epoxy resins. J. Electron Micros. 19:191-192.

17. Lipetz, J., and V.J. Cristofalo. 1972. Ultrastructural changes accompanying the aging of human diploid cells in culture. J. Ultrastruc. 39:43-46.

18. Porter, K.R., and M.E. Sterns. 1981. Stereo-microscopy of whole cells, pp. 53-56. In: J.N. Turner, ed. Methods in Cell biology. Vol 22. Three Dimentional Ultrastructure in Biology. Academic Press, New York.

19. Brunk, U., V.P. Collins, and E. Arro. 1980. The fixation, dehydration, drying and coating of cultured cells for SEM. J. Micros. 123 (2): 121-131.

F. QUESTIONS:

1. In the initial step, how would preparation of cells in suspension or scraped monolayer cells differ from the preparation of a tissue block? What purpose does this material serve?

2. Describe some of the differences in the appearance or orientation of cellular organelles when observed in different planes cut from cells embedded *in situ*.

3. Why must one take care to control carefully the temperature of baking during *in situ* embedment?

4. Is the concentration of solutions important? Why or why not?

5. How is ion concentration or tonicity measured (units), and what is the appropriate physiological condition (i.e. what concentration would be "isotonic" to blood or culture media)?

6. Which is more important, the concentration of the buffer vehicle or that of fixative?

7. Questions to look up:

What specific components do various fixatives preserve:

a) aldehydes

b) osmium tetroxide

c) uranyl acetate

APPENDIX A

Recipes

1. <u>2% Purified Agar</u>:

   a. Add 1 g agar to 50 ml water.

   b. Boil in water bath to melt (about 95^0C).

   c. Dispense in 1 ml aliquots into small vials.

   d. The vials can then be autoclaved and stored in the refrigerator indefinitely if sealed to prevent microbial growth and dehydration.

   If an autoclave is not available, use clean vials and heat agar thoroughly.

NOTE: WE USE SMALL DISPOSABLE TEST TUBES AND SEAL WITH PARAFILM; THE PREPARATION USUALLY LASTS FOR MONTHS WITHOUT STERILIZATION.

   e) When ready for use, remelt the agar in a boiling water bath.

NOTE: MELTING OVER A FLAME IS NOT ADVISABLE BECAUSE THE AGAR MAY SPUE OUT OVER THE WHOLE LAB.

   Allow the agar to cool to about 40^0C. Maintain this temperature in a water bath or beaker with warm water.

2. <u>2X Fixative for Cells in Suspension</u>:

   2% glutaraldehyde in 0.1 M PO_4, 1.8% sucrose, pH 7.4, 500 mOsm:

   2 10 ml vials 10% glutaraldehyde, EM grade
   40 ml H_2O
   50 ml 0.2 <u>M</u> PO_4 buffer
   1.8 g sucrose ($C_{12}H_{22}O_{11}$)

   It has been shown that the fewest artifacts occur when the vehicle (buffer) is at the same osmolarity as that of the cells, regardless of the concentration of the fixative (1).

   In this case, the buffer plus the sucrose is at 300 mOsm.

Each percent of glutaraldehyde adds about 100 mOsm yielding a fixative of 500 mOsm.

When the fixative and media are mixed in equal amounts the total solution is 400 mOsm but that contributed by the buffer medium still 300 mOsm.

3. Stock 0.2 M Sodium Phosphate Buffer

Solution A (monobasic sodium phosphate in water).

27.6 g NaH_2PO_4-H_2O/liter

Solution B (dibasic sodium phosphate in water).

28.4 g Na_2HPO_4/liter (anhydrous) or
53.65 g Na_2HPO_4-$7H_2O$/liter or
71.7 g Na_2HPO_4-$12H_2O$/liter

pH desired	ml A	ml B
8.0	5.3	94.7
7.8	8.5	91.15
7.6	13.0	87.0
7.4	19.0	81.0
7.2	28.0	72.0
7.0	39.0	61.0
6.8	51.0	49.0
6.4	73.5	26.5

NOTE: DEPENDING ON THE ACCURACY OF THE WEIGHING PROCEDURE, AMOUNTS MAY VARY SLIGHTLY.

TO ACHIEVE THE EXACT pH DESIRED, POUR THE LARGER VOLUME REQUIRED (e.g. FOR pH 7.4 USE 81 ml SOLUTION B) INTO A BEAKER WITH A STIR BAR AND PLACE ON MAGNETIC STIRRER.

INSERT A pH ELECTRODE AND TURN ON METER. MEASURE SMALLER AMOUNT NECESSARY (e.g. 19 ml SOLUTION A FOR pH 7.4) IN A GRADUATED CYLINDER.

ADD TO SOLUTION SLOWLY WHILE STIRRING AND MONITORING pH. THE AMOUNT MAY BE SLIGHTLY MORE OR LESS THAN LISTED.

PHOSPHATE BUFFERS EASILY SUPPORT FUNGAL GROWTH. IF LARGE BATCHES ARE TO BE MADE AND KEPT FOR LONG PERIODS OF TIME, CONTAMINATION CAN BE PREVENTED BY ADDING 0.01% SODIUM AZIDE (NaN_3) (0.01 g/liter).

4. Phosphate - Sucrose (PS) Buffer (0.1 M PO_4 buffer with 1.8% sucrose):

100 ml stock 0.2 M PO_4 buffer pH 7.4
100 ml H_2O
2.6 g sucrose ($C_{12}H_{22}O_{11}$)

5. 1% Osmium Tetroxide in 0.1 M Phosphate Buffer, pH 7.4:

At least 24 hr before needed, in a fume hood place a 1 g ampoule of OsO_4 into a heavy glass bottle.

Crush with a metal or thick glass rod, being careful not to break the bottle.

Add 100 ml 0.1 M PO_4 buffer (no sucrose) and stir with a magnetic stirrer overnight.

Maintain stock without sucrose, adding it to a final concentration of 1.8% a few hours before use.

If OsO_4 is to be used by others in a different concentration or buffer, a 4% aqueous solution can be kept for dilution and immediate use in the desired system.

NOTE: STORE ALL OSMIUM SOLUTIONS REFRIGERATED AND PROTECTED FROM LIGHT (COVER BOTTLE WITH FOIL).

BEFORE USE, MAKE SURE THE SOLUTION IS CLEAR OR A FAINT YELLOW.

DISCARD IF IT HAS TURNED PURPLE.

6. 2% Aqueous Uranyl Acetate:

Add 2 g uranyl acetate to 100 ml H_2O. Stir overnight and allow any undissolved material to settle to the bottom. (This solution is also used undiluted for poststaining). Uranyl acetate is not soluble in phosphate or cacodylate buffer. We use it in veronal acetate buffer. Others have used aqueous solutions, assuming the cells are fairly well fixed by then and thus resistant to osmolarity changes.

7. Veronal Acetate (VA) Buffer:

This buffer does not keep well for long periods as it may crystalize out or support fungal growth.

It is best made fresh from stock solution which can be stored in the cold.

Stock veronal acetate

19.43 g sodium acetate ($NaC_2H_3O_2$)
29.43 g sodium barbital ($NaC_8H_{11}N_2O_3$)
34.0 g sodium chloride (NaCl)

Boiled, distilled water to make 1 liter.

Stock 0.1 M HCl

8.5 ml concentrated HCl
991.5 ml H_2O

Stock 1.0 M Calcium Chloride ($CaCl_2$)

110.9 g $CaCl_2$

Water to 1 liter.

Working 0.22 M VA buffer

5 ml stock VA buffer
13 ml distilled H_2O
0.25 ml 1.0 M $CaCl_2$

Place into beaker with stir bar and on a magnetic stirrer.

Insert pH electrode and add 5-7 ml 0.1 M HCl while stirring to reach pH 7.4.

Use 0.11 M VA buffer for washing.

Dilute 0.22 M 1:1 with distilled water.

8. <u>1% Uranyl Acetate in 0.11 M Veronal Acetate Buffer</u>:

Dilute stock 2% aqueous uranyl acetate 1:1 with 0.22 M veronal acetate buffer.

9. <u>Lead Aspartate en bloc stain (4)</u>:

NOTE: USE BOILED, GLASS-DISTILLED H_2O (CO_2-FREE).

2 ml 0.03 M L-aspartic acid (FW = 133.1)
(Stock: Add H_2O to 3.99 g to make 100 ml)

2 ml 0.02 M lead nitrate (FW = 331.2)
(Stock: Add H_2O to 6.62 g to make 100 ml)

pH to 6 with 1 M KOH

add 10 ml 0.02 M HEPES buffer (pH 7.0)
(Stock: Dissolve 0.48 g N-2-Hydroxyethyl-piperazine-N-2-ethanesulfonic acid (FW = 238.3) in 90 ml H_2O; pH to 7.0 with 1 m KOH; add H_2O to make 100 ml).

final pH about 6.5

10. Ethanol Solutions:

An easy way to make ethanol dilutions is first to note the percent you are starting with (e.g. 95%) and then measure an amount in millimeters equal to the percent you wish to end up with (e.g. 50% or 50 ml).

Add water to give a final volume equal to the original concentration (95% or 95 ml).

Thus, in this case you would measure 50 ml of 95% ethanol and add 45 ml of water, to give 95 ml of 50% ethanol.

Make up a graded series of 30%, 50% and 70%. Store together with bottles of 95% and absolute ethanol.

Maintain the 100% ethanol anhydrous by adding molecular sieves to 500 ml ethanol.

Shake and allow beads to settle.

When ethanol is used up, remove beads.

Bake to dry, and reuse over and over.

11. Embedding Resins:

Much ado has been made about equivalent weights and accurate measurements.

We have never been unable to section blocks after measuring the ingredients by weight instead of volume as long as the tissue has been properly infiltrated.

On a top-loading balance, weigh into a disposable polypropylene beaker:

Epon 812, (EMbed 812, Poly/Bed 812)

EMbed 812	16.2 g
DDSA (Dodecenyl Succinic Anhdride)	10.0 g
NMA (Nadic Methyl Anhdride)	8.9 g

or

Spurr Low Viscosity Resin

ERL 4206	(Vinyl Cyclohexene Dioxide)	10.0 g
DER 736	(Diglycidyl Ether of Polypropylene Glycol)	6.0 g
NSA	(Nonenyl Succinic Anhydride)	26.0 g

(Experienced technicians can pour directly from the stock bottle measuring to the nearest 0.01 g. Beginning students should start by using a Pasteur pipette.

NOTE: IF THE COMPONENT IS THICK, CAREFULLY BREAK OFF THE TIP OF THE PIPETTE TO ALLOW FASTER FLOW.

Mix with a stir bar on a magnetic stirrer slowly to minimize bubble formation.

After about 10 min, examine the mixture against a solid background for the disappearance of interfaces between components.

When the resin appears homogenous, measure the accelerator in a 1 ml disposable pipette (0.6 ml DMP-30 for Epon, or 0.4 ml DMAE for Spurr).

Add it dropwise while continuing stirring, mix for another 5-10 min.

NOTE: MAKE SURE INTERFACES HAVE DISAPPEARED.

Unlike Epon 812, Embed 812 turns an amber color after this addition, which a tech-rep from EMS has assured us is normal and not an indication of wetness.

This method is certainly as accurate as measuring with a graduated cylinder, since the viscosity of the components prevents removal of the "last drop(s)" and it is by far less messy.

REFERENCES:

Brunk, U., V.P. Collins, and E. Arro. 1980. The fixation, dehydration, drying and coating of cultured cells for SEM. J. Micros. 123:(2):121-131.

Hyatt, M.A. 1970. Principles and Techniques of Electron Microscopy. Biological Applications, Vol. 1. Van Nostrand Reinhold Co., New York.

Kellenberger, E., A. Ryter, and J. Sechaud. 1958. Electron Microscope Study of DNA containing plasms. II Vegetative and mature phage DNA, as compared with normal bacterial nucleoids in different physiological states. J. Biophys. Biochem. Cytol 4:671-678.

APPENDIX B

Suppliers

Cell Scrapers:

Bellco Glass, Inc.
P.O. Box 13
Vineland, NJ 08360

Costar
205 Broadway
Cambridge, MA 02139

Chamber Slides:

Lab-Tek Products
Div. Miles Laboratories, Inc.
Naperville, IL 60540

Culture Dishes and Flasks:

Corning Glass Works
Corning, NY 14830
(607) 974-9000

Linbro Scientific, Inc.
Flow Laboratories, Inc.
Hamden, CT 06517

Fisher Scientific
711 Forbes Avenue
Pittsburgh, PA 15219
(Most branches have toll-free numbers; call information operator)

Film-Lined Culture Dishes:

Becton Dickinson Labware
1950 Williams Drive
Oxnard, CA 93030
(800) 235-5953 or in California
Collect (805) 485-8711

NOTES

EXERCISE NUMBER 35

RAPID PROCESSING OF TISSUES FOR DIAGNOSTIC ELECTRON MICROSCOPY

Sara E. Miller, Ph.D.

A. OBJECTIVES:

1. To process tissues for EM in a very short time in order to aid in the diagnosis of disease

2. To view tissues within at least 2 hours with little sacrifice of quality

B. MATERIALS:

(SEE RECIPES IN APPENDIX A OF EXERCISE NUMBER 34).

1. a) Biopsy tissue

   b) Single edge razor blades

2. a) Monolayer cells

   b) Cell Scraper

3. 2% Glutaraldehyde in PS buffer

4. Rocker or shaker if possible

5. Phosphate-sucrose (PS) buffer

6. 1% Osmium tetroxide in 0.1 M phosphate buffer

7. Veronal acetate (VA) buffer

8. 1% Uranyl acetate in 0.11 M veronal acetate buffer

9. Lead aspartate

10. Graded ethanols

11. Spurr resin

12. Oven

13. Drierite or other dessicant

14. Freezer

C. RATIONALE AND DISCUSSION:

It has been shown that tissue can be rapidly processed when time is of the essence, such as in detection of possible viral infections (1). The cell ultrastructure is not ideally preserved, but is acceptable, and virus structures are identifiable if present in sufficient quantity. Another method helpful in diagnosis is amplification of virus numbers in tissue culture.

The efficacy in treating any disease is improved by rapid diagnosis and treatment with the proper agent. Since viral infections can run rampant, and also with the advent of antiviral agents, rapid viral identification is becoming of utmost importance.

Direct examination of biopsy material may be negative if viruses are present only in low numbers. The amount of virus can be amplified by inoculating the suspected material into tissue culture, a procedure performed by many viral diagnostic labs. But then diagnosis is usually based on cell cytopathology in the light microscope, and several days to a couple of weeks may pass before any cytopathological effect (CPE) is evident.

By electron microscopy, even immature virions are recognizable, thus only one or less than one complete replication cycle may be necessary for identification. On the other hand, if the inoculum is low, the few infected cells in the culture may be difficult to locate in the electron microscope, and more than one generation may be required to spread the virus to other cells. Thus, several plates should be inoculated for harvest at different times. In any case, EM is definitive, while light microscopy CPE may be equivocal until a late stage of infection.

Furthermore, preparation time for EM can be reduced to that required for fluorescence microscopy (about 2 hr). Also use of a lead stain (2) during embedding enhances virions over cell cytoplasm and eliminates the need for poststaining. Rapid processing requires constant attention as handling is frequent.

D. PROCEDURE:

I. Tissue

1. As soon as possible after removal, place biopsy into warm (37^0C) 2% glutaraldehyde in PS buffer in a flat dish such as a petri dish.

   (See Appendix A in Exercise Number 34 for Preparation of PS buffer).

2. With scalpel or new (clean) razor blade (not scissors) cut tissue into slivers 0.5 mm x 0.5 mm x 1 mm.

   Slice tissue by drawing cutting edge over it while holding it gently with EM forceps.

NOTE: DO NOT MASH THE TISSUE.

3. Fix for 15 min, or if fixative has become bloody or cloudy during slicing, transfer slivers to fresh fixative for another 5 min.

   Place specimen on a shaker if available; if not, agitate manually every 2-3 min.

   Continue agitation throughout procedure.

4. Turn on oven to 95^0C and place a jar of Drierite into freezer for use in #14.

5. Wash in 3 changes of PS buffer (2 min each change).

   Use at least 10 ml per wash for each 2-4 pieces and drain thoroughly after decanting each wash.

NOTE: SLIVERS CAN EASILY BE LOST WHILE USING AN ASPIRATOR AS DESCRIBED IN "EM OF TISSUE CULTURE", OR WHILE POURING. BE CAREFUL NOT TO DO SO.

   Remove all excess solution or resin at each stage, removing the last few drops with a Pasteur pipette.

6. Fix 15 min in 1% OsO_4 in PS buffer

7. Rinse briefly in PS buffer.

8. Wash once for 5 min in VA buffer (See "Appendix A of Exercise Number 34).

9. Stain 5 min in 1% UA/VA.

10. Rinse briefly in distilled water.

11. Dehydrate in graded ethanols:

    50% 10 min
    95% 10 min
    100% 2 min ea, 5 changes

12. Infiltrate:

    50% ethanol - 50% Spurr, 15 min
    100% Spurr, 5 min ea, 5 changes

13. Bake 30 min. at 95^0C.

NOTE: CAPSULES WILL MELT IF TEMPERATURE GOES OVER 97^0C.

14. Place into Drierite in freezer for 5 min.

15. Cut sections.

16. View at 60 KV to increase contrast of nonpoststained sections.

II. <u>Tissue Culture</u>:

1. Decant media.

NOTE: IF CELLS CONTAIN OR MAY CONTAIN AN INFECTIOUS AGENT, POUR MEDIA INTO A BEAKER WITH 10-20 ml CLOROX OR OTHER DISINFECTANT.

Rinse briefly with PS buffer.

2. Fix in 2% glutaraldehyde in PS buffer for 10 min.

3. Wash in PS buffer 5 times, 1 min each.

4. Postfix in 4% OsO_4 in PS buffer.

5. Rinse briefly 2 times with PS buffer.

6. Rinse briefly 2 times with VA buffer.

7. Rinse briefly in boiled, glass distilled H_2O 4 times.

8. Stain 10 min with lead aspartate (See "Appendix A of Exercise Number 34).

   Meanwhile, do number 10.

9. Wash 4 times in 4 min with boiled H_2O.

10. Make up Spurr resin increasing catalyst to 3X --3 times what the recipe calls for.

11. Wash 4 times in 2 minutes with water.

12. Dehydrate:

50% ethanol,	4 changes,	30 sec each
80% ethanol,	2 changes,	30 sec each
100% ethanol,	4 changes,	1 min each

13. Add 50% ethanol-50% Spurr (3X catalyst).

14. Scrape cells off plate with a rubber policeman or cell scraper.

15. Centrifuge at 6000 g 7 min.

16. Pipette off 50% resin.

17. Place a jar of Drierite into the freezer for use in #19.

18. Infiltrate with 100% Spurr (3X catalyst).

    Change 3 times in 15 min.

    Place pellet into a capsule.

19. Bake at 97-98^{0}C for 25 min.

NOTE: DO NOT USE A HIGHER TEMPERATURE, OR THE CAPSULES WILL MELT.

20. Cool in the freezer in a bottle of Drierite 5 min.

21. Trim, section, view without poststaining and at 60 KV (for increased contrast).

E. REFERENCES:

1. Miller, S.E., and D.J. Lang. 1982. Rapid diagnosis of herpes simplex infection: Amplification for electron microscopy by short-term in vitro replication. J. Infec. 4:37-41.

2. Walton, J.R. 1978. Lead aspartate: A new stain for electron microscopy. pp. 34-35. In: G.W. Bailey, ed. Electron Microscopy. Claitor's Publishing Division, Baton Rouge, LA.

F. QUESTIONS:

1. Why might it be necessary to rush an embedding procedure even if ultrastructural preservation is not optimal?

2. Approximately how much time would be required from the fresh tissue stage to have sections _in the microscope_ for examination?

3. How can you increase contrast in addition to the use of osmium, uranium, and poststains? (2 ways)

4. What is Clorox used for in this procedure?

NOTES

EXERCISE NUMBER 36A

SPECIAL TECHNIQUES FOR PROBLEM TISSUES

BLOOD CELLS

John P. Caulfield, M.D.

A. OBJECTIVES:

1. To prepare cells obtained from washings of the rat peritoneal cavity for electron microscopy

2. To test the ability of these cells to endocytose heat killed yeast (zymosan)

B. MATERIALS:

1. 2% glutaraldehyde in 0.1 M cacodylate buffer, pH 7.2

2. 1% OsO_4 in acetate veronal buffer

3. Uranyl acetate in acetate veronal

4. Zymosan solution

5. Two 300 gram rats

6. Scissors, forceps, dissecting board, pins, razor blades

7. Tyrode's buffer with heparin 5 U/ml

8. Disposable plastic, Beral-type pipets

9. Glass vials

10. 15 cc conical centrifuge tubes

11. Microfuge and 400 µl microfuge tubes

12. Flat embedding molds

NOTE: SEE END OF EXERCISE FOR PREPARATION OF SOLUTIONS.

C. RATIONALE AND DISCUSSION:

The rat peritoneum contains a variety of cell types that are also found in the blood, i.e., neutrophils, eosinophils, monocytes, macrophages and lymphocytes.

In addition, mast cells, which are not found in the blood but are located at "portals of entry" beneath the skin, gastrointestinal, and urogenital epithelia, are present in large numbers. This mixed population of cells is very effective in protecting the peritoneum from infection.

There are two overall goals of the experiment. First, you will prepare these cells for microscopy and learn to distinguish them from one another. Secondly, you will inject a mock pathogen into the peritoneum and see how the cells behave in response to it.

The fixation of these and other isolated cells offers a technical problem to the electron microscopist. Should the cells be handled in suspension or be put in a solid form? We have chosen the latter approach. The cells will be isolated, washed and fixed in aldehyde.

They will then be pelleted in a microfuge and the resultant pellet removed from the centrifuge tube. This pellet will then be treated as a piece of tissue through postfixation, staining in block, dehydration, and embedding.

The identification of the various cell types should be carried out on thick (0.2 µm) and thin (600 A^{o}) sections at the light and electron microscopic level, respectively.

In brief, the cells will or will not contain granules. The cells without granules are lymphocytes and monocytes. The cells with granules are either mononuclear or polynuclear.

The mononuclear cells with granules are either mast cells that have many large granules or macrophages that contain a variable number of smaller granules.

The polymorphonuclear leukocytes are either eosinophils or neutrophils. The eosinophils have large granules with a central core that appears crystalline in favorable planes of section.

The neutrophils have smaller granules than the eosinophils and they are of two types, azurophil and specific. For more details on the identity of these cells, consult references 1 and 4.

In the experiment with the zymosan you will inject heat-killed yeast into the peritoneum. After an hour, you will prepare the peritoneal cells for microscopy, as described below, and see how the cells respond to the yeast. In particular, you will see if any cells have attached to or internalized the yeast.

D. PROCEDURE:

1. The experiment will require two 300 gram rats. One hour before sacrifice inject one of the rats intra-peritoneally with 0.1 cc of zymosan solution.

   Hold the rat behind the head and by the tail. Push the needle into the abdomen.

   After aspirating to make sure you are not in a blood vessel or a loop of bowel, inject the zymosan.

   All subsequent steps are to be done in parallel for both the zymosan injected and uninjected rats.

2. Decapitate the rat.

NOTE: THE FOLLOWING STEPS SHOULD BE DONE AS RAPIDLY AS POSSIBLE TO PREVENT CLOTTING IN THE PERITONEAL CAVITY.

3. Quickly pin all four extremities to a cork board and tilt the cephalic end upward.

4. Make a midline incision through the skin but not the muscle.

5. With scissors, make a small opening in the muscle and inject 10 cc of Tyrode's buffer with heparin.

   One member of the group should elevate the sides of the opening in the muscle with forceps.

6. The abdomen should be massaged gently for one minute.

7. After massaging, enlarge the muscle incision to 4-5 cm and take care not to cut any loops of bowel.

8. Remove the injected fluid with a plastic pipet.

NOTE: IN THIS AND ALL SUBSEQUENT STEPS, USE DISPOSABLE PLASTIC PASTEUR PIPETS SUCH AS THOSE MANUFACTURED BY BERAL.

9. Inject another 6 cc of Tyrode's buffer into the abdomen, massage, and remove.

10. Pool both abdominal washings in a 15 cc conical centrifuge tube and centrifuge at 1,000xg for 3 minutes and withdraw all but 0.7 cc of the supernatant.

    Resuspend the cells and add an equal volume (0.7 cc) of glutaraldehyde fixative to the suspension and mix well with a pipet.

11. After 5 minutes, fill four 400 µl microfuge tubes with the fixed cell suspension.

    However, the exact volume of cell suspension required for each tube and the total number of pellets obtained from each rat will vary depending on the efficiency of the initial recovery of cells from the peritoneum.

NOTE: EACH CELL PELLET SHOULD HAVE A TOTAL VOLUME OF 4 µl

    Centrifuge for 5 minutes at 10,000xg in a microfuge, such as the Beckman Microfuge B.

12. Remove the pellets from the microfuge tube by cutting off the tip of the tube with a razor blade and pushing on the bottom of the pellet with the corner of the razor blade.

    This will dislodge the pellet from the bottom of the tube and it will sink into the supernatant when the tube is inverted.

13. Cut the tube in half and allow the pellet to flow out with the supernatant into a glass vial or onto a plastic dish whence it can be

transferred to the glass vial with a plastic pipet.

The vial should contain 1 cc of glutaraldehyde fixative. Only 1 vial is needed/rat.

14. After all the pellets are placed in the vial, remove the aldehyde fixative and replace with 1 cc of acetate veronal buffered OsO_4.

    Incubate for 1 hour at 4^oC on ice.

15. Remove the OsO_4 solution. Replace with 1 cc uranyl acetate in acetate veronal.

    Allow to incubate for 2 hrs at room temperature.

16. Remove the uranyl acetate. Dehydrate with the following sequence:

    95% ETOH, 1 min

    100% ETOH, 7 min (X2)

    propylene oxide, 7 minute (X2)

    1:1 EPON: propylene oxide overnight at (4^oC).

    3:1 EPON: propylene oxide (1 hour, 25^oC)

    Pure EPON (3 hours, at 25^oC)

17. Embed in EPON in flat embedding molds.

    The pellet should be oriented so that the plane of sectioning will be the same as the long axis of the centrifuge tube, i.e., so the pellet will be cut from top to bottom.

NOTE: THIS IS NECESSARY TO APPRECIATE ANY GRADIENTS WHICH FORM DURING CENTRIFUGATION.

18. After polymerization (48 hours at 60^oC), thick (0.2 μm) sections should be cut and stained with azure II-methylene blue or toluidine blue.

    Thin sections should be stained on grids with uranyl acetate and lead citrate.

E. MODIFICATIONS:

a. Other embedding media such as Spurr's or Araldite may be substituted for EPON.

b. Other salt solutions or culture media can be used instead of Tyrode's but heparin should be included.

c. The experiment is most logically interrupted, if necessary, during infiltration.

However, time on the day of experiment can be saved by having made all solutions except the OsO_4 and the plastic on the previous day.

If absolutely necessary, the pellets can be stored overnight at 4^oC in 0.1 M cacodylate, pH 7.2, after aldehyde fixation.

d. An optional control is to mix 5 ul of the zymosan with aldehyde fixed cells and pellet this mixture in the microfuge.

This will demonstrate the surface of the yeast before injection into the peritoneum and before ingestion by cells.

e. A further optional experiment is to add 5 µl of the zymosan solution to 1×10^6 cells isolated from the peritoneum and to incubate the cells and zymosan together for 1 hr at 37^oC.

The suspension can be fixed and processed as in steps 10 to 18 of the procedure.

F. PREPARATION OF SOLUTIONS:

1. <u>Zymosan suspension</u>

Use Schwartz-Mann zymosan, Catalogue Number 812714 or its equivalent. 25 mg/ml will equal 4×10^9 particles/ml. Approximately 9 rats can be injected per ml.

To make the solution, put the zymosan in normal saline or in phosphate buffered saline in a test tube and place the test tube in a beaker of boiling water for 20 minutes.

Wash the zymosan 3 times by centrifuging in a table top centrifuge, removing the supernatant, and resuspending in fresh saline.

The final concentration should be 25 mg/ml. The solution may be frozen indefinitely.

2. 0.2 M cacodylate stock

Put 4.28 g of sodium cacodylate into solution in 80 cc distilled water. Put in a 100 cc volumetric flash and bring to 100 cc. Take pH which should be 8 but do not adjust.

3. 2% glutaraldehyde in 0.1 M cacodylate, pH 7.2.

For 50 cc, add 10 cc of 10% glutaraldehyde to 25 cc of 0.2 M cacodylate stock and 10 cc of distilled water.

Adjust pH to 7.2 and bring final volume to 50 cc with distilled water.

4. Acetate veronal stock

Dissolve 1.94 g Na Acetate-3H_2O (MW = 136) and 2.94 g Na barbiturate (Veronal) (MW = 206) in distilled H_2O and bring to a final volume of 100 cc in a volumetric flask.

5. 1% OsO_4 in acetate veronal

For 20 ml add 4 cc of acetate veronal stock, 4 cc of 0.1 N HCl and 7 cc of distilled water.

Adjust pH to 7.6. Add 5 cc of 4% OsO_4 stock in distilled H_2O.

6. Uranyl acetate in acetate veronal

For 25 cc, add 5 cc acetate veronal stock, 7 cc of 0.1 N HCl and 13 cc of distilled H_2O.

Adjust pH to 6.0 and add 125 mg of uranyl acetate. Dissolve the uranyl acetate on a magnetic stirrer.

7. Tyrode's buffer

To make 1 l of stock that can be diluted to 5 l, add 5 g dextrose, 5 g Na bicarbonate, 1 g KCl, 0.40 g NaCl, 0.282 g Na H_2PO_4 H_2O, 5 ml of 20% $CaCl_2$, 2 ml of 25% $MgCl_2$ to distilled water and bring to 1 l in a volumetric flask.

Before using, gas with 95% CO_2 and 5% O_2 until the pH falls to 7.2.

The buffering capacity may be enhanced with 10 mM HEPES or MOPS in the final solution.

Add 30 mg of DNAase/1 and 5U/ml of Pauheparin to the final solution before use.

G. REFERENCES:

1. Zucker-Franklin, D., Greaves, M.F., Grossi, C.E., and Marmout, A.M. Atlas of Blood Cells: Function and Pathology. Lea and Febriger, Philadelphia, 1981.

2. Padawer, J. Phagocytosis of Particulate Substances by Mast Cells. Lab. Invest. 25:320, 1971.

3. Sher, A., Hein, A., Moser, G., and Caulfield, J.P. Complement Receptors Promote the Phagocytosis of Bacteria by Rat Peritoneal Mast Cells. Lab. invest. 41:490, 1979.

4. Bainton, D.F., Nichols, B.A., and Farquhar, M.G. Primary Lysosomes of Blood Leukocytes. In: Lysosomes in Biology and Pathology 5. Dingle, J.T. and Dean, R.T., eds. Elsevier, New York, p. 3, 1976.

H. QUESTIONS:

1. Describe fully monocytes, lymphocytes, neutrophils, eosinophils and mast cells.

2. Which cells endocytosed zymosan? What cells were the most effective ingestors? Can you quantitate the percentage of zymosan in various cell types?

3. What happens to the zymosan within the various cell types? Draw a diagram of the internalization of zymosan and its interaction with cytoplasmic organelles.

4. If the zymosan were a living microbe, how could it be neutralized or killed by cells?

NOTES

EXERCISE NUMBER 36B

SPECIAL TECHNIQUES FOR PROBLEM TISSUES

BONE AND COLLAGEN AND RELATED NON-CALCIFIED AND CALCIFIED TISSUES, INCLUDING ENAMEL

William J. Landis, Ph.D.

I. Conventional Processing of Specimens fixed with Glutaraldehyde and Osmium Tetroxide

A. OBJECTIVES:

1. To preserve the ultrastructural appearance of the cellular and extracellular organic components and compartments of the normally uncalcified and collagen-containing vertebrate tissues such as skin, tendon, ligament, cornea and cartilage

2. To preserve the ultrastructural appearance of the cellular and extracellular organic components and compartments of the normally calcified vertebrate tissues such as bone, calcifying cartilage and tendon, dentin, cementum, and enamel

3. To enhance the electron density of ultrastructural features of these vertebrate tissues without additional staining with such agents as uranyl acetate and lead citrate

4. To stabilize thin sections of some calcified vertebrate tissues, notably enamel, against electron beam damage on microscopic examination

NOTE: THE TERMS "CALCIFIED" AND "MINERALIZED" ARE USED HERE INTERCHANGEABLY.

B. MATERIALS FOR DISSECTION AND FIXATION:

1. Cold (4^0C) 0.1 M cacodylate buffer solution pH 7.4

2. 2.5% glutaraldehyde in 0.1 M cacodylate buffer

3. Bottle (4^0C) of physiological saline solution

4. Four ethanol solutions (solvent of double-distilled, deionized water) of concentrations 70,80,95, and 100%

5. Freshly prepared 1% osmium tetroxide in 0.1 M cacodylate buffer, stored in a dark, glass-stoppered bottle at 4^0C

6. Animal tissue of interest

7. Dissecting kit including sharp scissors and #11 and 15 scalpel blades and handles

8. Razor blades cleaned with absolute ethanol

9. Pasteur pipettes with rubber bulbs

10. Ice and ice bucket

11. Ice-filled glass or disposable-plastic Petri dishes

12. Small glass specimen bottles such as scintillation vials (capacity 10-15 cc), capped and labeled

13. Dissecting microscope

NOTE: BOTH GLUTARALDEHYDE AND OSMIUM TETROXIDE ARE VOLATILE AND TOXIC. ALL PROCEDURES SHOULD BE CARRIED OUT IN A FUME HOOD. AVOID FUME CONTACT WITH EYES, MOUTH AND THROAT. AVOID SPILLS ON SKIN. WEAR SURGICAL MASK AND DISPOSABLE GLOVES.

C. RATIONALE AND DISCUSSION:

The methods to be described below are appropriate for the broad group of tissues which contain collagen as their principal protein component. These include in vertebrates the normally unmineralized collagenous tissues such as ligament, cornea, and the like, and the normally mineralized collagenous tissues including bone, dentin, and cementum.

Cartilage and tendon may be found both non-calcified and calcified. The preparative methods outlined are also appropriate to a fifth normally calcified vertebrate tissue, enamel, which contains no collagen.

A few characteristics of these particular tissues must be mentioned and understood to appreciate the special procedures applied in preparation of these samples for electron microscopy:

1. In general, over 90% of the extracellular matrix of these tissues (except enamel) consists of the highly cross-linked, fibrous, tough, densely-packed structural protein, collagen.

2. The extracellular mineralized component of these tissues is a calcium phosphate crystal formed within the collagen matrix to produce a reinforced two-phase system having tremendous mechanical strength and providing great structural support.

3. Some cells, cellular components, and extracellular constituents are located within the mineralized matrices of the tissues, nearly entirely surrounded by mineral in many cases.

4. There are large differences in tissue densities between the cells, uncalcified collagen and the extracellular organic matrix, and the calcified (collagen-mineral) regions of the specimen.

Because of the characteristic close-packing, fibrous nature of collagen in these tissues, as well as the hard and highly dense inorganic phases of those tissues which calcify (including enamel although not a collagenous-based tissue), it is relatively more difficult - compared to other non-collagenous, non-mineralized samples - for fixatives to diffuse and penetrate these specimens and for solvents to infiltrate and embed.

Sectioning and collecting the specimens are also more tedious. Staining presents a certain problem since the intrinsic electron density of the mineral, itself, in a sample is often indistinguishable from the density provided by the heavy metal stain of uranium, lead, or osmium, for example; in this case the precise identification of the calcium phosphate comprising the mineral can be very confusing.

As a result of these considerations, it is a common practice during electron microscopic preparation of these tissues to allow longer infiltration times and extended hardening of epoxy or polyester blocks, to minimize the sizes of tissue and blocks, and to section almost exclusively with diamond knives.

In addition it may sometimes be necessary to examine completely unstained (and unosmicated) tissue to identify the mineral phase. The following method for conventional processing of enamel and collagenous tissues, both normally non-calcifying and calcified, has been developed to achieve the objectives specified in (1 and 2) previously.

D. PROCEDURE:

1. <u>Fixation</u>

a. All solutions, fixatives and covered glass vials should be placed and kept cold in an ice bucket ready for use.

All procedures requiring glutaraldehyde and osmium must be carried out under a fume hood.

b. After an animal has been sacrificed (with out anaesthesia if possible) as quickly and humanely as possible, obtain the tissue of interest rapidly, removing excess and unwanted material.

If long times are required to obtain the tissue, one may apply cold stock glutaraldehyde (2.5%, see end of exercise for preparation of fixatives and buffer) using a pasteur pipette.

c. Place the excised tissue on the cover of an ice-filled petri dish in a few drops of cold stock glutaraldehyde (2.5%) under a fume hood.

d. Dissect a precise region of the tissue required with a clean, alcohol-washed razor blade or dissecting tools.

Specimens should be as small as possible with at least one dimension smaller than 0.5 mm and total volume about 1 mm^3.

e. Transfer specimens to labeled, clean glass scintillation vials, covered with aluminum foil to prevent light from entering vials.

f. Add 5-7 ml of cold (4^0C) glutaraldehyde-osmium tetroxide fixative to each glass vial using a pipette.

The addition of fixative should be made quickly since light exposure will degrade osmium.

The fixative consists of the addition of 1 part 2.5% glutaraldehyde in 0.1 M cacodylate buffer to 2 parts 1% osmium tetroxide in 0.1 M cacodylate buffer, pH 7.4.

g. Cap the vials and replace them in the ice bucket. Cover the ice bucket and fix tissues for 3 hours at 4^0C.

h. Following fixation, pour or pipette off the fixative solutions and rinse tissues twice for 10 minutes each in 5-7 ml cold (4^0C) physiological saline or cold cacodylate buffer.

If necessary tissues may be left overnight in saline at 4^0C (in ice bucket). After fixation, tissues should appear black.

2. Dehydration and Infiltration

a. Pour last rinse out of each vial or remove rinse solutions with a pipette. Begin dehydration sequence with cold (4^0C) 70% ethanol for 45 minutes. Complete the dehydration series with 80% ethanol, 95% ethanol, and two changes of 100% ethanol, all for 45 minutes each at 4^0C.

b. Infiltrate the tissues at room temperature. The sequence is the following:

(a) If using Spurr's medium, follow the last absolute ethanol change during dehydration with a 1:1 mixture of 100% ethanol and Spurr's for 1 hour, then 1 hour in a 1:3 mixture of 100% ethanol and Spurr's, and overnight in Spurr's alone.

(b) If using Epon, replace last absolute ethanol change during dehydration with propylene oxide for 10 minutes, then change the propylene oxide for an additional 10 minutes, and finally replace the propylene oxide with a 1:1 mixture of propylene oxide-Epon for one week.

Infiltration should be carried out with all specimen vials placed on a rotator.

3. Embedding

a. After appropriate infiltration period, fill embedding molds, Beem capsules, or other suitable sample holders with the embedding medium (Spurr's or Epon, for example).

Place individual identification labels in each mold or capsule making certain these will not interfere with subsequent sectioning of the tissue.

b. Transfer one tissue specimen to each mold or capsule.

c. Orient the specimens under a dissecting microscope using the narrow cut end of an applicator stick or a fine dissecting needle.

d. Place molds or capsules in an oven.

For Epon embedment, set these at 60^0C for 12 hours, then at 37^0C for one week.

For Spurr's embedment, set samples overnight at 37^0C only.

4. Sectioning

a. Apply usual methods to thin-section embedded and trimmed tissue blocks.

For bone and collagen and related non-calcified tissues, trim block faces to small sizes, with edges no greater than 0.5 mm long.

Remove as much excess Epon or Spurr's resin as possible.

b. Use glass knives to face blocks and diamond knives for final sectioning.

c. Use sectioning speeds of 0.5-2 mm/sec set on the ultramicrotome.

d. Be aware of the effects of the microtome trough fluid on the ultrastructural appearance of the resultant thin sections.

For example, usual aqueous buffer solutions or distilled water will decalcify mineralized tissue sections within minutes or even a shorter time.

Such decalcification results in sections in which normally dense and dark mineral zones appear relatively electron-transparent when viewed under the electron microscope.

To circumvent this artifact, trough fluids which are not aqueous in nature may be used.

Organic solvents such as 100% ethylene glycol minimize section decalcification (See Exercise Number 36C for more detailed notes).

If decalcification is not a concern (as with skin, cornea, and the other normally uncalcified tissues, of course), suitable trough fluids are distilled water, veronal acetate or cacodylate solutions.

Aqueous phosphate buffer may also be used, but be aware that the phosphate ion may artifactually affect the phosphate of the mineral in a section and that the water will decalcify the section.

5. Staining

a. Stain sections for 10 minutes each with 8% uranyl acetate in 50% ethanol-distilled water and 0.2% aqueous lead citrate.

It must be realized that the uranyl acetate as prepared has an acid pH which will also rapidly decalcify mineralized tissue sections.

To preserve the mineral, use an 8% solution of uranyl acetate in 100% ethylene glycol, absolute ethanol, or methanol

(See Exercise Number 36C for more details).

b. Stain sections on clean parafilm or wax plates in closed petri dishes.

Grids are floated face down on single drops of the stains.

Use sodium hydroxide pellets around the inside of the petri dish containing lead citrate.

This practice reduces the possible formation of lead carbonate precipitates on the tissue sections.

c. Rinse grids after each staining by repeatedly dipping them into a beaker of fresh solvent (15-30 seconds).

6. Preparation of Solutions

a. Cacodylate buffer (0.1 M)

1. Cacodylic acid 6.9 gm

2. Distilled water 500 ml

3. Adjust pH to 7.4 with 5 N NaOH

If pH exceeds 7.4, readjust with dilute acetic acid.

b. Glutaraldehyde fixative (2.5%)

1. Cacodylate buffer (0.1 M) 22 ml

2. 8% glutaraldehyde 10 ml

c. Osmium tetroxide solution (1%)

1. Crystalline OsO_4 (Osmic acid anhydride, purified) 1 gm

2. Cacodylate buffer (0.1 M) or distilled water (100 ml)

7. Decalcification of samples prior to fixation

In some instances, fully decalcified specimens may be needed for microscopic examination. In these situations,

a. Wash excised tissue in two changes 0.1 M cacodylate buffer or physiological saline, 10 minutes each.

b. Place samples in decalcifying solution from 1 day to 4 weeks or more, depending on the age of the animal and type and density of the mineralized tissue.

The decalcifying solution is follows:

1. Cacodylate buffer (0.1 M) 22 ml

2. 8% Glutaraldehyde 10 ml

3. Sodium ethylenediaminetetraacetic acid (EDTA) 2.4 gm

4. Add EDTA to the mixture of glutaraldehyde and buffer, place on magnetic stirrer until EDTA dissolves.

c. Change decalcifying solution every three days.

d. Following decalcification period, wash samples 3 times, 15 minutes each, with 0.2 M cacodylate buffer or saline.

e. Dehydrate, infiltrate and embed as usual.

E. QUESTIONS:

1. What difficulties should be anticipated in the processing of calcified tissues for electron microscopy?

2. What is the molecular structure and organization of collagen and how do these characteristics affect an approach to processing collagenous tissues for electron microscopy?

3. Describe some of the physical and chemical features of the inorganic mineral phase of vertebrate calcified tissues which provide for their important function in musculoskeletal strength and support?

EXERCISE NUMBER 36C

SPECIAL TECHNIQUES FOR PROBLEM TISSUES

BONE AND COLLAGEN AND RELATED NON-CALCIFIED AND CALCIFIED TISSUES, INCLUDING ENAMEL

William J. Landis, Ph.D.

II. Processing of Calcified Tissues with Anhydrous Ethylene Glycol

A. OBJECTIVES:

1. To maintain the integrity and ultrastructural features of the intra- and extracellular inorganic mineral of calcified vertebrate tissues

2. To allow adequate preservation of organic components and compartments of calcified tissues so that ultrastructural interrelations between inorganic and organic matrix constituents may be observed

3. To preserve the intrinsic electron density of inorganic mineral components of the calcified tissues without additional section staining

B. MATERIALS FOR DISSECTION AND TISSUE TREATMENT:

1. See items 6-13 described in Exercise Number 36B.

2. Reagent grade 100% ethylene glycol (< 0.1% water)

3. Cellosolve, the monoethyl ether of ethylene glycol

4. A small vacuum desiccator

5. A mechanical pump

C. RATIONALE AND DISCUSSION:

The inorganic (mineral) components of the normally calcified vertebrate tissues--bone, calcified cartilage, tendon, dentin, enamel, and cementum--may undergo serious changes during their preparation for electron microscopy.

Such changes include artifactual mineral phase transformations to hydroxyapatite, the most mature form of calcium-phosphate found in these tissues; translocation (movement) of diffusible ions such as calcium and phosphorus; dissolution and loss of mineral salts; and mineral reprecipitation.

The alterations will significantly influence the size, shape, location, distribution, and organization of the mineral phase particles in the tissues and any ultrastructural results or interpretation of these characteristics may be compromised if these changes occur.

The principal factor contributing to the possible changes given above is exposure of the tissue, and hence the mineral phase, to aqueous solvents during microscopic preparation.

Aqueous solvents in the form of fixatives, dehydrating agents, microtome trough fluids, or electron microscopic stains, are to be avoided in order to maintain optimally the state of the mineral phase as it may exist *in vivo*.

One of the methods of circumventing water and aqueous solvents for microscopic sample preparation utilizes organic solvents such as 100% ethylene glycol.

Studies have shown that this and other anhydrous reagents are suitable for minimizing or eliminating the potential artifacts in inorganic tissue components attributable to aqueous solvents. Such reagents adequately preserve the ultrastructural features of the organic tissue matrices as well.

There are, however, a few limitations to the application of these organic solvents. First, they do not provide the same quality of ultrastructural preservation to *organic* tissue components as that imparted by aqueous fixatives.

With ethylene glycol treatment, for example, the nuclei of bone and other calcified tissue cells may become distorted in shape and some regions of the tissue thin sections may be punctuated by holes as a possible result of osmotic pressure imbalances during specimen processing.

Second, the cytomembranes of the cells, including the nuclear envelope, cristae and other mitochondrial membranes, and the endoplasmic reticulum appear quite faint on microscopic examination, even after staining.

It is thought that the membrane constituents have either been extracted in the organic solvents or altered in a manner which affects their interaction with electron-dense stains.

Finally, the use of these organic solvents is technically more demanding than that of conventional aqueous solvents.

Ethylene glycol is more viscous than water and this characteristic leads to some problems occurring especially during sectioning - frequent wetting of block faces, difficulties with section flotation and the formation of ribbons may be common. Collection of the sections may also be a tedious exercise. Despite these disadvantages of the method, treatment of calcified tissues with anhydrous ethylene glycol is a superior approach for preserving the inorganic portion of the specimens.

The ethylene glycol method should be considered whenever any question concerning the physical or chemical nature of the mineral in biological sciences is to be investigated.

D. PROCEDURE:

1. Fixation and Dehydration

a. Sacrificing an animal and dissection of tissues are carried out as described in Exercise Number 36B (a-e) of the procedure, except that 100% ethylene glycol is used as tissue fixative.

   Glycol is at room temperature although dissection is carried out on the cover of an ice-filled Petri dish.

   It is important that specimens are cut as small as possible with at least one dimension < 1.0 mm.

b. Specimens are placed in labeled, clean scintillation vials (label both vial and its cap).

c. Place uncapped specimen vials, filled with 5-7 ml of 100% ethylene glycol, into a vacuum desiccator connected to a mechanical pump and keep samples under vacuum for 3 hours.

   Disconnect the mechanical pump at that time, and maintain vacuum overnight in the desiccator.

d. Pour or pipette off glycol and replace with cellosolve at room temperature and pressure.

e. Cap the vials, place them on an agitator or rotator.

f. After 12 hours, change cellosolve solution.

g. After an additional 12 hours, replace cellosolve with absolute ethanol, 45 minutes at room temperature and pressure.

   Vials should continue to be agitated or rotated.

h. Replace first ethanol wash with a second, 45 minutes at room temperature and pressure.

2. Infiltration and Embedding

   a. Infiltrate the tissues at room temperature and pressure.

   b. All procedures utilizing Spurr's or Epon resin should be followed as described in Exercise Number 36B, parts 2b and 3a-d.

3. Sectioning

   Microtomy requires patience with ethylene glycol-treated tissue. The usual methods are applicable with a few special technical hints as follows:

a. Blocks are trimmed to widths no larger than 0.5 mm.

b. Sections may be best obtained by utilizing a diamond knife filled first with double distilled water.

   Once a few sections are cut, sectioning is to be stopped, and the water to be replaced by 100% ethylene glycol.

   With a pipette, carefully fill the diamond knife with glycol to the point where the level of liquid does not quite reach the knife edge.

   The meniscus of glycol is brought to the knife edge using an eyelash. Restart sectioning.

c. As new sections are obtained, and these are usually compressed, add glycol drop by drop until satisfactory sections, now flat, are cut.

   The optimum glycol level is principally determined through trial and error.

d. Wetting of specimen block face or the back face of the knife occurs commonly.

   In this circumstance, stop sectioning, remove the knife, rinse it with distilled water and dry it.

   The face of the specimen block should be rinsed with 100% ethanol and dried.

   Sectioning is resumed as in part 3b above.

e. Compressed sections may result despite all efforts to adjust glycol levels, change the knife angle, and so on.

   In such cases, it may be best to retrim or reorient the block face or simply begin sectioning anew under different sectioning parameters.

f. Use sectioning speeds of 0.5-2 mm/sec.

4. Section collection

a. The most practical means of collecting sections are touching grids to the glycol trough surface from above the liquid.

b. When sections are collected, there is usually a drop of residual glycol remaining with them.

Excess glycol is removed first with the corner of a piece of clean filter paper and then by placing the grids under vacuum for a few hours.

NOTE: THE PRINCIPAL REASON FOR UTILIZING THE ETHYLENE GLYCOL TREATMENT IS TO MINIMIZE MINERAL CHANGES IN THE CALCIFIED TISSUES.

LOSSES OF THE MINERAL AND OTHER ARTIFACTS MAY YET OCCUR DURING THE PROCEDURE AND A PARTICULARLY SENSITIVE TIME IS THAT PERIOD OF SECTION FLOTATION ON GLYCOL.

It is very important, then, to collect sections quickly (less than 1 minute later) and efficiently once they are sectioned and floating on the trough.

5. Staining

a. For determining certain features of the inorganic components of the calcified tissues, sections may be examined directly without additional staining.

The mineral, itself, in tissue sections prepared in the manner above, should be quite electron dense under the microscope.

b. Anhydrous staining may be used with an 8% solution of uranyl acetate in 100% ethylene glycol, absolute ethanol, or methanol.

Stain grids face down on single drops of filtered solution for 10 minutes to 1 hour at 37°C or at room temperature.

The exact conditions are determined best by experience.

Methanol is a solvent for parlodion (collodion) coated grids.

For methanolic uranyl solutions use formvar coated grids.

Lead salts are not used to stain sections anhydrously.

c. Stain sections on clean parafilm or wax plates in closed petri dishes.

Rinse grids by rapid and repeated dippings in a beaker of the appropriate solvent (glycol, absolute ethanol, or methanol).

d. Dry grids by touching them to filter paper and, for ethylene glycol, placing them under vacuum for a short time.

e. Conventional staining may be followed for 5 minutes each with an 8% solution of uranyl acetate in 50% ethanol-water and an aqueous solution of 2% lead citrate.

NOTE: IT SHOULD BE REALIZED THAT THIS STEP (e) WILL DEMINERALIZE THE THIN SECTIONS.

E. QUESTIONS:

1. If one wishes to compare the ultrastructural features in a sample of bone tissue, what would be the most suitable approach to preserve both organic and inorganic matrices of the specimen?

2. What is the physical-chemical basis for ethylene glycol fixation?

3. Could the anhydrous method of processing vertebrate tissues be useful for treatment of invertebrates? What basic differences exist between the two types of tissues as considerations for microscopy?

EXERCISE NUMBER 36D

SPECIAL TECHNIQUES FOR PROBLEM TISSUES

LIPIDS

E. Tessa Hedley-Whyte, M.D.

A. OBJECTIVES:

1. To preserve lipids in tissues
2. To immobilize and preserve the structure of cells
3. To obtain sections of tissues
4. To obtain sections of tissues which can be stained
5. To obtain sections which are stable under the electron beam

B. RATIONALE AND DISCUSSION:

Lipids are extracted from tissues by organic solvents. There are times when one wants to preserve the lipids either for direct examination or for examination with autoradiography or other labeling procedures.

This exercise avoids all contact of the tissues with organic solvents. The method is finicky and is rendered hopeless if the carbohydrazide is impure.

However, although glutaraldehyde urea is easier to prepare it is more difficult to section.

C. MATERIALS:

1. Glutaraldehyde, 50%
2. Carbohydrazide (CH)

NOTE: PURITY OF CARBOHYDRAZIDE IS CRITICAL. THE COMPANIES WE HAVE HAD GOOD LUCK WITH ARE POLYSCIENCES AND EASTMAN.

3. Barium carbonate

4. Sodium hydroxide, 0.1 M

5. Minced tissues that have been fixed in glutaraldehyde and postfixed in osmium tetroxide in buffer

6. Rubber molds

7. Crushed ice

8. Polymerized epoxy blocks

9. Unpolymerized epoxy

10. Applicator sticks

11. Mechanical paddle stirrer

12. Suction device

13. Whatman #3 filter paper

14. pH meter

15. Balance

16. Salt-ice water bath

17. Disposable syringe

18. Cold room

19. Mechanical rotator

D. PROCEDURE:

<u>Neutralization of glutaraldehyde</u>

NOTE: THE PH OF THE GLUTARALDEHYDE AT THE TIME OF CARBOHYDRAZIDE ADDITION IS IMPORTANT.

IF PH VARIES SIGNIFICANTLY FROM 7.0, POLYMERIZATION WILL BE INITIATED BEFORE MIXING IS ACCOMPLISHED, EVEN AT 0^0C.

1. Add a small amount of $BaCO_3$ to 50% glutaraldehyde (1 g/100 ml).

2. Stir vigorously for 10 minutes.

3. Filter with suction through Whatman #3 paper.

NOTE: THE BARIUM SALT COMPLEXES WITH ACID DECOMPOSITION PRODUCTS PRESENT IN GLUTARALDEHYDE OF HIGH CONCENTRATION.

4. Determine pH and adjust to 7.0 with NaOH if necessary.

5. Weigh out and keep chilled 150 mg carbonhydrazide/ml neutralized glutaraldehyde.

6. Pulverize all lumps to facilitate dissolving.

7. Place beaker of cold neutralized glutaraldehyde in a salt water-ice bath.

8. Stir vigorously with a mechanical stirring apparatus.

NOTE: A MAGNETIC STIRRER IS NOT POWERFUL ENOUGH FOR THIS THICK MIXTURE.

9. Gradually add about 1/3 of the weighed cold carbohydrazide and let stir for 20 minutes.

10. Add 1/2 of remaining CH and stir 20 minutes.

11. Add rest of CH and stir 20 minutes.

NOTE: REPLENISH ICE IN SALT-ICE BATH AS NECESSARY TO KEEP MIXTURE VERY COLD.

12. Pour glutaraldehyde-carbohydrazide (GACH) into chilled disposable syringes.

13. Wrap in foil and store at -20^{0}C.

    The embedment will keep for at least one month if frozen, or for up to 12 hours at 4^{0}C.

<u>GACH TISSUE PROCESSING</u>

Fixation and staining en bloc:

1. Nervous tissue: Fix in situ with 5% glutaraldehyde in cacodylate buffer.

2. Remove selected areas of fresh fixative for 1 hour.

3. Wash several times Sabatini's buffered sucrose.

4. Postfix tiny pieces of tissue in Dalton's chrome osmium 1 hour.

5. Rinse in Sabatini's solution.

6. Stain in 0.25% uranyl acetate in 0.01 M acetate buffer 1 hour.

7. Rinse several times in Glass Distilled Water (GDW).

NOTE: BUFFER SALTS INTERFERE WITH POLYMERIZATION OF GACH AND MUST BE WASHED OUT.

THIS STEP IS TO BE DONE IN THE COLDROOM SO THAT VIALS ARE THOROUGHLY CHILLED.

Infiltration:

NOTE: TO AVOID PREMATURE POLYMERIZATION OF GACH, ITS VARIOUS SOLUTIONS MUST BE PREPARED AND USED AT 0-4^{0}C.

ALL (DISPOSABLE) GLASSWARE TO BE IN CONTACT WITH GACH AND WATER FOR DILUTION MUST BE <u>COLD</u>.

GELATION OCCURS IN A MATTER OF MINUTES IF THE GACH IS WARMED SLIGHTLY.

1. Prepare 20% GACH in coldroom, using chilled GDW to dilute.

2. Add to chilled vials after removing GDW rinse.

3. Keep overnight, 0-4^{0}C.

4. 50% GACH, 2-4 hours on rotator at 40 rpm in coldroom.

5. 80% GACH, 2-4 hours on rotator at 40 rpm in coldroom.

6. 100% GACH, 1-2 hours on rotator at 40 rpm in coldroom.

7. To embed, place each block of tissue in a <u>tiny</u> drop of 100% GACH on a smooth surface such as the back of an embedding mold.

NOTE: BE CERTAIN TO KEEP REMAINING VIALS CHILLED WHILE HANDLING OTHERS.

8. Put in a 37^0C oven overnight.

   Droplets will crack extensively while curing, and polymerized GACH will be extremely hard and brittle.

9. Carefully remove blocks from support and glue onto epoxy rods for sectioning.

GACH SECTIONING

1. Trim very carefully with sharp razor blades.

NOTE: DO NOT ATTEMPT TO CUT THICK SLICES, AS BLOCKS CRACK AND CHIP OUT EASILY.

2. Use 38^0 angle knives for sectioning.

   These are prepared on a LKB Knifemaker by breaking trapezoids as shown in Figure 36D-a.

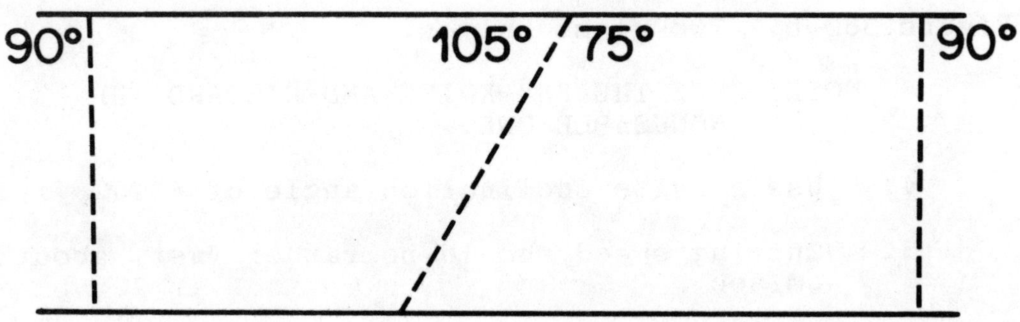

Figure 36D-a. Positioning of glass on knifemaker to obtain proper angle.

3. Use the knife from one half which has a 38^0 edge and a 90^0 base (See Figure 36D-b).

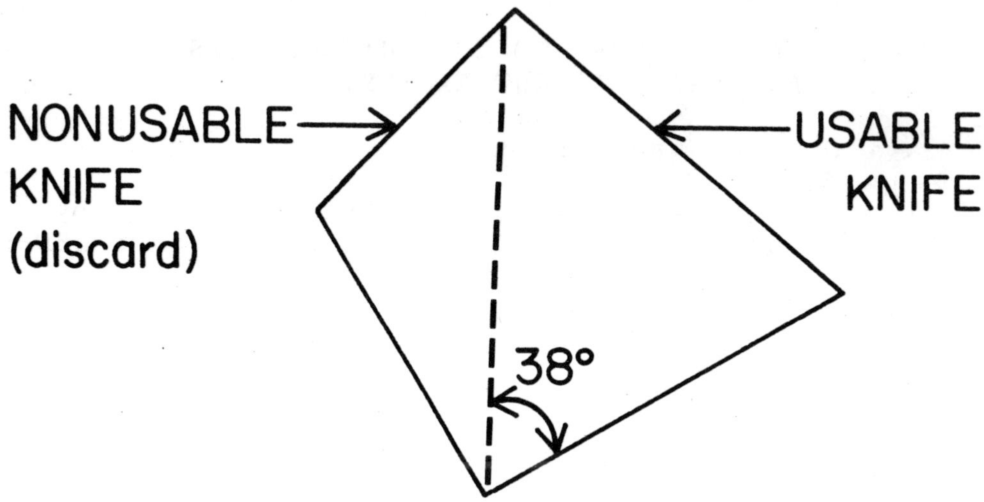

Figure 36D-b. Two glass knives.

NOTE: USE THE 38^0 KNIFE AND DISCARD THE NONUSABLE ONE.

4. Use a knife inclination angle of 4^0.

5. Cutting speed should be rather fast, about 0.6 cm/sec.

6. Keep water level low.

7. Do not cut sections thicker than about 0.5 μ or large cracks will develop during drying.

8. Dry sections and observe with phase optics or Paragon multiple stain for LM.

9. Stain EM grids with lead citrate.

NOTE: EM CONTRAST IS LOW.

MYELINATED TISSUES HAVE NUMEROUS SMALL CRACKS ALONG STRUCTURAL EDGES.

E. RESULTS:

1. 0.5 micron sections can be cut for orientation.

   On the whole they are less satisfactory than ultrathin sections.

2. Dark gold sections are approximately 100 nanometers thick as opposed to light gold sections with epoxy.

3. Tissue preservation is not as good as epoxy embedment but is adequate for specific purposes.

4. Contrast is less than with traditional embedment.

F. EVALUATION:

PRESENT A MICROGRAPH TO THE INSTRUCTOR AS A RESULT OF GACH TISSUE PROCESSING.

G. QUESTIONS:

1. What are the advantages of water miscible non-organic solvent tissue processing?

2. What are the disadvantages?

3. What effect does the various sizes of droplets have on polymerization and quality of the block?

EXERCISE NUMBER 37

PREPARATION OF PLANT TISSUES FOR SCANNING AND TRANSMISSION ELECTRON MICROSCOPY

Patricianne M. Hurd, PH.D.

A. OBJECTIVES:

1. To become familiar with the proper handling of plant tissues for electron microscopy

2. To preserve the structure of plant tissues using conventional fixatives

3. To become familiar with specialized fixation procedures for some selective plant tissue types

4. To study plant cell fine structure in sectioned material

5. To study plant cell fine structure by examining three-dimensional surface topography images

B. MATERIALS:

1. 3% glutaraldehyde in Michaelis buffer, pH 6.0

2. Osmium tetroxide (OsO_4) in Michaelis buffer, pH 6.0

3. 4% glutaraldehyde in cacodylate-HCl buffer, pH 6.8

4. 1% OsO_4 in cacodylate-HCl buffer pH 6.8

5. 8% paraformaldehyde-8% glutaraldehyde in cacodylate-HCl buffer, pH 7.2

6. 1% OsO_4 in cacodylate buffer, pH 7.2

7. 1% methylene blue

8. 1% noble agar

9. Water bath

10. Calcium chloride
11. Petri dishes
12. Disposable pipettes with rubber bulbs
13. Dark amber bottles
14. Glass vials with caps
15. Applicator sticks
16. new ethanol cleaned razor blades
17. Forceps
18. Lead citrate
19. Uranyl acetate
20. Ethanol
21. Acetone
22. Sodium acetate
23. Sodium veronal
24. Sodium chloride
25. Grids
26. Centrifuge
27. Spatula

NOTE: PREPARATION OF SOLUTIONS IS INCLUDED AT THE END OF THIS EXERCISE.

C. RATIONALE AND DISCUSSION:

The importance of fixation is discussed in exercise number 11 of this manual. As is the case with animal tissues, glutaraldehyde and osmium tetroxide are generally used for preparation of plant tissues for EM.

The presence of cellulose, chitin, tannins and lignins as plant cell components makes fixation difficult. Particular attention must be given to the pH of buffers and fixatives.

Plant cells that are highly vacuolated may also present difficulties if modifications are not made with the fixative solutions. Disruption of the cell membrane and dispersion of the cytoplasm are frequently observed.

Plant cells are more susceptible to various shifts in the pH than animal cells. This is because mature plant cells contain a large vacuole or vacuoles surrounded by a thin layer of cytoplasm. The osmotic pressure inside plant cells is high. Therefore a pH of 6.0–7.2 is suggested for the fixation.

This exercise is primarily written to expose the student to fixation procedures of some selective plant tissue types. The student has a choice of preparing blue-green algae, fungi or higher plant parts such as roots, petioles, stems and leaves for TEM and SEM.

D. PROCEDURE:

Blue-Green Algae and Other Microorganisms

1. Assemble materials under hood.
2. Fix specimens in 3% glutaraldehyde in Michaelis buffer, pH 6.0 for 1 hour at room temperature.
3. Centrifuge at 5000 g for five minutes.
4. Decant supernate and replace with buffer.
5. Resuspend pellet in buffer and let stand for 2 minutes.
6. Recentrifuge suspension for 3 minutes.
7. Decant supernate and replace with buffer.
8. Recentrifuge again for 3 minutes.
9. Repeat steps 7 and 8.
10. Suspend pellet in OsO_4 in Michaelis buffer.
11. Fix for 1-2 hours at 4^0C.
12. Prepare 1% nobel agar.

13. Allow agar to cool at room temperature slightly above solidification (*ca* 45-50 ^{0}C).

14. Keep in water bath until postfixation is completed.

15. Following osmication, place pellet in Michaelis buffer.

16. Allow to stand for 2 minutes.

17. Centrifuge (3X) for 3 minutes each decanting supernate each time and replacing with buffer.

18. Mix pellet with 1-2 ml of 1% noble agar and 1 drop of 1% methylene blue.

19. Quickly pour mixture on a glass slide and allow to solidify.

20. Cut the mixture into small square pieces.

21. Using a spatula, transfer agar blocks to several vials.

22. Dehydrate blocks in 25, 50, 70, and 100% ethanol, 5 minutes each and again in 100% (2X), 5 minutes each.

23. Infiltrate in propylene oxide and spurr low viscosity resin in ratios of 2:1, 1:2 and 1:1, 30 minutes each.

24. Leave in full strength spurr resin overnight.

25. Embed tissues in beem capsules and polymerize in oven for 24-48 hours at 70 ^{0}C.

26. Trim blocks.

27. Cut 0.25-1 micron sections for light microscopy.

28. Stain with methylene blue.

29. Cut ultrathin sections (500-800 A^{0}).

30. Stain in lead citrate and uranyl acetate.

Fungi

1. Fix specimens in 4% glutaraldehyde in cacodylate-HCl buffer at pH 6.8, 2 hours at room temperature.

NOTE: LEAVE SPECIMENS UNDER VACUUM FOR 10-20 MINUTES AT THE BEGINNING OF FIXATION.

THIS WILL FACILITATE BETTER PENETRATION OF THE FIXATIVE IF CHITIN IS PRESENT IN THE CELL WALL.

2. Rinse specimens in cacodylate-HCl (3X) 5 minutes each.

3. Postfix in 1% OsO_4 in cacodylate buffer for 1 hour at 4^0C.

4. Dehydrate in acetone (15, 30, 70, 95, and 100%) (3 changes of 100%, 5 minutes each).

5. Infiltrate in spurr-acetone mixture in ratios of 2:1, 1:2, and 1:1 for 15, 30, and 60 minutes respectively.

6. Repeat steps 25-30 of procedure.

Higher plants (roots, petioles, stems and leaves)

NOTE: TO PREVENT BLOCKAGE IN THE PLANT CELLS, THE BRANCH OF THE PLANT FROM WHICH THE PLANT PART IS OBTAINED SHOULD BE RE-CUT UNDER WATER.

1. Place cacodylate-HCl buffer in petri dish.

2. Cut plant parts into quarter pieces.

3. Remove buffer and fix in 8% glutaraldehyde in cacodylate-HCl buffer for 1.5 hours, pH 7.2 at room temperature.

4. Rinse specimen in the same buffer for 1 hour.

5. Postfix in 1% OsO_4 for 2 hours at 4^0C.

6. Repeat previously described steps for fungi (4-6).

Scanning Electron Microscopy of Plant Tissues

NOTE: FOR PREPARATION OF PLANT TISSUES FOR SEM SEE EXERCISE NUMBERS 20A, B AND C OF THIS MANUAL.

Preparation of Solutions:

1. 3% glutaraldehyde in Michaelis buffer

   25% glutaraldehyde, 12 ml Michaelis buffer (pH 6.0) to make 100 ml

2. Osmium Tetroxide

   Solution A

   5 ml stock Michaelis buffer (pH 6.0)
   7 ml of 0.1 N HCl
   13 ml of distilled water
   0.25 ml 1 M $CaCl_2$

   Solution B

   Osmium tetroxide 0.5 gm
   double distilled or deionized water, 25 ml

   Working Solution

   Dilute solution A 1:1 with solution B

3. 4% Glutaraldehyde in cacodylate-HCl buffer

   25% Glutaraldehyde 16 ml
   Cacodylate-HCl at (pH 6.8)
   to make 100 ml

4. 1% OsO_4 in cacodylate-HCl buffer

   Osmium tetroxide 1 gm
   Cacodylate-HCl buffer (pH 6.8) 100 ml

5. 8% paraformaldehyde - 8% glutaraldehyde in cacodylate buffer

   8% paraformaldehyde 25 ml
   8% glutaraldehyde 25 ml
   0.2 M cacodylate buffer, (pH 7.2), 25 ml
   distilled water, 25 ml

   1% OsO_4 in cacodylate buffer
   osmium tetroxide 1 gm
   0.2 M cacodylate 100 ml
   buffer (pH 7.2)

NOTE: PREPARE THE FIXATIVE IN A DARK BOTTLE.

6. Methylene Blue 1 gm

   distilled water 100 ml

7. 1 M calcium chloride ($CaCl_2$)

   $CaCl_2$ 11 g
   boiled distilled water 100 ml

8. 1% noble agar

   noble agar 1 gm
   distilled water 100 ml
   autoclave media

9. Michaelis buffer

   1.94 gm sodium acetate
   2.94 gm sodium veronal (barbital)
   3.4 gm NaCl
   add water to 100 ml
   adjust pH to 6.0

10. Cacodylate-HCl (0.2 M)

   Solution A

   Sodium cacodylate 42.8 gm
   $Na(CH_3)_2 AsO_2 \cdot 3H_2O$
   Distilled water 1000 ml

   Solution B

   0.2 M HCl
   conc. HCl (36-38%) 10 ml
   distilled water 603 ml

   Working Solutions:

   a) To obtain a pH of 6.0, add 50 ml of solution A and 29.6 ml of solution B and dilute to a total volume of 200 ml.

   b) To obtain a pH of 6.8, add 50 ml of solution A to 9.3 ml of solution B and dilute to a total volume of 200 ml.

   c) To obtain a pH of 7.2, add 50 ml of solution A to 4.2 ml of HCl.

   Dilute to 200 ml.

E. EVALUATION:

SELECT ONE PROCEDURE OF YOUR CHOICE FOR PREPARATION OF PLANT TISSUES.

CARRY OUT IN THE LABORATORY AND PRESENT ONE MICROGRAPH TO THE INSTRUCTOR FOR EVALUATION.

F. QUESTIONS:

1. Why are plant cells more susceptible to pH shifts than animal cells?

2. How may changes in the pH of the buffer and fixative affect the fine structure of the plant cell?

3. Draw and label a typical plant cell and its components.

EXERCISE NUMBER 38

PREPARATION OF NON-BIOLOGICAL SPECIMENS FOR TRANSMISSION ELECTRON MICROSCOPY

Anthony J. Garratt-Reed, Ph.D.

A. INTRODUCTION:

The remainder of this manual has taken the user, by way of a series of practical demonstrations, through the range of procedures commonly used in the microscopy of biological materials.

In this exercise we will address the problems of preparing specimens of non-biological materials for transmission electron microscopy.

However, as this field is quite as large as that for biological materials, we will adopt a different format for the exercise. This section will take the form of a narrative rather than a detailed guide through the procedures. It is not expected that the reader will as a result of reading this exercise, be able to produce specimens of the material of interest.

Rather, it is intended to make the reader aware of the philosophy of the various techniques available, so that should it be desired to perform microscopy on non-biological materials, the user will know at least the sort of detailed advice that will be required.

B. OBJECTIVES:

The objective of any specimen preparation technique for transmission electron microscopy (of biological or non-biological materials) is to produce an electron transparent section of the specimen of interest, without introducing artifacts.

This can, in fact, be quite demanding, for in some cases it implies that a specimen must be cut from the bulk sample and thinned to a few tens of a nanometer, without ever introducing even the slightest bend.

It is also necessary that the section be small enough to fit into the specimen holder, which is usually 3.05 mm in diameter. While this is rarely a problem for the biologist, it is rarely not so for the materials scientist.

Even laboratory samples are usually several centimeters across and it is common to perform microscopy on specimens as diverse as lunar rock, parts of nuclear reactors, parts of automobiles, etc.

Thus in general, preparation of specimens proceed in two steps:

a) Forming of small piece of material from the bulk

b) Final thinning to produce electron transparent regions

C. INITIAL FORMING OF SPECIMENS:

It is usual when performing microscopy of larger specimens to form a disc 3.0 mm in diameter and 0.1-0.5 mm thick. The central portion of this disc will later be subjected to a final thinning, resulting in a self-supporting specimen which can be placed in the microscope without the need for supporting grids.

Before selecting the method of cutting, it is necessary to consider the problem under investigation. For instance, if dislocation structures are important, then no deformation of the specimen can be allowed.

Grinding the specimen on 600-grade paper (very fine) will cause deformation typically to 0.2 mm from the surface.

Thus final grinding must leave the specimen 0.5 mm thick to allow for chemical polishing to remove the deformed layer from either side and to leave undeformed material for study.

If on the other hand, only the grain size on precipitate distribution is to be studied, then the presence of a few spurious dislocations will not affect the results, and it is possible to grind the specimen much thinner.

It is also necessary to consider the effect of the specimen on the microscope. In general, non-magnetic specimens will not cause a problem.

However, magnetic specimens will severely distort the magnetic field of the objective lens, producing extreme astigmatism and beam tilt, and on some occasions the force on the specimen may overcome the restraining system, and pull the specimen out of the holder.

Not only will the user loose the specimen, but the microscope maintenance personnel and other users will be very unhappy, for the microscope will have to be dismantled to remove the specimen!

Both these difficulties can be mitigated by using less magnetic material.

Some ways that 3.0 mm diameter disks can be made from metal samples are:

a) Form a rod 3.0 mm diameter by careful toning on a lathe.

Cut slices using a diamond slitting wheel or a spark erosion machine with a wire-cutting attachment.

b) Cut thin slices from a larger specimen using a diamond slitting wheel.

Punch out 3.0 mm disks using a good quality mechanical punch, or a spark erosion machine with a 3.0 mm ID tube as a tool, or an ultrasonic drill with a similar tool.

c) Form thin slices by grinding.

Hold the specimen in a jig or glue to a block.

Using shims, grind to a thickness of about 0.5 mm.

Unglue and re-glue inverted.

Grind to final desired thickness. Prepare disks as in case (b) above.

If the sample is ceramic, then probably only careful grinding will be possible. Forming into discs will be done with an ultrasonic drill.

D. FINAL THINNING:

By far the majority of specimens will be thinned using either a jet electropolisher or an ion-beam milling machine.

Before describing these processes in more detail, however, we will list other processes which could be applied to prepare thin specimens:

a) Collect thin samples directly, eg. dust particles in smoke, or use finely powdered samples.

b) Microtome samples. Applicable mainly to polymers, or to powders embedded in resin.

c) Chemical solution.

The sample is suspended in a suitable chemical polish and simply dissolves slowly away.

If a depression is first formed in the center of one of the faces, (by, for example, using a spark machine with a hemispherical tool) then eventually a perforation will occur.

Polishing is terminated, and the area round the perforation is examined.

d) It sometimes is the case that the sample as made has suitable thin areas.

No other preparation whatever is required - simply place the specimen in the microscope!

NOTE: OFTEN GRAPHITE FIBERS FALL INTO THIS CATEGORY.

E. JET-POLISHING:

The basic scheme for a jet-electropolishing system is illustrated in Figure 38-a. Exact details of course, vary from system to system, but always the specimen is held in a plastic (usually teflon or similar material) holder in between two opposed jets of electrolyte.

A current is passed between platinum electrodes in the jets (the cathodes) and the specimen (the anode). Electro-chemical attack, assisted by the flow of electrolyte causes dishing and eventual perforation of the specimen.

Suitable choice of electrolyte, temperature, voltage and jet velocity allows successful polishing of many metallic specimens to be undertaken.

Problems may well be encountered with multi-phase systems, where it will be found that differential polishing on local electrochemical attack produces poor local uniformity.

Jet-electropolishing clearly cannot be applied to non-conductive samples, although sometimes it can be possible for jet chemical polishing to be useful.

F. ION-BEAM MILLING:

If ions of moderate energy ($\approx$3-10 KeV) inpact on a solid, then in a high proportion of cases, one or more of the atoms of the surface will be dislodged and removed, in a process known as sputtering.

The ion-beam milling machine exploits this process. A beam typically of argon ions, is shown at a moderately oblique angle onto the surfaces of the specimen, which is mounted on a rotating stage (See Figure 39-b).

The oblique angle enhances the sputtering yield, and the rotation of the stage promotes uniform thinning.

With most commercial machines, thinning rates typically of 1-5 μm/hr will be obtained (depending upon the material), so to avoid extremely prolonged milling, initial thicknesses of $<$ 100 μm should be aimed for.

Ion-beam milling will work on almost all specimens, but because it is slow, it is usually only when other methods are not available.

Good uniformity is obtained locally (eg. from phase to phase), but usually the large-scale uniformity of the specimen (over distances of the order of several microns) is rather poor.

As examples of the considerations we have listed, we shall conclude by listing for two specific examples, the actual processes aimed out to prepare specimens.

1) The sample is a portion of 2" diameter steam pipe which failed in service.

   The aim is to prepare samples enabling us to study precipitation on the grain boundaries.

   a) Using hacksaw, cut piece of pipe about 1" long.

   b) Again using hacksaw, cut longitudinal slice roughly 1/4" wide from the 1" long portion.

c) Glue this slice to a brass, steel or other metal block (about 2" x 2" x 1/2") using Duco or similar cement, concave side towards block.

d) On rotary grinder, grind so that exposed surface of pipe is flat.

e) Dissolve glue in acetone, invert sample and re-glue.

f) Grind again until exposed surface is flat. You should now have a flat slice about 1" x 1/4" x ~0.04". Dissolve glue.

g) Using a good quality mechanical punch, diameter 0.120", punch out four or five samples.

h) Glue these, again using Duco back to the block.

i) Using suitable shims, grind specimens to a thickness of 0.010 $\pm$ 0.002, finishing on 600-grade SiC paper.

j) Dissolve the glue, invert the specimens and re-glue.

k) Again using suitable shims (2 sheets of paper, for example) grind specimens to 0.005 $\pm$ 0.001, again finishing on 600-grade paper.

l) Dissolve the glue, wash specimens well in acetone.

m) Electropolish using 20% Perchloric acid in Methanol at -30^0C.

NOTE: PERCHLORIC ACID SOLUTIONS ARE POTENTIALLY EXPLOSIVE.

DO NOT USE THIS SOLUTION UNLESS YOU HAVE RECEIVED ADEQUATE INSTRUCTION ON THE USE AND HAZARDS OF PERCHLORIC ACID.

n) Wash well in reagent grade methanol, dry using hot-air blower.

o) Examine in the electron microscope.

2. The sample is magnesia doped with trace levels of scandium. It is in the form of a disk 1 cm in diameter, 1 mm thick.

The aim is to study scandium segregation to the grain boundaries.

a) Carefully break the disk into smaller pieces.

Select several whose dimensions do not exceed 3.0 mm, but which are as large as possible otherwise.

b) Fix these specimens to a brass block or jig from an automatic polishing machine using hard wax.

c) Using diamond paste, polish the specimens until their thickness is approximately halved.

d) Melt the wax, invert the specimens and re-fix.

e) Continue to polish until the specimens near the edges of the block begin to break-up.

f) Very carefully melt the wax, wash the specimens in acetone carefully.

g) Working under a binocular microscope, fix specimens using epoxy adhesive to slotted copper grids.

h) Ion-mill samples to perforation, using argon ions at 6.0 KV, at an incident angle of 15^0.

This process will probably take about 1 day.

i) Using a vacuum evaporator, deposit a thin (5-10 mm) layer of carbon on one surface of the sample.

The carbon bleeds away the charge that would otherwise build up on the insulating specimens.

j) Examine in a scanning transmission electron microscope.

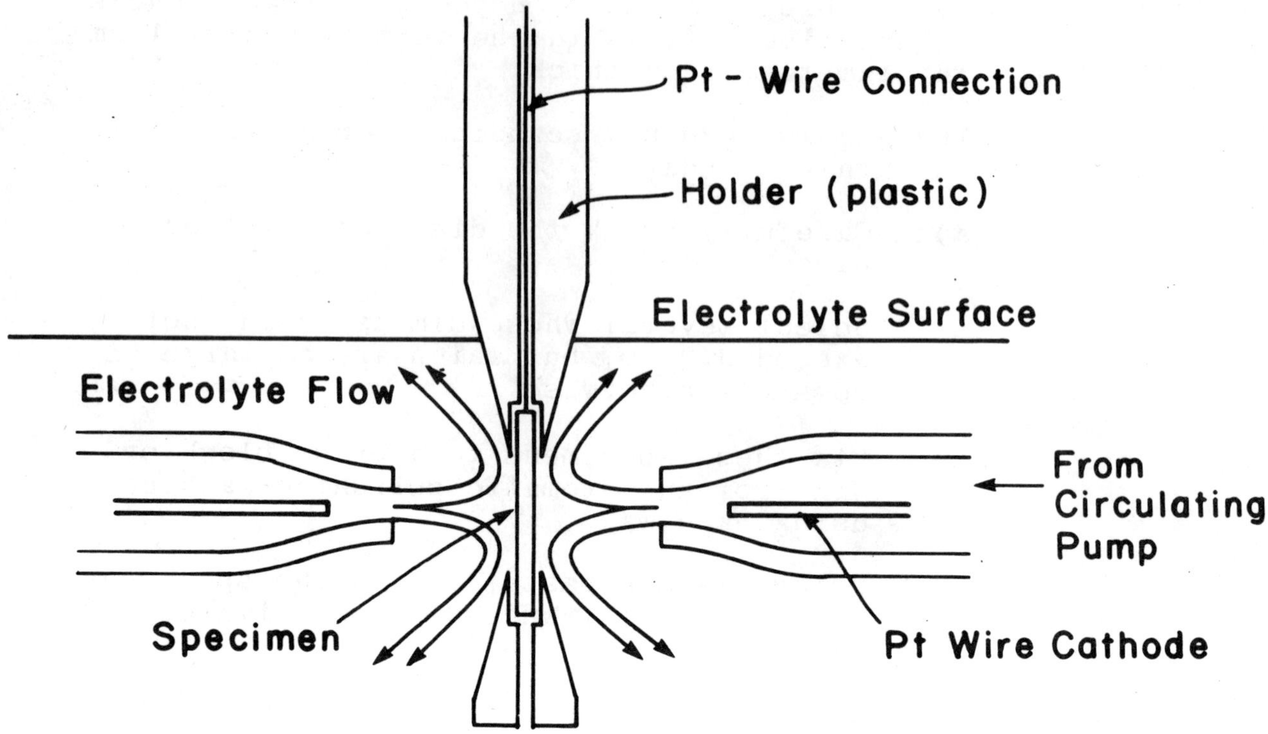

FIGURE 38-a. Basic arrangement of jet-electro-polisher.

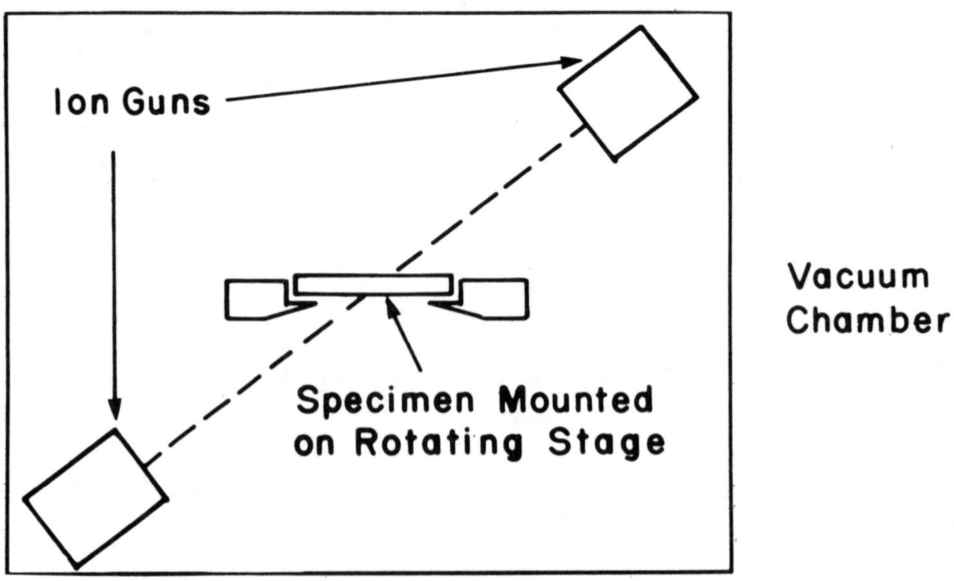

FIGURE 38-b. Basic arrangement of Ion-beam miller.

G. QUESTIONS:

1. What type of information might be sought from samples of (a) metals, (b) minerals, (c) polymers?

2. What constitutes a "satisfactory" electron microscopy specimen of typical inorganic material?

3. When planning a preparative procedure, for a specimen, what factors must be taken into consideration?

4. How does an ion mill work? What type of specimens are typically used for it?

5. How does a jet-electropolishing system work? What type of specimens are typically used for it?

EXERCISE NUMBER 39

QUANTITATIVE ANALYSIS OF ELECTRON MICROGRAPHS

Brenda R. Eisenberg, Ph.D.

A. OBJECTIVES:

1. To estimate the volume of a cell organelle as a fraction of the cell volume (volume fraction)

2. To estimate the surface area of a cell organelle per unit volume of the cell (surface density)

B. MATERIALS:

1. 2 clear plastic sheets
2. Ruler
3. Felt tip pen
4. 1 cm graph paper
5. Calculator
6. Tape
7. String
8. Paper clips
9. 3 Micrographs are provided for students to measure
10. Optional: A Laboratory Tally Counter will speed up counting
11. Optional: Student to provide a set of 10 Micrographs suitable to measure

C. RATIONALE AND DISCUSSION:

Electron Micrographs provide visual information about cells and organelles and these can be readily described in qualitative terms. We often wish to compare cell composition in response to some experimental intervention. Descriptive words allow us to say an organelle is larger or more abundant than before.

But we need better methods to give us numerical values in order to make quantitative statements about the extent of the change. The scientific discipline of stereology, or morphometry enables us to make meaningful measurements from micrographs.

A cell is a solid three-dimensional object which contains other three-dimensional organelles inside. When we cut an ultrathin section and view it in the electron microscope we see flat two-dimensional profiles. The outer profile comes from the cell and the numerous small, irregular shaped profiles come from the organelles.

We view many sections through many cells and gain experience from which to reconstruct a three-dimensional model of how the solid organelles are packed within a typical cell. One micrograph cannot give us enough information.

In quantitative measurements, we face the same problem of choosing enough micrographs to provide a good representative sample of a cell. Measurement of these few micrographs will then produce a good statistical estimate of the composition of the population of cells from which they were taken.

Unfortunately the choice of "enough representative" micrographs varies from one project to the next and it is difficult to suggest a universal method in a laboratory exercise. The most important point is that the selection of the micrographs must be made in an unbiased way. If all the micrographs are centered on the nucleus, for example, the amount of nucleus in the cell will be overestimated and the amount of cytoplasm will be underestimated.

Therefore, we propose some simple rules to generate the first group of micrographs from one section. In practice, this is by photographing at the corners of the copper EM grid (Figure 1).

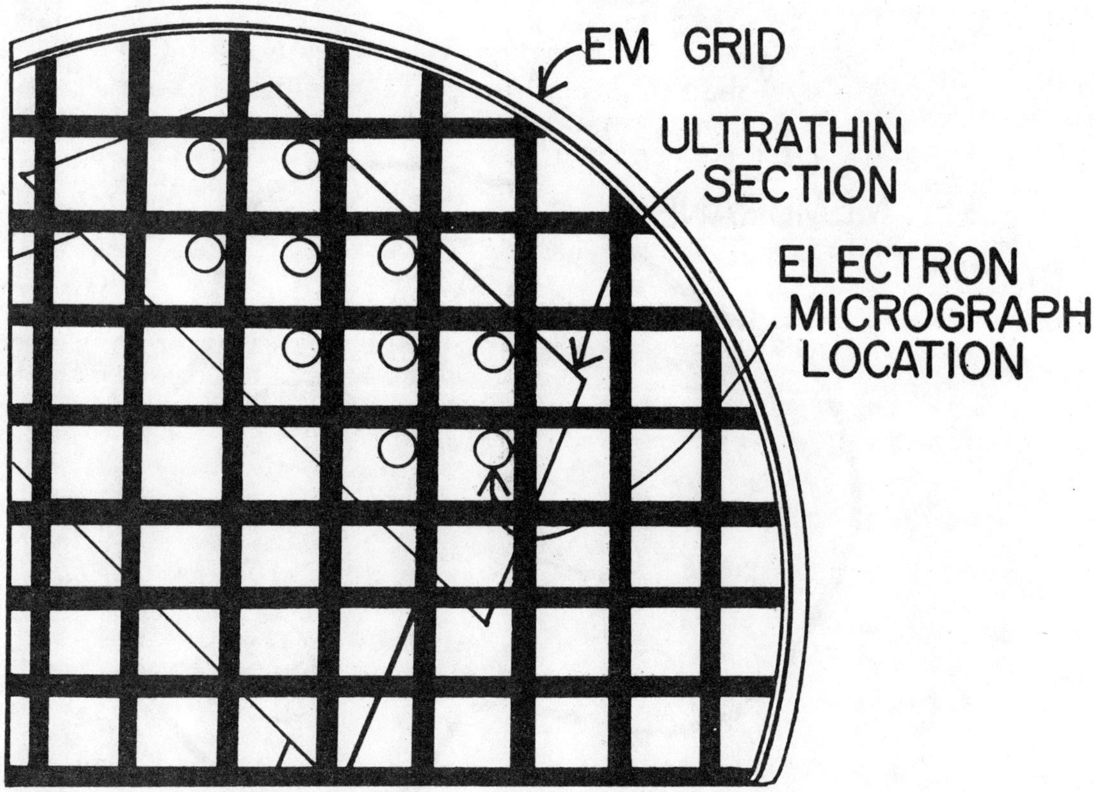

Figure 1. One section on an EM grid to show unbiased selection of 10 fields for electron micrographs at the upper left corners of grid spaces.

The volume density or volume fraction is defined as the volume of the organelle, V_o, per unit volume of the cell, V_{cell}. The volume density can be written as a fraction of V_o/V_{cell} which has no units. The volume density allows us to say how much of the cell volume is occupied by an organelle.

For example, if the nuclear volume occupies one tenth of the cell volume, then $V_{nucleus}/V_{cell} = 0.1$; or if the mitochondrial volume occupies one quarter of the cell volume it is written as $V_{mitochondria}/V_{cell} = 0.25$.

The electronmicrographs have flat 2-dimensional profiles, not real 3-dimensional volumes. The two dimensional areas contained within the profiles can be mathematically related to the 3-dimensional bodies from which they were sectioned (Figure 2).

If the area of the cell is A_{cell} and the area of all the organelle profiles within the cell is A_o then a simple equation relates the volume ratios to the areal ratios:

$$\frac{V_o}{V_{cell}} = \frac{A_o}{A_{cell}} \quad (1)$$

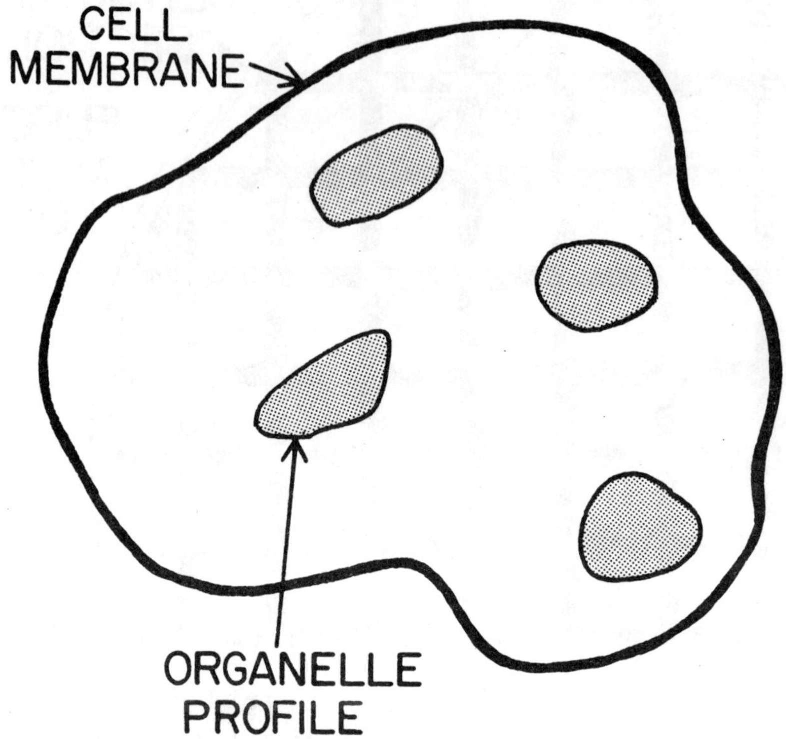

Figure 2. The cell membrane encloses a 2-dimensional area of cell, A_{cell}. The four organelle profiles summed together enclose the organelle are, A_o.

This means we need a method to measure areas and we can then calculate the volume fraction from equation 1. Expensive methods include digital planimeters to trace areas.

Cheap methods include cutting out the area and weighing them (but you end with confetti!). We suggest that the best, cheap and efficient method to estimate the area is by point counting.

Points are drawn on a plastic sheet which is then placed over a micrograph. It is practical to use points located at the intersection of a square test lattice (Figure 3).

A thick line can be used and the location just inside the upper left corner of each cross hair is taken as the test point. The distance between the points is d in cm of the plastic sheet which is placed over the micrograph.

Every point P has a little square area, d^2, associated with it. The actual distance between the points, D (units of cm of tissue), will depend on the magnification of the micrograph, M, where M is given by

$$M = \frac{d\ (1\ \text{cm of micrograph})}{D\ (1\ \text{cm of tissue})} = \frac{d\ (1\ \text{cm of micrograph})}{D\ (10^4\ \mu\text{m of tissue})} \qquad (2)$$

and 1 cm is 10^4 μm.

For example, 1 cm of micrograph is 1 μm of tissue, if m = 10,000 x. an actual area of tissue a is given by

$$a = pd^2 = pd^2/m^2 \text{ in units of cm}^2 \text{ of tissue.} \qquad (3)$$

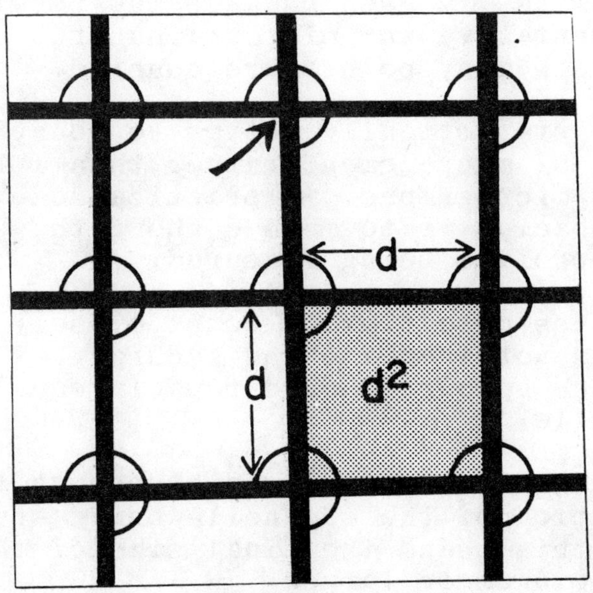

Figure 3. A square test lattice with 9 points, P, at corners. Every point has an area, d^2 associated with it.

Arrow indicates exact location of one test point.

The test points are placed over the micrograph (Figure 4) and we count the number of points, P_o, lying on top of the organelles and the number of points, P_{cell}, lying on top of the cell. Then $A_o = d^2 P_o$ and $A_{cell} = d^2 P_{cell}$,

so
$$\frac{V_o}{V_{cell}} = \frac{A_o}{A_{cell}} = \frac{d^2 P_o}{d^2 P_{cell}} \tag{4}$$

The area of the little square, d^2, cancels which gives

$$\frac{V_o}{V_{cell}} = \frac{P_o}{P_{cell}} \tag{5}$$

Therefore, we can estimate the volume of the organelle as a fraction of the cell volume, simply by taking the ratio of the point counts (Figure 4). The word "estimate" is used because the result depends on how representative the micrographs are, how many are taken and how many points are counted.

It is statistically better to count a few points from a lot of micrographs rather than a lot of points from a few micrographs. A practical choice of the point to point spacing is to make d about the same size as the organelle profile being measured.

It takes more time to count closely spaced points but it does not increase the accuracy of the method. Test lattice spacing should roughly equal the size of the organelle.

The magnification should be enough to make a clear identification of the organelle being studied. In practice, this means the final size of the organelle should be 0.5 cm or larger.

Nuclei are about 5 μm or 5×10^{-4} cm in diameter and need 1000 x to make a size on the micrograph of 0.5 cm (using equation 2). Similarly, mitochondria of about 0.5 μm diameter need 10,000 x and endoplasmic reticulum of about 0.2 μm in width needs 25,000 x.

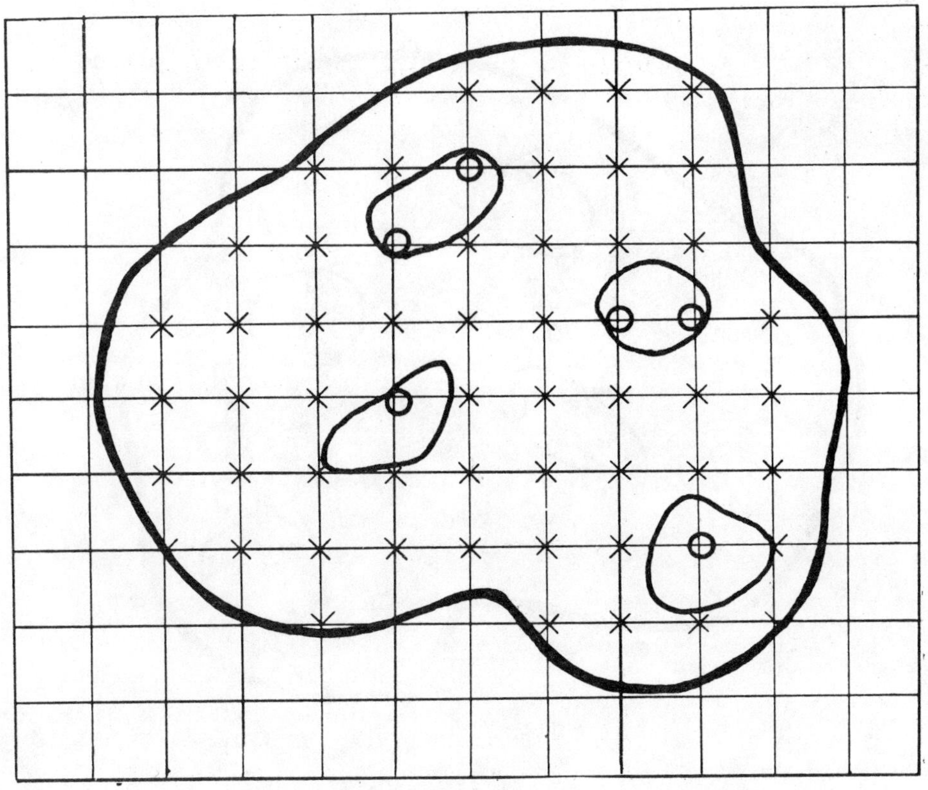

Figure 4. Point counting method to estimate areas. Six points circled fall over the organelles, P_o.

The 52 points marked with x fall over the cytoplasm, P_{cyt}. The points over the cell, P_{cell}, are $P_o + P_{cyt} = 58$. From equation 5, the volume density is

$$\frac{V_o}{V_{cell}} = \frac{P_o}{P_{cell}} = \frac{6}{58} = 0.103,$$

or the organelles occupy 10.3% of the cell volume.

The <u>surface density</u> is defined as the surface area of the organelles, S_o, per unit volume of the cell, V_{cell}.

The surface density can be written as a fraction S_o/V_{cell} and has the units of area/volume or cm^2/cm^3, or cm^{-1}.

This quantity is the total surface area of numerous small organelles in cm^2 referred to a unit volume of cell in cm^3.

This ratio is not an obvious idea. Very large areas of intracellular membranes can be folded up inside a unit volume of a cell. Most of us think in terms of a surface wrapped around its own volume, such as S_o/V_o, where V_o is the volume of the organelle. This surface-to-volume ratio can also be calculated.

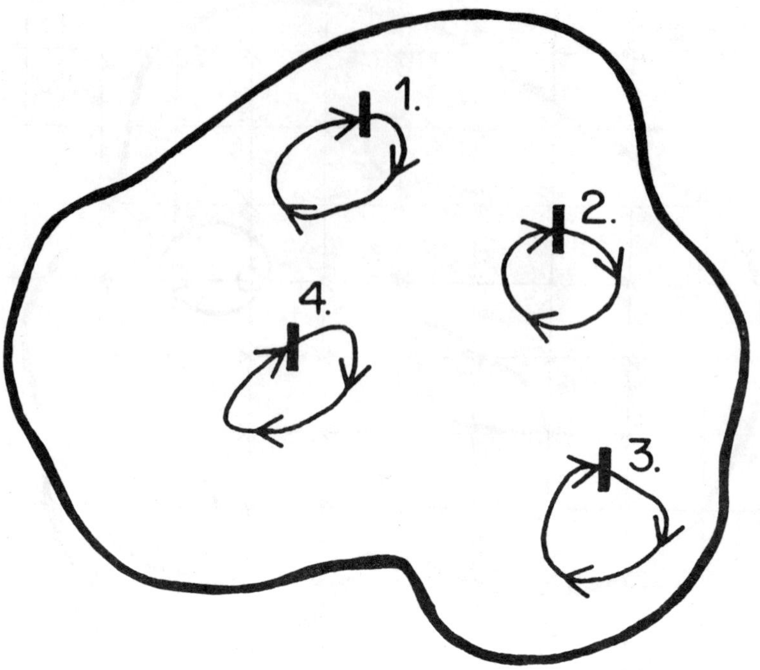

Figure 5. The boundary length, B_1, B_2..., of every profile is measured and summed. The total boundary length of the organelles, B_o is given by $B_1 + B_2 + B_3 + B_4$. String, a map wheel, or an electronic tracer is used to measure B_o.

The electron micrograph traced in (Figure 5) shows profiles or organelles in an area of cell. We wish to know which property in this 2D micrograph corresponds to the surface area of the organelle in 3D.

It can be mathematically shown that the required property is the total boundary length, B, traced around the perimeters of all the organelle profiles. For randomly oriented, uniformly distributed organelles the equation relating the surface density to the boundary lengths is

$$\frac{S_o}{V_{cell}} = \frac{4\ B_o}{\pi A_{cell}} \qquad (6)$$

where the constant $4/\pi = 1.27$ is derived from geometric probability theory.

This equation (6) means that if we can find a method to measure the boundary length, B, and then use the point counting method to measure A discussed above, we can calculate the surface density.

Expensive methods to measure B include digital tracers. Cheap methods include map wheels and segments of strings stretched over each profile which you can try in an exercise.

We suggest that a better, cheap and more efficient method is intersection counting.

An intersection, I, is counted when a test line intersects with the boundary of a profile, (Figure 6). It should be obvious that "I" will be bigger if the boundary length is longer or if more total length of the parallel test line (L_{cell}) is placed over the micrograph.

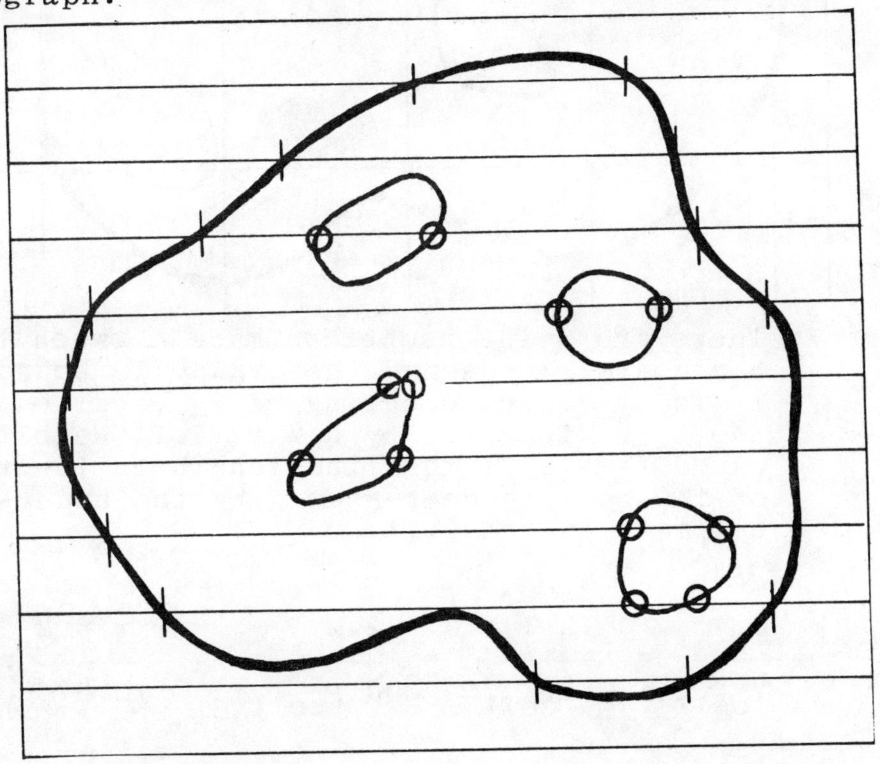

Figure 6. Intersections, I_o, are made between the test line and the organelle boundaries. $I_o = 12$. The total test line length over the cell, L_{cell}, marked by vertical lines at the cell membrane, can be measured by a ruler (approx 62 cm of print).

It is difficult to use geometric probability theory to derive the equation relating the surface density S_o/V_{cell} to I and L even for randomly distributed organelles. Fortunately, the equation for calculating surface density is simple and easy to use:

$$\frac{S_o}{V_{cell}} = \frac{2\ I_o}{L_{cell}} \tag{7}$$

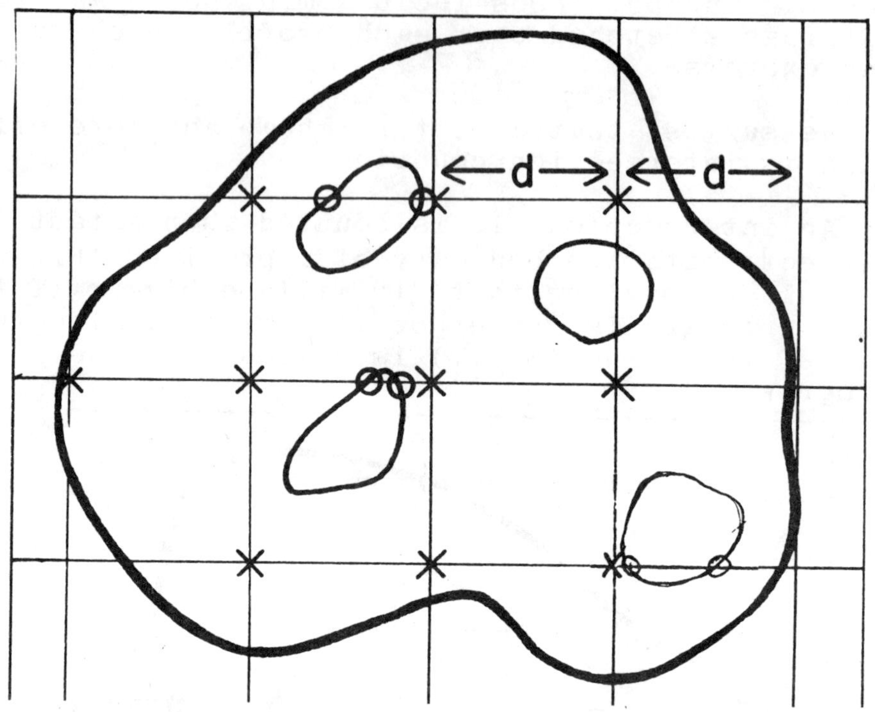

Figure 7. There are 6 intersections made between the horizontal lines and the organelle boundaries. I_o is 6. Point counting of P_{cell} gives 10, so the L_{cell} is given by $d/M\ P_{cell}$. Note the overestimate of the line length on the right of the cell is compensated by the underestimate on the left.

Then, $$\frac{S_o}{V_{cell}} = \frac{2\ I_o}{L_{cell}} = \frac{2\ M\ I_o}{d\ P_{cell}} = \frac{2 \times 6\ M}{10\ d}\ cm^2/cm^3$$

The length of test line falling over the cell L_{cell} could be measured in cm with a ruler but this is slow and inaccurate. In practice the "I" counting is done using the horizontal lines of a square test lattice with test segment length, d in cm of print, (Figure 7). The lattice spacing itself can be used as a ruler with $L_{cell} = d/M\ P_{cell}$, in units of cm of tissue. Therefore, substituting for L_{cell}, we find

$$\frac{S_o}{V_{cell}} = \frac{2MI_o}{d\ P_{cell}} \tag{8}$$

The units will be cm^2 of surface area per cm^3 of cell volume or cm^{-1}.

We count I_o and P_{cell} and we know the magnification M and test lattice space d in cm of print. This equation (8) is used to estimate the total surface area of the organelles per unit volume of the cell. Other surface-to-volume ratios can be calculated by counting the intersection of test lines with other structures.

D. PROCEDURE:

1. Make a 2 cm square test lattice. Tape the plastic sheet over the 1 cm graph paper and line up carefully.

   Use felt tip pen to draw horizontal and vertical lines at 2 cm spacing in both directions to create squares.

2. Make a 1 cm square test lattice. Set aside until Step 7.

3. Use paper clips to attach 2 cm square test lattice firmly over exercise micrograph #1.

   Avoid having the outer margins of the micrograph exactly covered by the test points.

   Move the test lattice up and sideways a bit to prevent this.

4. Volume Fractions. Use 2 cm lattice to make point counts of the nucleus, P_n, the mitochondria, P_m, the cytoplasm, P_{cyt}, and the extracellular space, P_{ecs}.

   The test point is just inside the corner (See Figure 3). If you cannot decide what it is, score 1/2 P for each adjacent organelle.

   Enter P counts onto Table I in RESULTS.

5. Use 2 cm test lattice and repeat the P counts on micrographs #2 and #3.

6. Add up the P counts in each column and use the summed P counts to calculate the ratios of volume fractions on the RESULTS pages.

7. Repeat Steps 4-6 using the 1 cm test lattice and make calculations on Table II.

8. <u>Surface Density.</u> Use horizontal lines from 2 cm test lattice to make intersection counts of mitochondria, I_m, and the cell membrane, I_{cell}.

   Transfer the points counted in Step 4, Table I of the mitochondria, P_m, and the cell, P_{cell}.

   Use all 3 exercise micrographs and enter I and P counts into Table III.

9. Sum the counts and calculate surface density from the equations given under Table III.

10. Repeat Steps 8-9 using the 1 cm test lattice.

11. Place string carefully over the length of the cell membranes in micrograph #1 to measure total boundary, B_{cell}.

    Record the length in cm of string on Table III.

    Repeat on micrographs #2, #3. Calculate surface density using the equations given. Take the total P_{cell} from Table I.

12. Optional: Students can obtain their own micrographs at the EM grid corner locations as in (Figure 1).

    Make an estimate of volume of mitochondria per unit volume of cell using a 2 cm test lattice.

Micrograph #1

Magnification M = 7500 times

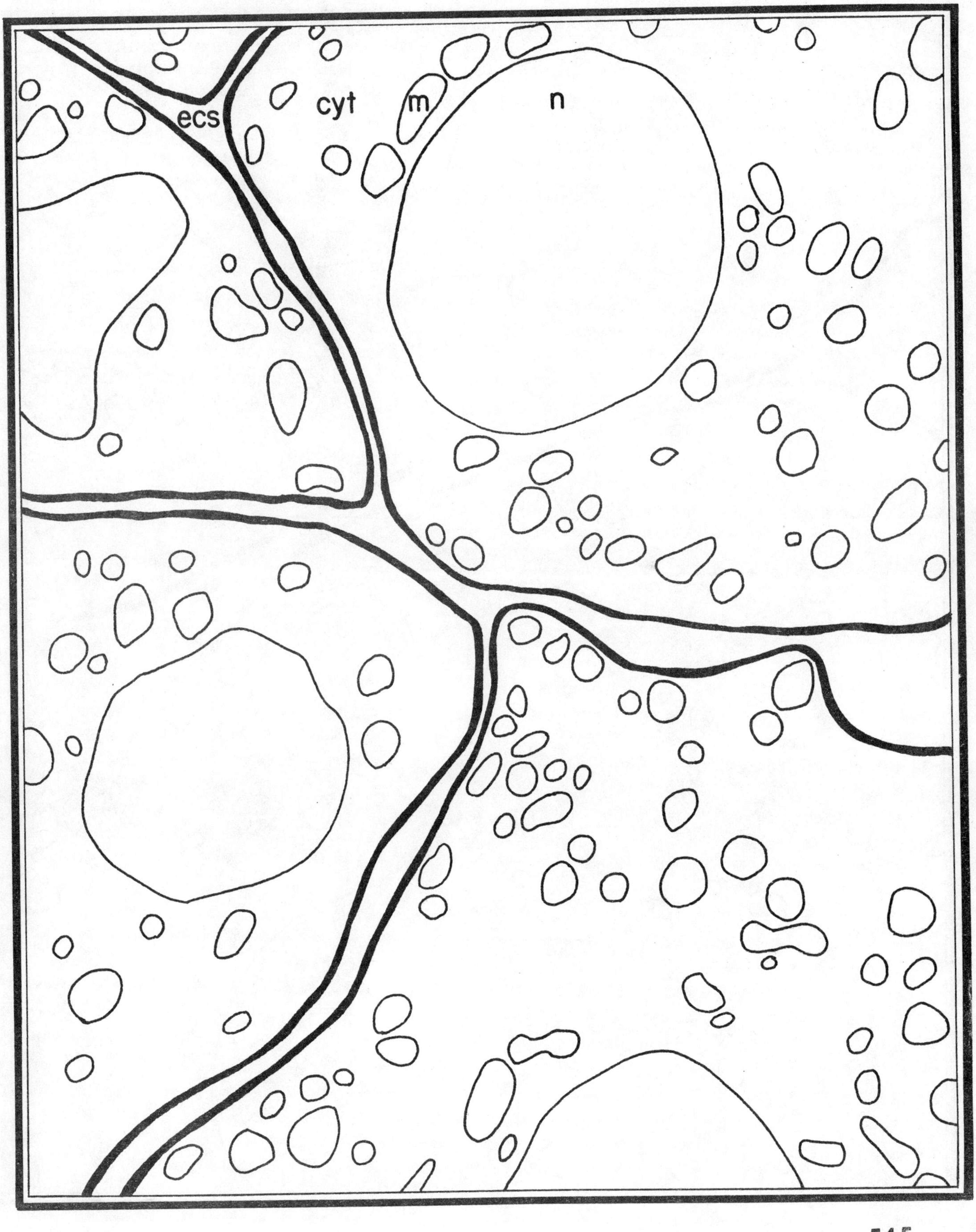

Micrograph #2 Magnification M = 7500 times

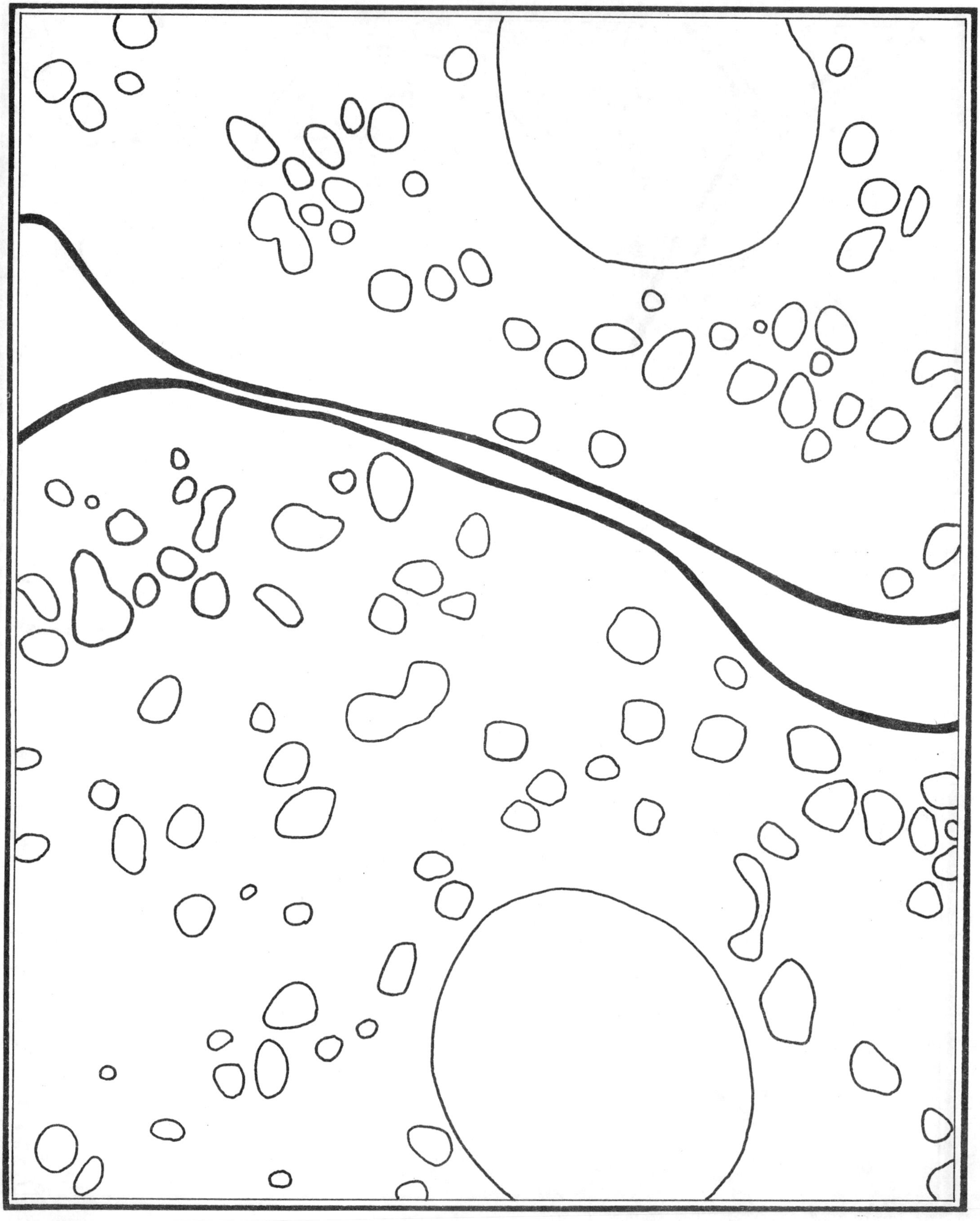

Micrograph #3

Magnification M = 7500 times

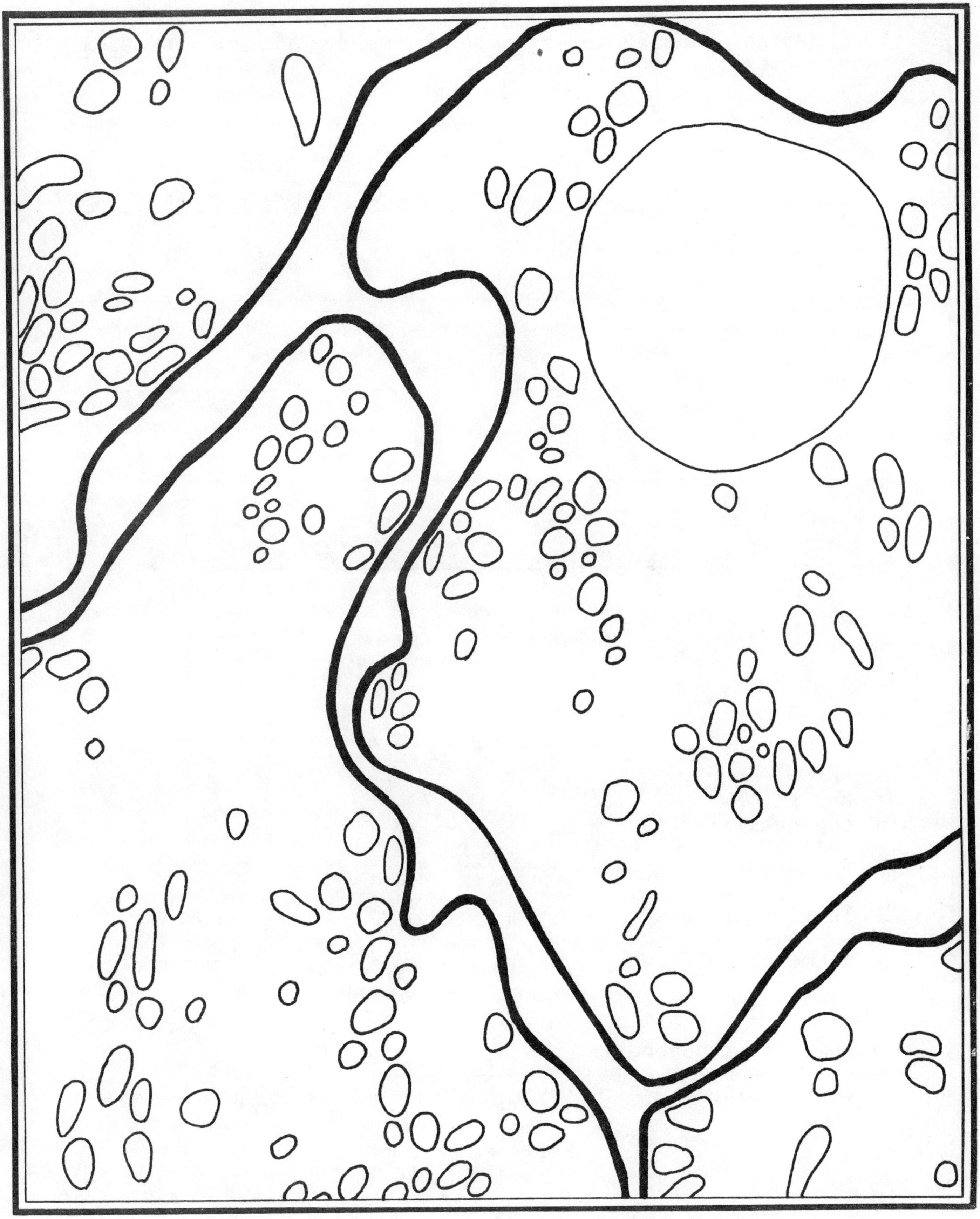

E. RESULTS:

TABLE I

Point counting to estimate volume fractions. Use 2 cm square test lattice.

Micrograph #	P_n	P_m	P_{cyt}	P_{ecs}
1				
2				
3				
Total counts from all 3 micrographs				

$$P_{tissue} = P_n + P_m + P_{cyt} + P_{ecs} = \_\_\_\_\_\_\_\_$$

$$P_{cell} = P_n + P_m + P_{cyt} = \_\_\_\_\_\_\_\_$$

Using equation (5), calculate:

$$\frac{\text{Volume of nucleus}}{\text{Volume of cell}} = \frac{V_n}{V_{cell}} = \frac{P_n}{P_n + P_m + P_{cyt}} =$$

$$\frac{\text{Volume of mitochondria}}{\text{Volume of cell}} = \frac{V_m}{V_{cell}} = \frac{P_m}{P_n + P_m + P_{cyt}} = \_\_\_\_\_\_\_\_$$

$$\frac{\text{Volume of mitochondria}}{\text{Volume of cytoplasm}} = \frac{V_m}{V_{cyt}} = \frac{P_m}{P_{cyt}} = \underline{\hspace{3cm}}$$

$$\frac{\text{Volume of mitochondria}}{\text{Volume of tissue}} = \frac{V_m}{V_{tissue}} = \frac{P_m}{P_n + P_m + P_{cyt} + P_{ecs}} =$$

TABLE II

Point counting to estimate volume density. Use 1 cm test lattice.

Micrograph #	P_n	P_m	P_{cyt}	P_{ecs}
1				
2				
3				
Total counts from all 3 a micrographs				

$$P_{tissue} = P_n + P_m + P_{cyt} + P_{ecs} = \underline{\hspace{3cm}}$$

$$P_{cell} = P_n + P_m + P_{cyt} = \underline{\hspace{3cm}}$$

Using equation (5), calculate:

$$\frac{\text{Volume of nucleus}}{\text{Volume of cell}} = \frac{V_n}{V_{cell}} = \frac{P_n}{P_n + P_m + P_{cyt}} = \_\_\_\_\_\_\_\_\_\_$$

$$\frac{\text{Volume of mitochondria}}{\text{Volume of cell}} = \frac{V_m}{V_{cell}} = \frac{P_m}{P_n + P_m + P_{cyt}} = \_\_\_\_\_\_\_\_\_\_$$

$$\frac{\text{Volume of mitochondria}}{\text{Volume of cytoplasm}} = \frac{V_m}{V_{cyt}} = \frac{P_m}{P_{cyt}} = \_\_\_\_\_\_\_\_\_\_$$

$$\frac{\text{Volume of mitochondria}}{\text{Volume of tissue}} = \frac{V_m}{V_{tissue}} = \frac{P_m}{P_n + P_m + P_{cyt} + P_{ecs}} =$$

Compare your answers counted from the 1 cm and 2 cm test lattice. Do you think they should differ? Why?

Which test lattice took less time to count? ______

TABLE III

Intersection counting to estimate surface density. Use 2 cm test lattice. Make counts from horizontal direction.

Micrograph #	I_m	I_{cell}	P_{cell}	P_m
1				
2				
3				
Total from all 3 micrographs				

Use equation (8) with lattice space, d = 2 cm and M = 7,500 x

$$\frac{\text{Surface area of mitochondria}}{\text{Volume of cell}} = \frac{2\ M\ I_m}{d\ P_{cell}} = \_\_\_\_\_\_\_\_\ \_\_\_\_\_\_\_\_ \text{ units}$$

$$\frac{\text{Surface area of mitochondria}}{\text{Volume of mitochondria}} = \frac{2\ M\ I_m}{d\ P_m} = \_\_\_\_\_\_\_\_\ \_\_\_\_\_\_\_\_ \text{ units}$$

$$\frac{\text{Surface area of cell}}{\text{Volume of cell}} = \frac{2\ M\ I_{cell}}{d\ P_{cell}} = \_\_\_\_\_\_\_\_\ \_\_\_\_\_\_\_\_ \text{ units}$$

Boundary length of all cell membrane: #1 ____ in cm of string
#2 ____
#3 ____

Total boundary length B_{cell}=#1 +#2 +#3 = ____ in cm of string

and = B_{cell}/M = ____ in cm of tissue

Use Equations (3) and (6)

$$\frac{\text{Surface area of cell}}{\text{Volume of cell}} = \frac{4B_{cell}}{\pi A_{cell}} = \frac{4M_{cell}}{\pi d^2 P_{cell}} = \underline{\qquad} \ \underline{\qquad} \text{ units}$$

Compare this answer with the intersection method above. Does it give a similar value? ______

Would you like to try the string method on the mitochondria now? ______

TABLE IV

Intersection counting to estimate surface density. Use the 1 cm test lattice in the horizontal direction.

Micrograph #	I_m	I_{cell}	P_{cell}	P_m
1				
2				
3				
Total from all 3 micrographs				

Use equation (8) with lattice space, d = 1 cm and M = 7,500 x.

$$\frac{\text{Surface area of mitochondria}}{\text{Volume of cell}} = \frac{2\ M\ I_m}{d\ P_{cell}} = \underline{\qquad} \ \underline{\qquad} \text{ units}$$

$$\frac{\text{Surface area of mitochondria}}{\text{Volume of mitochondria}} = \frac{2\,M\,I_m}{d\,P_m} = \frac{\_\_\_\_\_\_\_\_}{} \quad \frac{\_\_\_\_\_\_\_\_}{\text{units}}$$

$$\frac{\text{Surface area of cell}}{\text{Volume of cell}} = \frac{2\,M\,I_{cell}}{d\,P_{cell}} = \frac{\_\_\_\_\_\_\_\_}{} \quad \frac{\_\_\_\_\_\_\_\_}{\text{units}}$$

Compare your answers counted from the 1 cm and 2 cm test lattice. Do they differ? Why?

F. EVALUATION:

Take your calculated answers to the instructor for corrections. Be prepared to explain what you have calculated.

Answers: Total P_{tissue} is approximately 290 with the 2 cm lattice and P_{tissue} is approximately 1160 with the 1 cm lattice, or 4 x more.

Volume densities are approx:

$$V_n/V_{cell} = 0.15,\ V_m/V_{cell} = 0.12,$$

$$V_m/V_{cyt} = 0.17,\ V_m/V_{tissue} = 0.11.$$

Total intersection counts of I_m is about 500 and I_{cell} is about 140 with the 1 cm lattice. The 2 cm lattice gives half these counts.

Surface densities are approx:

$$S_m/V_{cell} = 7{,}000\ cm^{-1},\ S_m/V_m = 55{,}000\ cm^{-1},$$

$$S_{cell}/V_{cell} = 2{,}000\ cm^{-1}.$$

The total length of the string is about 225 in cm of string and this gives the same value for S_{cell}/V_{cell} of 2,000 cm^{-1}.

The surface-to-volume, S/V, ratio of a sphere is known. S for a sphere is $4\pi r^2$, where r is the radius. V is $4/3\ \pi r^3$. Therefore,

$$\frac{S}{V} = \frac{4\pi r^2}{4/3\pi r^3} = \frac{3}{r}$$

We can use this to check our answers. S_m/V_m = 55,000 = 3/r cm^{-1}. So, the radius of the mitochondrion is r = 0.00005 cm, or 0.5 μm. This is reasonable. The cell radius is 15 μm.

G. IMPORTANT HINTS:

Real projects require a lot of work and so it is important to choose an efficient method. Biological variation is often large and a sample of 5 animals is used, with 5 blocks from each, one section per block and 10 micrographs per section.

This produces 5 x 5 x 1 x 10 = 250 micrographs. Choice of the test lattice and magnification is very important too. Consult books for further information.

A most useful elementary reference is Weibel's Chapter 6 in Hayat's "Principles and Techniques of EM" Volume 3, Van Nostrand Reinhold, 1973. An advanced text is Weibel's "Stereological Methods" Vol 1, Academic Press, 1979.

NOTE: REMEMBER THE MEASURING ITSELF IS EASY BUT THE SELECTION OF WHAT IS MEASURED IS HARD.

SECTION III

EXERCISE NUMBER 40

DEVELOPMENT OF A RESEARCH PROTOCOL USING ELECTRON MICROSCOPY

Raymond L. Hayes, Ph.D.

and

Betty R. Jones, Ph.D.

A. OBJECTIVES:

1. To identify a particular research problem in one's field of interest
2. To ask specific questions directed toward solving the problem
3. To follow the scientific method in seeking a solution to the problem
4. To develop a precise, detailed protocol using electron microscopy (EM) techniques
5. To understand and determine why and how EM should be the method of choice for solving the problem
6. To search for several alternative methods other than EM for solving the problem
7. To be able to show how a combination of approaches can be used
8. To show how SEM and TEM can serve as a valuable tool in revealing data on the fine external and internal ultrastructure of cells, tissues and organs
9. To develop a protocol that will yield reproducible results
10. To have knowledge of pertinent research papers

that have been published in the area

See Figure 40-a for major headings for writing a research paper after data is interpreted and conclusions are drawn.

B. MATERIALS:

1. Background knowledge of research area

2. Available resource materials

C. RATIONALE AND DISCUSSION:

Once having identified a problem area for research, the student must pose a question or questions directed toward resolution of that problem or some component thereof.

In most instances several experimental methodologies might be pertinent to the questions asked. These methodologies might require the accumulation of preliminary data or sequential acquisition of information.

That is, before pursuing a question related to the function or chemistry of an organ, one might wish to become familiar with the histology and/or cytology of that organ.

Depending upon the degree of detail required, light or electron microscopy would be the chosen methodology for the investigation.

The choice of an experimental organism requires consideration of several factors. Availability, cost, ease of accommodation within or near the laboratory and standardization are valid concerns.

Having selected the experimental organism, however, the development of a research protocol must precede any laboratory effort.

The use of in vivo or in vitro methodology is a primary decision and the determination of suitable controls for experimental manipulations must be determined.

In most instances, control tissues and experimental tissues are dissected from the organism using a parallel procedure and these tissues are then processed similarly to assure valid comparisons of derived data.

Surgical procedures must not introduce artifact within tissues and stabilization of morphological features of tissues must be initiated promptly and efficiently.

To assure adequate penetration of fixative solutions as rapidly as possible, tissues in question must be exposed to the fixative as completely and comprehensively as possible.

This might require dripping the fixative directly on the tissue prior to its removal from the organism.

Also, this might require perfusion through the blood vascular tree of the organ. In most instances and as a first approximation, however, the tissue may be surgically removed and immersed in the fixative solution.

Trimming away unnecessary tissue masses or surface tissues which may have been stressed during surgical exposure is performed to core down the tissue of interest.

Shaping the tissue may be done to preserve orientation in the intact organ, if significant for later reference to other experimentals or to controls.

Optimal penetration of fixatives and other solutions requires that the tissue be minced into small units.

One millimeter cubed pieces or one millimeter squared strips are preferred for rapid and thorough absorption and exchange of solutions as well as for embedment in supportive plastic matrices.

D. PROCEDURE:

NOTE: WRITE OUT A DETAILED PROTOCOL FOR SOLVING A RESEARCH PROBLEM UTILIZING EM TECHNIQUES.

MAKE SURE THE PROBLEM IS STATED WITH CLARITY AND THE METHODS OF APPROACH ARE PRECISELY DEFINED.

PRESENT A DISCUSSION ON THE SIGNIFICANCE OF THE EXPECTED OUTCOME.

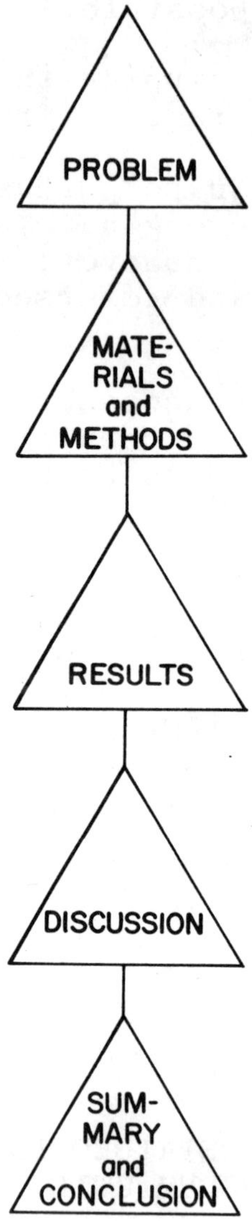

FIGURE 40-a. Sequence of steps for developing a research paper.

E. EVALUATION:

EACH STUDENT WILL BE ASKED BY THE INSTRUCTOR TO MAKE AN ORAL PRESENTATION ON THEIR RESEARCH PROTOCOL TO THE CLASS.

THEIR WRITTEN MATERIAL WILL ALSO BE CHECKED BY THE INSTRUCTOR.

F. QUESTIONS:

1. Discuss the significance of diagnostic EM in medicine.

2. Discuss the role and impact EM has on the biological and physical sciences with respect to solving research problems.

3. Discuss the first reported data garned from SEM and TEM techniques.

SECTION IV

EXERCISE NUMBER 41

ELECTRON MICROSCOPY TRAINING IN THE FORM OF SHORT INTENSIVE LABORATORY COURSES

Morton D. Maser, Ph.D.

A. INTRODUCTION:

Most of the descriptions and exercises in the previous sections have pertained particularly to educational programs intended to match conventional university calendars; the courses are conceived as extending over a semester or quarter and meeting a few times each week with opportunity between class sessions for study, reflection, and practice.

Usually, a university course in electron microscopy (EM) is only one of several being taken simultaneously by the student who is forced to apportion his available time judiciously among competing interests.

There is another format used for training in EM, as old as the discipline itself, and having special characteristics that make it particularly appropriate for certain circumstances. I refer to the so-called Short Courses by which I mean full-time intensive training programs, emphasizing laboratory work, and extending for a few weeks at most, but more commonly for five days.

With the right organization and execution, the accomplishments of students in a five-day Short Course can be equivalent to those in a conventional course lasting a quarter, for which two or three credits are earned.

Educational principles such as the setting of purposes, goals, and motivations are no different for Short Courses than for other programs but compressing the time scale allows any deficiencies or failures to be observed with embarassing ease. By the same token, success is apparent and the rewards immediate.

It has been my good fortune to be closely associated, as teacher or developer, with more than a dozen Short Courses which have been offered nearly 100 times in the aggregate at several different institutions.

Although the courses have dealt with various biochemical, microscopical, and analytical techniques, both elementary and advanced, they all have in common an approach and style I will exemplify by describing in some detail a particular Short Course on conventional biological transmission EM that is descended from a course taught principally by Dr. Blair Bowers of NIH and myself since 1964.

The course begins on a Sunday afternoon and ends the following Friday afternoon. Instruction is scheduled from 8:30 a.m. until 11:30 p.m. with breaks for lunch and dinner.

The efficiency resulting from this rigorous exclusive schedule more than compensates for the absence of opportunity for reflective study and independent practice; students have time for nothing else but EM and are not distracted by extraneous matters.

The course is intended primarily for persons having little or no experience in EM but who require it in their work or research. It is <u>not</u> intended for students wishing to learn EM as a potentially useful tool or as part of their general education, but rather for those who have an immediate practical vocational motivation.

The goals are for each student (1) to produce and evaluate technically acceptable electron micrographs of biological specimens he has produced by ultrathin sectioning and (2) to be able to do the same thing in his own laboratory within a short time after the course.

Given the availability of two electron microscopes, four ultramicrotomes, and between three and five expert faculty, as well as customary (but less crucial) ancillary equipment, 12 students can be taught.

We consider the most challenging technical task in the chain that begins with a living specimen and ends with a micrograph to be the production of ultrathin sections and we select that as the primary focus of accomplishment.

Although instruction in microscope operation and evaluation of its importance is a very important component of the course, we approach microscopy, in the curricular sense, as an assay necessary for critical assessment of sections.

The other key instructional emphasis is to provide the students with a readily achievable standard method of specimen preparation that they can modify for their own purposes but which will serve as a reliable control procedure should difficulties occur with the modifications. Because short courses are severely constrained by the clock and calendar, we strive continually for efficiency and maximum use of the students' time. Laboratory setups are prepared in advance so that valuable instructional time is not spent in searching for implements and supplies.

Students in general do not watch things being done; rather, there is enough equipment for them all to be doing those things. Laboratory instruction is repeated as many as three times to different small groups of students so that they are not limited by equipment availability. For example, since only four students can be working on microtomes at one time, instruction on sectioning is given only to those four while the other eight are doing microscopy or trimming blocks or making knives or doing some other task.

During the next laboratory period the microtomy instruction is repeated for another four students, etc. Laboratory instructional periods usually range from 1.5 to 2 hours. As a general rule, in this kind of short course, the fundamental structured teaching should be completed by the end of the third day, and the final two days should be devoted in large part to individual consultation and practice dwelling on specific applications of the basic techniques to the student's own situation.

In summary, a short course of this type requires:

Highly motivated students with similar vocationally oriented goals.

Ample equipment and faculty to employ the student's time maximally.

Emphasis on the most difficult technical task.

Practical laboratory success.

Examinations are neither desired nor necessary. A sample annotated schedule implementing the foregoing ideas follows.

B. SAMPLE ELECTRON MICROSCOPY COURSE SCHEDULE

SUNDAY

3:00 - 5:00 PM LECTURE: Orientation; introduction to EMs.

Purpose, theory, limitations, and advantages of EMs with particular reference to ensuing lab sessions; history;preparation for lab

7:00 - 11:00 PM LAB: Trim at least two blocks for sectioning; make at least four glass knives with boats; prepare fixative solutions.

Untrimmed blocks are provided since students will not produce their own for several days. Four trimming setups are provided with examples of partially and completely trimmed blocks available at a fifth station.

We feel strongly that breaking glass for knives by hand, rather than with a machine, is an excellent exercise in the self-discipline and attention to detail required for success in the course.

Knifemakers are not introduced until later in the week. Cleaned glass, pliers, and other materials sufficient for at least four students to work at once are provided.

The blocks and knives will be used tomorrow by the students. In anticipation of tomorrow morning's fixation, students - in pairs - prepare glutaraldehyde and osmic acid solutions.

		Handling osmium is likely to be a new experience for many, but making buffers is not, so buffers are provided already mixed.

MONDAY

8:30 - 9:30 AM	LAB:	Dissection and fixation. Each student pair sacrifices a small animal and fixes several types of tissue, including some "easy" ones such as liver or muscle. Continuation of fixation, dehydration, and embedding schedules is done throughout the day and evening.
9:30 - Noon	LECTURE:	Fixation and dehydration.
1:00 - 5:30 PM	LAB:	Microscope operation; semi-thin sectioning; dehydration. Students in pairs learn to change specimens, manipulate electron beam, and recognize out-of-focus Fresnel fringes. They learn microtome operation and cut and stain several one-micron sections from blocks they trimmed using knives they made.
7:00 - 8:00 PM	LAB:	Preparation of embedding media. (Students weigh and mix epoxy media components and begin infiltration of tissue)
8:00 - 10:00 PM	LECTURE:	Evaluation of electron microscope performance.

(Critical focus, astigmatism, contamination, dirt, magnification calibration, and resolution checking. Tomorrow's parallel exercises are outlined).

TUESDAY

8:30 - 9:00 PM	LAB:	Transfer tissue to fresh resin mixture.
9:00 - Noon	LECTURE:	Embedding; ultramicrotomy; staining.
1:00 - 5:30 and 7:00 - 11:30 PM	LAB:	Finish embedding; EM evaluation exercises; ultrathin sectioning. (Students perform astigmatism correction, drift and contamination tests, magnification calibration, learn EM photographic system. They produce ultrathin sections and stain them for EM observation tomorrow. About half the students succeed in producing decent sections during this session).

WEDNESDAY

9:00 - 10:00 AM	LAB:	Analysis of EM performance tests. (Students measure drift, contamination, magnification, and critically evaluate test micrographs from yesterday).
10:00 - Noon	LECTURE:	EM photography; diamond knives; instrument maintenance; safety; miscellaneous topics.

1:00 - 5:30 and 7:00 - 11:30 PM	LAB:	EM operation; ultrathin sectioning; photographic printing; knifemakers. Student's own sections are studied and photographed if available; if not, prepared sections are used. Remainder of class usually succeeds in ultrathin sectioning. Printing is taught, but not emphasized, since most participants can acquire and practice this skill elsewhere.

THURSDAY

9:00 - Noon	LECTURE:	Non-sectioning specimen preparation methods; scanning electron microscopy. Negative staining, vacuum evaporation, replica methods, freeze-etching, etc. An overview of SEM is presented only for information i.e. there is no accompanying lab, and only if there is consensual interest.
1:00 - 5:00 and 7:00 - 11:00 PM	LAB:	EM operation; ultrathin sectioning; demonstrations and practice of non-sectioning methods; substrate preparation. Carbon and platinum evaporation are demonstrated;students all make formvar-coated grids; they practice with vacuum evaporators if interested. Students learn operation of the other available microscope and use it individually to study their own sections.

FRIDAY

9:00 - 11:00 AM LECTURE:	Artifacts and interpretation. Discussion of projected micrographs with emphasis on defects produced during various stages of preparation
11:00 - AM - 5:00 PM	LAB AND DISCUSSION. Practice with tissue blocks students fixed and embedded earlier in the week; microscope maintenance; lab design and management; individual consultation on specific student applications.

Students are asked to leave with the faculty at least one micrograph they have taken of a thin section produced in the course.

The surprisingly high quality of these pictures is evidence of the efficacy of the intensive Short Course format for instruction in specific technical topics.

These programs are an important supplement to conventional University courses and are to be preferred in many instances where immediate professional need for the technical knowledge is present.

APPENDIX

APPENDIX C

LIST OF COMMERICAL SUPPLIERS

American Optical Company
1150 Avenue of the Americas
New York, New York 10036

Amray Corporation
160 Middlesex Turnpike
Bedford, Massachusetts 01730

Applied Research Lab.
(Bausch and Lomb)
9545 Wentworth Street
Sunland, California 91040

Balzers Corporation
8 Sagamore Park Road
Hudson, New Hampshire 03051

Bausch and Lomb
Instruments and Systems Division
42 East Avenue
Rochester, New York 14603

Better Equipment for Electron Microscopy, Inc. (BEEM)
1745 Williamsbridge Road
Bronx, New York 10461

Bowmar Company
P.O. Box 225
Tacoma, Washington 98401

Cambridge Instruments, Inc.
40 Robert Pitt Drive
Monsey, New York 10952

Carl Zeiss, Inc.
444 Fifth Avenue
New York, New York 10018

Colonial Metals, Inc.
P.O. Box 726
Elkton, Maryland 21921

Cooke Vacuum Products, Inc.
13 Merritt Street
South Norwalk, Connecticut 06854

Coolwell
(Technics International)
96 Woodland Avenue
Westwood, New Jersey 07675

Degussa Corporation
Technical Metals
P.O. Box 2004
Rte. 46 at Hollister Road
Teterboro, New Jersey

Denton Vacuum, Inc.
8 Fellowship Road
Cherry Hill, New Jersey 08034

D.J. Bleu Engineering Company
1201 El Curtola Boulevard
Walnut Creek, California 94596

Dupont Company/Sorvall Microtomy
Dupont Instruments
Room X38606
Wilmington, Delaware 19898

Ebtec Corporation
120 Shoemaker Lane
P.O. Box 465
Agawam, Massachusetts 01001

EDAX International
P.O. Box 135
Praire View, Illinois 60069

Efjelds Company
3 A Street
Burlington, Massachusetts 01803

Electron Microscopy 13 Sciences
Box 251
Fort Washington, Pennsylvania 19034

LIST OF COMMERCIAL SUPPLIERS - CONT.

Ernest F. Fullam, Inc.
P.O. Box 44
Schenectady, New York 12301

FTS Systems, Inc.
Route 209
Stone Ridge, New York 12494

Gatan, Inc.
780 Commonwealth Drive
Warrendale, Pennsylvania
15086

G.W. Electronics, Inc.
22 Perimeter Park, Suite 101
Atlanta, Georgia 30341

Hitachi Scientific
Instruments
Nissei Sangyo America, Ltd.
460 E. Middlefield Road
Mountainview, California
94043

International Equipment
Trading Ltd.
3005 Commercial Ave.
Northbrook, Illinois 60062

International Scientific
Instruments, Inc. (ISI)
3255-6C Scott Boulevard
Santa Clara, California
95051

Jack Moorlag Associates
Incorporation
3607 Huron
Kalamazoo, Michigan 49007

JCPDS
International Centre for
Diffraction Data
1601 Park Lane
Swarthmore, Pennsylvania
19081

JEOL, Inc.
11 Dearborn Road
Peabody, Massachusetts
01960

Kevex Corporation
1101 Chess Drive
P.O. Box 4050
Foster City, California
94404

Kimball Physics, Inc.
Wilton, New Hampshire
03086

Ladd Research Industries
P.O. Box 901
Burlington, Vermont 05401

Lemont Scientific
2011 Pine Hall Drive
State College, Pennsylvania
16801

Link Systems, Inc.
11735 Bowman Green Drive
Reston, Virginia 22090

LKB Instruments, Inc.
12221 Parklawn Drive
Rockville, Maryland 20902

Marcon Instrument Company
P.O. Box 375
Norwood, Massachusetts
02062

M.E. Taylor Engineering,
Inc.
11506 Highview Avenue
Wheaton, Maryland 20902

Micron, Inc.
P.O. Box 3536
Wilmington, Delaware 19807

APPENDIX C

LIST OF COMMERCIAL SUPPLIERS - CONT.

Microscopy Laboratories
P.O. Box 338
Red Bank, New Jersey 07701

M.J.O. - Diatome Company
P.O. Box 125
Fort Washington, Pennsylvania
19034

Normco, Inc.
1717 Elton Road, Suite 206
Silverspring, Maryland 20903

Ortec/EGG
100 Midland Road
Oak Ridge, Tennessee 37830

Pasat Research Associates
585 Royervista Drive, N.E.
Atlanta, Georgia 30342

Philips Electronic Instruments
85 McKee Drive
Mahwah, New Jersey 07430

Physimetrics, Inc.
6400 Powers Ferry Road
Atlanta, Georgia 30339

Perkin-Elmer Physical
Electronics Division
5 Progress Street
Edison, New Jersey 08820

Pesto, Inc.
Electron Microscope Services
11 Kaylor Court
Cold Spring Harbor, New York
11724

Polaron Instruments
Incorporation
4099 Landisville Road
Doylestown, Pennsylvania 18901

Polysciences, Inc.
Paul Valley Industrial Park
Warrington, Pennsylvania
18976

Princeton Gamma-Tech
1200 State Road
Princeton, New Jersey 08540

Scientific Instruments
and Sales Company
P.O. Box 11628
Atlanta, Georgia 30355

SPI (Division of Structure
Probe, Inc.)
P.O. Box 342
Westchester, Pennsylvania
19380

Technicon Instruments
Corporation
511 Benedict Avenue
Tarrytown, New York 10591

Technics Electron
Microscopy Systems, Inc.
7653 Fullerton Road
Springfield, Virginia 22153

Ted Pella Company
P.O. Box 510
Tustin, California 92680

Tousimis Research
Corporation
P.O. Box 2189
Rockville, Maryland 20852

Tracor Northern
3040 Holcomb Bridge Road
Suite F-1
Norcross, Georgia 30071

NOTE: THIS LIST IS BY NO MEANS EXHAUSTIVE, USERS OF THIS MANUAL MAY REFER TO THE APPENDIX OF EM TEXTBOOKS AND HANDBOOKS FOR OTHER SUPPLIERS.

APPENDIX D

BIBLIOGRAPHY

REFERENCE TEXTS:

Bloom, W. and Fawcett, D. 1975. Textbook of Histology. W.B. Saunders Company, Philadelphia, Pennsylvania.

Dawes, C.J. 1971. Biological Techniques in Electron Microscopy. Ladd Research Industries, Burlington, Vermont.

__________ 1979. Biological Techniques for Transmission and Scanning Electron Microscopy. Ladd Research Industries, Burlington, Vermont.

Fawcett, D. 1966. The Cell: An Atlas of Fine Structure. W.B. Saunders Company, Philadelphia, Pennsylvania.

__________ 1981. The Cell. W.B. Saunders Company, Philadelphia, Pennsylvania.

Glauert, A.M. 1974. Practical Methods in Electron Microscopy. Volumes 1-8, North-Holland Company, New York.

Griffin, R.L. 1972. Ultramicrotomy. Williams and Wilkins Company, Baltimore, Maryland.

Hayat, M.A. 1970. Principles and Techniques of Electron Microscopy: Biological Applications. Volume 1-5, Van Nostrand Reinhold Company, New York.

__________ 1970. Principles and Techniques of Scanning Electron Microscopy: Biological Applications. Van Nostrand Reinhold Company, New York.

__________ 1978. Introduction to Biological Scanning Electron Microscopy. University Park Press, Baltimore, Maryland.

APPENDIX D

BIBLIOGRAPHY

REFERENCE TEXTS - CONT.

_______ 1981. Fixation for Electron Microscopy. Academic Press, New York.

Kay, D.H. 1965. Techniques for Electron Microscopy. Blackwell Scientific Publications, Oxford, England.

Ladd, M.W. 1973. The Electron Microscope Handbook. Ladd Research Industries, Inc., Burlington, Vermont.

Meek, G. 1976. Practical Electron Microscopy for Biologists. John Wiley and Sons, Inc., New York.

Mercer, E.H. and Birbeck, M.S.C. 1966. Electron Microscopy: A Handbook for Biologists. F.A. Davis Company, Philadelphia, Pennsylvania.

Pease, D.C. 1964. Histological Techniques for Electron Microscopy. Academic Press, Inc., New York.

Porter, K.R. and Bonnerville, M.A. 1963. An Introduction to the Fine Structure of Cells and Tissues. Lea and Febiger, Philadelphia, Pennsylvania.

_______ 1973. Fine Structure of Cells and Tissues. Lea and Febiger, Philadelphia, Pennsylvania.

Porter, K.R. 1977. Microanatomy of Cell and Tissue Surfaces: An Atlas of Scanning Electron Microscopy. Lea and Febiger, Philadelphia, Pennsylvania.

Reith, E.J. and Ross, M.H. 1977. Atlas of Descriptive Histology. Harper and Row Company, New York.

APPENDIX D

BIBLIOGRAPHY

REFERENCE TEXTS - CONT.

Rhodin, J.A.G. 1963. Atlas of Ultrastructure. W.B. Saunders Company, Philadelphia, Pennsylvania.

Sanborn, E.B. 1970. Light and Electron Microscopy of Cells and Tissues. Academic Press, Inc., New York.

Sjostrand, F.S. 1967. Electron Microscopy of Cells and Tissues. Academic Press, Inc., New York.

Thin Sectioning and Associated Technics for Electron Microscopy 1967. Ivan Sorvall, Inc., Norwalk and Connecticut.

Toner, P.G. and Carr, K.E. 1971. Cell Structure. Williams and Wilkins Company, Baltimore, Maryland.

Trump, B.F. and Jones, R.T. 1980. Diagnostic Electron Microscopy. Volume 3, John Wiley and Sons, New York.

Weakley, B.S. 1972. A Beginner's Handbook in Biological Electron Microscopy. Churchill Livingstone Company, London, England.

____________ 1981. A Beginner's Handbook in Biological Transmission Electron Microscopy. Churchill Livingstone Company, London, England.

Wischnitzer, S. 1962. Introduction to Electron Microscopy. Pergamon Press, New York.